Motion Deblurring

A comprehensive guide to restoring images degraded by motion blur, bridging traditional approaches and emerging computational photography-based techniques, and bringing together a wide range of methods drawn from basic theory and cutting edge research. It encompasses both algorithms and architectures, providing detailed coverage of practical techniques by leading researchers.

From an algorithms perspective, blind and non-blind approaches are discussed, including the use of single or multiple images; projective motion blur model; image priors and parametric models; high dynamic range imaging in the irradiance domain; and recognition of blurred images. Performance limits for motion deblurring cameras are also presented.

From a systems perspective, hybrid frameworks combining low resolution high-speed and high resolution low-speed cameras are described, along with the use of inertial sensors and coded exposure cameras. An architecture exploiting compressive sensing for video recovery is also included.

This book will be a valuable resource for researchers and practitioners in computer vision, image processing, and related fields.

A. N. Rajagopalan is a Professor in the Department of Electrical Engineering at the Indian Institute of Technology, Madras. He co-authored the book *Depth From Defocus: A Real Aperture Imaging Approach* in 1998. He is a Fellow of the Alexander von Humboldt Foundation, Germany, a Fellow of the Indian National Academy of Engineering, and a Senior Member of the IEEE. He received the Outstanding Investigator Award from the Department of Atomic Energy, India, in 2012, and the VASVIK Award in 2013.

Rama Chellappa is Minta Martin Professor of Engineering and an Affiliate Professor of Computer Science at the University of Maryland, College Park. He is also affiliated with the Center for Automation Research and UMIACS, and is serving as Chair of the ECE department. He is a recipient of the K.S. Fu Prize from the International Association for Pattern Recognition (IAPR) and the Society, Technical Achievement, and Meritorious Service Awards from the IEEE Signal Processing Society. In 2010, he was recognized as an Outstanding ECE by Purdue University. He is a Fellow of the IEEE, IAPR, OSA, ACM, and AAAS, a Golden Core Member of the IEEE Computer Society, and has served as a Distinguished Lecturer of the IEEE Signal Processing Society, and the President of the IEEE Biometrics Council.

Motion Deblurring

Algorithms and Systems

Edited by

A. N. RAJAGOPALAN

Indian Institute of Technology, Madras

RAMA CHELLAPPA

University of Maryland, College Park

CAMBRIDGE
UNIVERSITY PRESS

CAMBRIDGE
UNIVERSITY PRESS

University Printing House, Cambridge CB2 8BS, United Kingdom

Cambridge University Press is part of the University of Cambridge.

It furthers the University's mission by disseminating knowledge in the pursuit of education, learning and research at the highest international levels of excellence.

www.cambridge.org
Information on this title: www.cambridge.org/9781107044364

First published 2014

Printed in the United Kingdom by Clays, St Ives plc

A catalog record for this publication is available from the British Library

Library of Congress Cataloging in Publication Data
Motion deblurring : algorithms and systems / edited by A. N. Rajagopalan, Indian Institute of Technology, Madras, Rama Chellappa, University of Maryland, College Park.
pages cm
ISBN 978-1-107-04436-4 (Hardback)
1. Image processing–Mathematics. 2. Image processing–Digital techniques. 3. Digital video–Editing–Mathematics. 4. Photography–Retouching–Data processing. 5. Motion–Mathematical models. I. Rajagopalan, A. N. II. Chellappa, Rama.
TA1637.5.M68 2014
006.6–dc23 2013043749

ISBN 978-1-107-04436-4 Hardback

Contents

	List of contributors	*page* ix
	Preface	xi
1	**Mathematical models and practical solvers for uniform motion deblurring** Jiaya Jia	1
	1.1 Non-blind deconvolution	1
	1.2 Blind deconvolution	12
2	**Spatially-varying image deblurring** Neel Joshi, Sing Bing Kang, and Richard Szeliski	31
	2.1 Review of image deblurring methods	31
	2.2 A unified camera-shake blur model	32
	2.3 Single image deblurring using motion density functions	35
	2.4 Image deblurring using inertial measurement sensors	36
	2.5 Generating sharp panoramas from motion-blurred videos	46
	2.6 Discussion	54
3	**Hybrid-imaging for motion deblurring** Moshe Ben-Ezra, Yu-Wing Tai, Michael S. Brown and Shree K. Nayar	57
	3.1 Introduction	57
	3.2 Fundamental resolution tradeoff	57
	3.3 Hybrid-imaging systems	59
	3.4 Shift-invariant PSF image deblurring	61
	3.5 Spatially-varying PSF image deblurring	67
	3.6 Moving object deblurring	70
	3.7 Discussion and summary	72
4	**Efficient, blind, spatially-variant deblurring for shaken images** Oliver Whyte, Josef Sivic, Andrew Zisserman and Jean Ponce	75
	4.1 Introduction	75
	4.2 Modelling spatially-variant camera-shake blur	76
	4.3 The computational model	82

4.4 Blind estimation of blur from a single image 83
4.5 Efficient computation of the spatially-variant model 87
4.6 Single-image deblurring results 94
4.7 Implementation 95
4.8 Conclusion 97

5 Removing camera shake in smartphones without hardware stabilization 100
Filip Šroubek and Jan Flusser

5.1 Introduction 100
5.2 Image acquisition model 100
5.3 Inverse problem 102
5.4 Pinhole camera model 109
5.5 Smartphone application 112
5.6 Evaluation 117
5.7 Conclusions 119

6 Multi-sensor fusion for motion deblurring 123
Jingyi Yu

6.1 Introduction 123
6.2 Hybrid-speed sensor 124
6.3 Motion deblurring 125
6.4 Depth map super-resolution 128
6.5 Extensions to low-light imaging 132
6.6 Discussion and summary 137

7 Motion deblurring using fluttered shutter 141
Amit Agrawal

7.1 Related work 141
7.2 Coded exposure photography 142
7.3 Image deconvolution 142
7.4 Code selection 144
7.5 Linear solution for deblurring 147
7.6 Resolution enhancement 150
7.7 Optimized codes for PSF estimation 151
7.8 Implementation 156
7.9 Analysis 156
7.10 Summary 159

8 Richardson–Lucy deblurring for scenes under a projective motion path 161
Yu-Wing Tai and Michael S. Brown

8.1 Introduction 161
8.2 Related work 163

8.3 The projective motion blur model 164
8.4 Projective motion Richardson–Lucy 165
8.5 Motion estimation 170
8.6 Experiment results 171
8.7 Discussion and conclusion 179

9 HDR imaging in the presence of motion blur 184
C. S. Vijay, C. Paramanand, and A. N. Rajagopalan

9.1 Introduction 184
9.2 Existing approaches to HDRI 186
9.3 CRF, irradiance estimation, and tone-mapping 189
9.4 HDR imaging under uniform blurring 191
9.5 HDRI for non-uniform blurring 192
9.6 Experimental results 199
9.7 Conclusions and discussions 203

10 Compressive video sensing to tackle motion blur 207
Ashok Veeraraghavan and Dikpal Reddy

10.1 Introduction 207
10.2 Related work 208
10.3 Imaging architecture 211
10.4 High-speed video recovery 213
10.5 Experimental results 216
10.6 Conclusions 219

11 Coded exposure motion deblurring for recognition 222
Scott McCloskey

11.1 Motion sensitivity of iris recognition 223
11.2 Coded exposure 225
11.3 Coded exposure performance on iris recognition 239
11.4 Barcodes 240
11.5 More general subject motion 241
11.6 Implications of computational imaging for recognition 242
11.7 Conclusion 244

12 Direct recognition of motion-blurred faces 246
Kaushik Mitra, Priyanka Vageeswaran, and Rama Chellappa

12.1 Introduction 246
12.2 The set of all motion-blurred images 249
12.3 Bank of classifiers approach for recognizing motion-blurred faces 251
12.4 Experimental evaluation 252
12.5 Discussion 255

13 Performance limits for motion deblurring cameras 258
Oliver Cossairt and Mohit Gupta

13.1 Introduction 258
13.2 Performance bounds for flutter shutter cameras 261
13.3 Performance bound for motion-invariant cameras 265
13.4 Simulations to verify performance bounds 269
13.5 Role of image priors 270
13.6 When to use computational imaging 273
13.7 Relationship to other computational imaging systems 274
13.8 Summary and discussion 278

Index 283

List of contributors

Amit Agrawal
Mitsubishi Electric Research Labs (MERL), USA

Moshe Ben-Ezra
Massachusetts Institute of Technology, USA

Michael S. Brown
National University of Singapore, Singapore

Rama Chellappa
University of Maryland, USA

Oliver Cossairt
Northwestern University, USA

Jan Flusser
Czech Academy of Sciences, Czech Republic

Mohit Gupta
Mitsubishi Electric Research Labs (MERL), USA

Jiaya Jia
The Chinese University of Hong Kong

Neel Joshi
Microsoft Research, USA

Sing Bing Kang
Microsoft Research, USA

Scott McCloskey
Honeywell ACS Labs, USA

Kaushik Mitra
Rice University, USA

Shree K. Nayar
Columbia University, USA

C. Paramanand
Indian Institute of Technology, Madras, India

Jean Ponce
INRIA, France

A. N. Rajagopalan
Indian Institute of Technology, Madras, India

Dikpal Reddy
Rice University, USA

Josef Sivic
INRIA, France

Filip Šroubek
Czech Academy of Sciences, Czech Republic

Richard Szeliski
Microsoft Research, USA

Yu-Wing Tai
Korea Advanced Institute of Science and Technology, Republic of Korea

Priyanka Vageeswaran
University of Maryland, USA

Ashok Veeraraghavan
Rice University, USA

C. S. Vijay
Indian Institute of Technology, Madras, India

Oliver Whyte
Microsoft Corporation, USA

Jingyi Yu
University of Delaware, USA

Andrew Zisserman
University of Oxford, UK

Preface

The computer vision community is witnessing a major resurgence in the area of motion deblurring spurred by the emerging ubiquity of portable imaging devices. Rapid strides are being made in handling motion blur both algorithmically and through tailor-made hardware-assisted technologies. The main goal of this book is to ensure a timely dissemination of recent findings in this very active research area. Given the flurry of activity in the last few years in tackling uniform as well as non-uniform motion blur resulting from incidental shake in hand-held consumer cameras as well as object motion, we felt that a compilation of recent and concerted efforts for restoring images degraded by motion blur was well overdue. Since no single compendium of the kind envisaged here exists, we believe that this is an opportune time for publishing a comprehensive collection of contributed chapters by leading researchers providing in-depth coverage of recently developed methodologies with excellent supporting experiments, encompassing both algorithms and architectures.

As is well known, the main cause of motion blur is the result of averaging of intensities due to relative motion between a camera and a scene during exposure time. Motion blur is normally considered a nuisance although one must not overlook the fact that some works have used blur for creating aesthetic appeal or exploited it as a valuable cue in depth recovery and image forensics. Early works were non-blind in the sense that the motion blur kernel (i.e. the point spread function (PSF)) was assumed to be of a simple form, such as those arising from uniform camera motion, and efforts were primarily directed at designing a stable estimate for the original image. Research in this area then surged towards PSF estimation to deal with arbitrarily-shaped blur kernels resulting from incidental shakes in hand-held cameras and due to object motion in real scenes. This led to the evolution of blind deconvolution algorithms in which the PSF as well as the latent image had to be estimated. Initial works dealt with space-invariant kernels, only to quickly pave the way for investigation of methods for dealing with space-variant blur situations which are actually more prevalent. Parallel efforts addressing the motion blur problem from an acquisition point of view too were also being developed. Traditional approaches to motion deblurring have separated the sensor from the scene being sensed in that the motion-blurred images are post-processed to mitigate blurring effects. Given the recent advances in computational photography, novel approaches are being developed that integrate the sensor design and motion deblurring aspects. These have ranged from integrating inertial sensor data, to proposing hybrid system architectures consisting of multiple cameras with different sensor characteristics, to suitably tailoring the nature of the PSF itself through coded exposure cameras.

Research continues to grow at an amazing pace as portable imaging devices are becoming commonplace. Commercial significance of the motion deblurring problem is amply evident from the plethora of software and hardware-based approaches that have been reported in recent years and this can only be expected to grow in the coming years. In an attempt to provide a comprehensive coverage of early as well as recent efforts in this area, the contents of this edited book are spread across *thirteen* self-contained chapters. These can be broadly categorized under two major headings, namely, *Algorithms* and *Systems* for tackling the motion deblurring problem. We have deliberately refrained from sequestering the chapters along these themes and instead allowed the chapters to seamlessly weave through both these topics.

The first chapter by Jiaya Jia deals with shift-invariant or uniform single image motion deblurring and provides a systematic survey that covers early as well as recent developments for single-input motion blur models and solvers. Representative methods such as regularized and iterative approaches for model design and solver construction, and recent advances including construction of natural priors for a latent image and variable splitting solver for sparse optimization are described for the non-blind deconvolution problem. For blind deconvolution, marginalized estimation and alternating minimization strategies along with edge prediction for very large PSFs are presented.

The chapter by Joshi *et al.* addresses the problem of spatially-varying blur and presents a unified model of camera shake that can represent space-variant blur as a function of camera motion. The discussions range from fully blind methods to hardware design for deblurring using sensor data. A method for generation of sharp panoramas from blurry sequences is also presented.

The chapter by Ben-Ezra *et al.* involves hybrid-imaging system design targeting image deblurring due to camera shake and, to some extent, moving objects in a scene. This is a hardware-assisted technique that combines a high resolution camera along with an auxiliary low resolution camera to effect deblurring. The secondary imager is principally used to acquire information about the spatially-invariant or spatially-variant discrete parametric 2D motion field. The flow field is then used to derive the PSF corresponding to the primary camera so as to arrive at a high resolution non-blurred image.

The chapter by Whyte *et al.* describes a compact global parameterization of camera-shake blur, based on the 3D rotation of the camera during the exposure. A model based on three-parameter homographies is used to connect camera motion to image motion and, by assigning weights to a set of these homographies, this formulation can be viewed as a generalization of the standard, spatially-invariant convolutional model of image blur. A scheme for blur estimation from a single image followed by restoration is presented. Different approximations are introduced for the global model to reduce computational complexity.

The chapter by Šroubek *et al.* presents an attractive semi-blind implementation for image deblurring on a smartphone device. It is shown that information from the inertial sensors such as accelerometers and gyroscopes, which are readily available in modern smartphones, is accurate enough to provide camera motion trajectory and, consequently, to estimate the blurring PSF. A simple yet effective space-variant implementation of the deblurring algorithm is given that can handle complex camera motion as well as rolling shutter issues. The method works in practical situations and is fast enough to be acceptable by end users.

The chapter by Jingyi Yu presents two multi-sensor fusion techniques for sufficient and low-light conditions, respectively, that combine the advantages of high speed and high resolution for reducing motion blur. The hybrid sensor consists of a pair of high-speed color (HS-C) cameras and a single high resolution color (HR-C) camera. The HS-C cameras capture fast-motion with little motion blur. They also form a stereo pair and provide a low resolution depth map. The motion flows in the HS-C cameras are estimated and warped using the depth map on to the HR-C camera as the PSFs for motion deblurring. The HR-C image, once deblurred, is then used to super-resolve the depth map. A hybrid sensor configuration for extension to low-light imaging conditions is also discussed.

The chapter by Amit Agrawal describes coded exposure photography in which the key idea is to assist motion deblurring by modifying the imaging process to avoid loss of high frequency information during acquisition. In addition to optimal temporal code design, the role of coded exposure in enabling resolution enhancement of blurred moving objects is also dwelt upon. Hardware and implementational considerations for consumer-grade cameras are also discussed.

The chapter by Tai *et al.* describes a projective motion path blur model which, in comparison to conventional methods based on space-invariant blur kernels, is more effective at modeling spatially-varying motion blur. The model is applicable to situations when a camera undergoes ego motion while observing a distant scene. The blurred image is modeled as an integration of the clear scene under a sequence of planar projective transformations that describes the camera's path. A modified Richardson–Lucy algorithm is proposed that incorporates this new blur model in the deblurring step. The algorithm's convergence properties and its robustness to noise are also studied.

The chapter by Vijay *et al.* describes a method that operates in the irradiance domain to estimate the high dynamic range irradiance of a static scene from a set of blurred and differently exposed observations captured with a hand-held camera. A two-step procedure is proposed in which the camera motion corresponding to each blurred image is derived first, followed by estimation of the latent scene irradiance. Camera motion estimation is performed elegantly by using locally derived PSFs as basic building blocks.

By reformulating motion deblurring as that of recovering video from an underdetermined set of linear observations, the chapter by Veeraraghavan *et al.* presents an alternate viewpoint for tackling motion blur. An imaging architecture is introduced in which each observed frame is a coded linear combination of the voxels of the underlying high-speed video frames which in turn are recovered by exploiting both temporal and spatial redundancies. The architecture can tackle complex scenarios, such as non-uniform motion, multiple independent motions and spatially variant PSFs, naturally without the need for explicit segmentation.

The chapter by Scott McCloskey dwells on deblurring in the context of improving the performance of image-based recognition, where motion blur may suppress key visual details. The development of a motion deblurring system based on coded exposure is discussed, and its utility in both iris recognition and barcode scanning is demonstrated. A method is introduced to generate near-optimal shutter sequences by incorporating the statistics of natural images. Also described are extensions to handle more general object motion that can even include acceleration.

The chapter by Mitra *et al.* addresses the problem of recognizing faces from motion-blurred images, which is especially relevant in the context of hand-held imaging. Based on the conditional convexity property associated with directional motion blurs, they propose a bank-of-classifiers approach for directly recognizing motion-blurred faces. The approach is discriminative and scales impressively with the number of face classes and training images per class.

The final chapter by Cossairt *et al.* derives performance bounds in terms of signal-to-noise ratio for various computational imaging-based motion deblurring approaches including coded-exposure and camera-motion based techniques and discusses the implications of these bounds for real-world scenarios. The results and conclusions of the study can be readily harnessed by practitioners to not only choose an imaging system, but also to design it.

The contents of this book will benefit theoreticians, researchers, and practitioners alike who work at the confluence of computer vision, image processing, computational photography, and graphics. It can also serve as general reference for students majoring in Electrical Engineering and Computer Science with special focus in these areas. It would be suitable as both a textbook as well as an advanced compendium on motion blur at graduate level. Given the impact that an application such as motion deblurring has on consumer cameras, the material covered herein will be of great relevance to the imaging industry too. In short, this book can serve as a one-stop resource for readers interested in motion blur.

This comprehensive guide to the restoration of images degraded by motion blur bridges traditional approaches and emerging computational techniques while bringing together a wide range of methods, from basic theory to cutting-edge research. We hope that readers will find the overall treatment to be in-depth and exhaustive, with a right balance between theory and practice.

A. N. Rajagopalan
Rama Chellappa

1 Mathematical models and practical solvers for uniform motion deblurring

Jiaya Jia

Recovering an un-blurred image from a single motion-blurred picture has long been a fundamental research problem. If one assumes that the blur kernel – or point spread function (PSF) – is shift invariant, the problem reduces to that of image deconvolution. Image deconvolution can be further categorized as non-blind and blind.

In non-blind deconvolution, the motion blur kernel is assumed to be known or computed elsewhere; the task is to estimate the un-blurred latent image. The general problems to address in non-blind deconvolution include reducing possible unpleasant ringing artifacts that appear near strong edges, suppressing noise, and saving computation. Traditional methods such as Wiener deconvolution (Wiener 1949) and the Richardson–Lucy (RL) method (Richardson 1972, Lucy 1974) were proposed decades ago and find many variants thanks to their simplicity and efficiency. Recent developments involve new models with sparse regularization and the proposal of effective linear and non-linear optimization to improve result quality and further reduce running time.

Blind deconvolution is a much more challenging problem, since both the blur kernel and the latent image are unknown. One can regard non-blind deconvolution as an inevitable step in blind deconvolution during the course of PSF estimation or after the PSF has been computed. Both blind and non-blind deconvolution are practically very useful; they are studied and employed in a variety of disciplines including, but not limited to, image processing, computer vision, medical and astronomic imaging, and digital communication.

This chapter discusses shift-invariant single image motion deblurring methods, which assume that the image is uniformly blurred with only one PSF, which may not be known a priori. This set of problems has a long history in theoretical and empirical research and has notably advanced in the last 5–10 years with a few remarkably effective models and solvers.

1.1 Non-blind deconvolution

Ideally, a blur observation is modeled as a linearly filtered version of the latent un-blurred signal. This process can be expressed as

$$\mathbf{b} = \mathbf{l} \otimes \mathbf{f}, \tag{1.1}$$

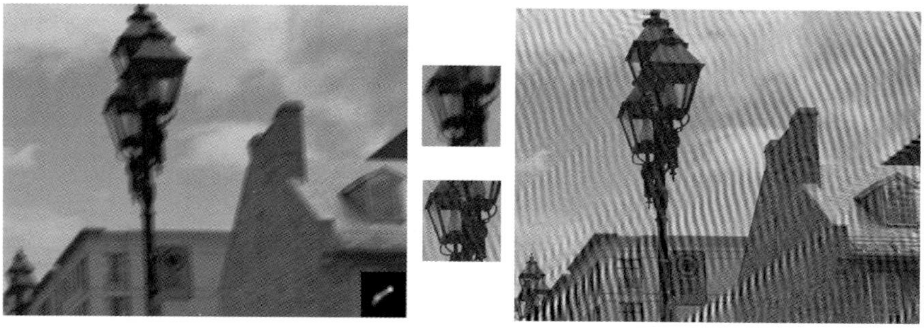

(a) (b) (c)

Figure 1.1 Inverse filtering problem. Visual artifacts caused by inverse filter. (a) Blurred image and PSF; (b) close-ups; (c) output of inverse filtering.

where **b**, **l** and **f** are the blurred image, latent un-blurred image, and PSF (or blur kernel), respectively. In the frequency domain,

$$\mathcal{F}(\mathbf{b}) = \mathcal{F}(\mathbf{l}) \cdot \mathcal{F}(\mathbf{f}), \tag{1.2}$$

where $\mathcal{F}$ is the Fourier transform.

If $\mathcal{F}(\mathbf{f})$ does not contain zero or very small values and the blurred image is noise-free, the latent image **l** can be obtained simply by inverting the convolution process using inverse filtering, the simplest method that solves for **l**. This process is expressed as

$$\mathcal{F}(\mathbf{l}) = \mathcal{F}(\mathbf{b})/\mathcal{F}(\mathbf{f}). \tag{1.3}$$

This strategy practically may produce severe visual artifacts, such as ringing, for the following reasons. First, the inversion of **f** may not exist, especially for low-pass filters. Second, motion PSFs caused by object or camera motion are typically band-limited and their spectra have zero or near-zero values at high frequency. Third, image formation causes problems including image noise, quantization error, color saturation, and non-linear camera response curve. They make blur violate the ideal convolution model and lead to a more flexible form

$$\mathbf{b} = \mathbf{l} \otimes \mathbf{f} + \mathbf{n}, \tag{1.4}$$

where **n** denotes error in the blurred image, which we call image noise in general. One deconvolution result by direct inverse filter is shown in Figure 1.1.

Development of more advanced non-blind deconvolution methods dates back to the 1970s. Early representative approaches include Wiener Deconvolution (Wiener 1949), Least Square Filtering (Miller 1970, Hunt 1973, Tikhonov & Arsenin 1977), Richardson–Lucy method (Richardson 1972, Lucy 1974), and recursive Kalman Filtering (Woods & Ingle 1981). Readers are referred to Andrews & Hunt (1977) for a review of these early approaches.

Simply put, many algorithms minimize an energy consisting of two terms, i.e. the *data* term E_{data} (corresponding to *likelihood* in probability) and *regularization* (also known as *prior*) E_{prior}. E_{data} measures the difference between the convolved image and the blur observation, and is written as

$$E_{\text{data}} = \Phi(\mathbf{l} \otimes \mathbf{f} - \mathbf{b}), \tag{1.5}$$

where Φ is a distance function. A common definition is $\Phi(\cdot) = \|\cdot\|^2$ (Wiener 1949), representing the L_2-norm of all elements. It is also called the Gaussian likelihood. E_{prior} is denoted as a function $\Psi(\mathbf{l})$, which has different specifications in existing approaches. Given E_{data} and E_{prior}, the latent image $\mathbf{l}$ can be estimated by minimizing the energy incorporating these two terms, expressed as

$$\min_{\mathbf{l}} \|\mathbf{l} \otimes \mathbf{f} - \mathbf{b}\|^2 + \lambda \Psi(\mathbf{l}), \tag{1.6}$$

where λ is a weight. In what follows, we discuss a few representative non-blind deconvolution methods with respect to model design and solver construction. Their respective strengths, disadvantages, and relations are also presented.

1.1.1 Regularized approaches

A number of early methods incorporated square regularization constraints. Two representative forms are $\Psi(\mathbf{l}) = \|\mathbf{l}\|^2$ and $\Psi(\mathbf{l}) = \|\nabla \mathbf{l}\|^2$, where ∇ is the gradient operator. They enforce smoothness on image values and image gradients, and are called Tikhonov and Gaussian regularizers, respectively. Substituting them into Eq. (1.6) yields

$$\min_{\mathbf{l}} \|\mathbf{l} \otimes \mathbf{f} - \mathbf{b}\|^2 + \lambda \|\mathbf{l}\|^2 \tag{1.7}$$

and

$$\min_{\mathbf{l}} \|\mathbf{l} \otimes \mathbf{f} - \mathbf{b}\|^2 + \lambda \|\nabla \mathbf{l}\|^2, \tag{1.8}$$

for overall energy minimization. Weight λ is typically a small value.

The main advantage of these constrained least square methods is in the simplicity of formation, which results in a solver similar to an inverse filter. Taking the Tikhonov method as an example, there exists a closed form solution $\mathbf{l}^*$ for Eq. (1.7) by setting its first order derivative to zero with respect to $\mathbf{l}$. Rearranging Eq. (1.7) in a matrix form and denoting by E the total energy yield

$$\begin{aligned} E &= \|F\nu(\mathbf{l}) - \nu(\mathbf{b})\|^2 + \lambda\|\nu(\mathbf{l})\|^2 \\ &= \nu(\mathbf{l})^T F^T F \nu(\mathbf{l}) - 2\nu(\mathbf{b})^T F \nu(\mathbf{l}) + \nu(\mathbf{b})^T \nu(\mathbf{b}) + \lambda \nu(\mathbf{l})^T \nu(\mathbf{l}), \end{aligned}$$

where F is a sparse convolution matrix generated from $\mathbf{f}$, and ν is the operator that transforms the image into its vector form. The partial derivative is

$$\frac{\partial E}{\partial \nu(\mathbf{l})} = 2F^T F \nu(\mathbf{l}) - 2F^T \nu(\mathbf{b}) + 2\lambda \nu(\mathbf{l}). \tag{1.9}$$

By setting the above equation to zero, the optimal solution $\mathbf{l}^*$ is

$$\nu(\mathbf{l}^*) = \frac{F^T}{F^T F + \lambda \Lambda} \nu(\mathbf{b}), \tag{1.10}$$

where Λ is an identity matrix, the same size as $F^T F$.

Regularization bias

If there is neither kernel error nor image noise and the kernel matrix F is invertible, the ground truth latent image $\hat{\mathbf{l}}$ is simply the reversion of convolution, expressed as

$$\nu(\hat{\mathbf{l}}) = F^{-1}\nu(\mathbf{b}) = \frac{F^T\nu(\mathbf{b})}{F^TF}. \tag{1.11}$$

The difference between Eqs. (1.10) and (1.11) makes it possible to analyze how the regularization term introduces bias in deconvolution in an ideal noise-free situation. It serves as guidance for future deconvolution model design.

We denote the error map of the recovered image as

$$\delta\mathbf{l} = \mathbf{l}^* - \hat{\mathbf{l}}, \tag{1.12}$$

where $\delta\mathbf{l}$ is the error introduced in deconvolution. Equations (1.11) and (1.12) together lead to

$$\nu(\delta\mathbf{l}) = \nu(\mathbf{l}^*) - \nu(\hat{\mathbf{l}}) = -\frac{\lambda\nu(\mathbf{b})}{(F^TF + \lambda\Lambda)F} = -\frac{\lambda}{F^TF + \lambda\Lambda}\nu(\hat{\mathbf{l}}). \tag{1.13}$$

Because $\lambda/(F^TF + \lambda\Lambda)$ can be regarded as a weight fixed by blur, this equation indicates that $\delta\mathbf{l}$ generally appears as a high frequency map dependent on image structures in $\nu(\hat{\mathbf{l}})$ as shown in Figure 1.2(d). Intuitively, a large λ makes the result lose detail.

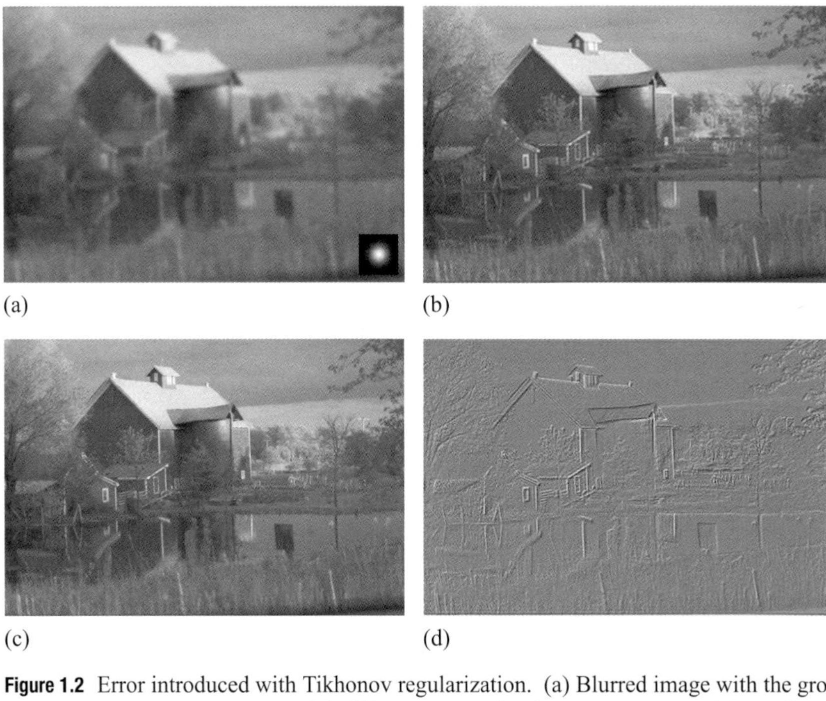

(a) (b) (c) (d)

Figure 1.2 Error introduced with Tikhonov regularization. (a) Blurred image with the ground truth PSF; (b) deblurred image with Tikhonov regularization; (c) ground truth latent image; (d) map of $\delta\mathbf{l}$ computed using Eq. (1.13).

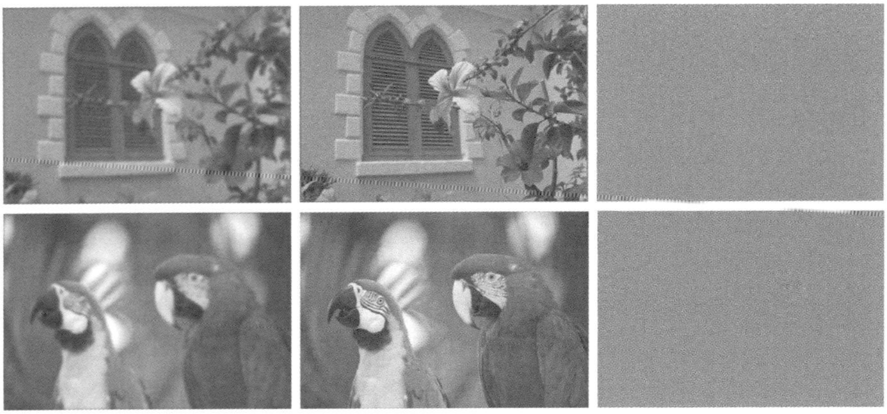

(a) (b) (c)

Figure 1.3 Influence of image noise in deconvolution using Tikhonov regularization. (a) Input images with additive noise; (b) deconvolution results; artifacts are primarily the amplified noise; (c) difference maps between (b) and the ground truth un-blurred noise-free images.

If we consider inevitable image noise and PSF error, the Tikhonov regularizer actually enhances the stability of deconvolution, discussed below.

Noise amplification

Now consider the case that image noise $\delta\mathbf{b}$ is present, which is common in natural images. With a derivation similar to Eq. (1.9), which takes derivatives and sets them to zero, the expression obtained is

$$\nu(\delta\mathbf{l}) = \frac{F^T\nu(\delta\mathbf{b})}{F^TF+\lambda\Lambda} + \frac{-\lambda\nu(\hat{\mathbf{l}})}{F^TF+\lambda\Lambda}. \tag{1.14}$$

We have explained the second term given the same expression in Eq. (1.13). It produces a map that in general contains high frequency structures.

In the first term, setting $\kappa = F^T/(F^TF+\lambda\Lambda)$ to represent a coefficient matrix, the expression simplifies to $\kappa\nu(\delta\mathbf{b})$. It actually functions as adding noise with a ratio κ, which yields noisy results. Summing up the effects of the two terms in Eq. (1.14), it can be concluded that the deconvolution results contain noise while lacking an amount of structural detail compared to the ground truth image. Two examples are shown in Figure 1.3.

Relation to Wiener deconvolution

Wiener filter is a method which has been widely used in non-blind deconvolution (Wiener 1949). Its specialty is in the use of image and noise power spectra to suppress noise, expressed as

$$\mathcal{F}(\mathbf{l}) = \frac{1}{\mathcal{F}(\mathbf{f})}\left(\frac{|\mathcal{F}(\mathbf{f})|^2}{|\mathcal{F}(\mathbf{f})|^2+\frac{1}{\mathrm{SNR}(\mathbf{f})}}\right)\cdot\mathcal{F}(\mathbf{b}), \tag{1.15}$$

where $\mathrm{SNR}(\mathbf{f})$ is the signal-to-noise ratio and $|\mathcal{F}(\mathbf{f})|^2$ denotes autocorrelation.

It can be proven that the Tikhonov regularized method is equivalent to the Wiener filter with a proper λ. First, Eq. (1.10) can be rewritten as

$$(F^T F + \lambda\Lambda)\nu(\mathbf{l}) = F^T \nu(\mathbf{b}). \tag{1.16}$$

Because it holds that

$$\begin{aligned} F\nu(\mathbf{l}) &= \nu(\mathbf{f} \otimes \mathbf{l}), \\ F^T\nu(\mathbf{l}) &= \nu(\mathbf{f} \oplus \mathbf{l}) = \nu(\mathbf{f}' \otimes \mathbf{l}), \\ \mathcal{F}(\mathbf{f}) \cdot \mathcal{F}(\mathbf{f}') &= |\mathcal{F}(\mathbf{f})|^2, \end{aligned}$$

where $\mathbf{f}'$ is the flipped version of $\mathbf{f}$, $\oplus$ denotes correlation, and $\cdot$ is an element-wise multiplication operator. Equation (1.16) finds the solution in image domain as

$$\nu(\mathbf{f}' \otimes (\mathbf{f} \otimes \mathbf{l})) + \lambda \mathbf{l} = \nu(\mathbf{f}' \otimes \mathbf{b}). \tag{1.17}$$

Taking the Fourier transform on both sides of Eq. (1.17) yields

$$\mathcal{F}(\mathbf{f}') \cdot \mathcal{F}(\mathbf{f}) \cdot \mathcal{F}(\mathbf{l}) + \lambda \mathcal{F}(\mathbf{l}) = \mathcal{F}(\mathbf{f}') \cdot \mathcal{F}(\mathbf{b}),$$

which can be further expressed as

$$\mathbf{l} = \mathcal{F}^{-1}\left(\frac{\mathcal{F}(\mathbf{f}') \cdot \mathcal{F}(\mathbf{b})}{\mathcal{F}(\mathbf{f}') \cdot \mathcal{F}(\mathbf{f}) + \lambda\Lambda}\right). \tag{1.18}$$

Equation (1.18) is the same as Eq. (1.15) when $\lambda\Lambda = 1/(\mathrm{SNR}(\mathbf{f}))$. This equivalence implies that Wiener deconvolution has similar noise amplification and structure information loss properties to Tikhonov regularized deconvolution.

1.1.2 Iterative approaches

Iterative computation was also used in several methods. The van Cittert (van Cittert 1931) solver can be applied to iteratively estimate the deconvolved image as

$$\mathbf{l}^{t+1} = \mathbf{l}^t + \beta(\mathbf{b} - \mathbf{l}^t \otimes \mathbf{f}), \tag{1.19}$$

where β is a parameter, adjustable automatically or manually, controlling the convergence speed while t and $t+1$ index iterations. Equation (1.19) converges ideally to a result close to that produced by the inverse filter expressed in Eq. (1.3), which does not incorporate any prior or regularization.

The widely employed Richardson–Lucy (RL) deconvolution (Richardson 1972, Lucy 1974) can be expressed as

$$\mathbf{l}^{t+1} = \mathbf{l}^t\left(\mathbf{f}' \otimes \left(\frac{\mathbf{b}}{\mathbf{l}^t \otimes \mathbf{f}}\right)\right), \tag{1.20}$$

where $\mathbf{f}'$ is the flipped version of $\mathbf{f}$, used in correlation instead of convolution. How Eq. (1.20) is constructed is explained in quite a number of papers and tutorials available online, and is thus omitted here. Different from direct inversion (Eq. (1.3)), RL deconvolution is iterative and can be stopped halfway, which *empirically* alleviates, in part, noise amplification. Performing it for many iterations or making it converge, contrarily, could yield less satisfactory results. The following derivation shows that the RL method is equivalent to the Poisson maximum likelihood, without imposing any image or kernel prior.

When assuming independent and identically distributed (i.i.d.) Gaussian noise $\mathbf{n} = \mathbf{b} - \mathbf{l} \otimes \mathbf{f}$, the maximum likelihood estimation of $\mathbf{l}$ is generally expressed as

$$p(\mathbf{b}|\mathbf{l}) \propto \prod_i \exp\left(-\frac{(\mathbf{b}_i - (\mathbf{l} \otimes \mathbf{f})_i)^2}{2\sigma^2}\right), \tag{1.21}$$

where $p(\mathbf{b}|\mathbf{l})$ is the conditional probability (also known as likelihood), i indexes pixels, and σ^2 is the Gaussian variance. Similarly, assuming that noise $\mathbf{n} = \mathbf{b} - \mathbf{l} \otimes \mathbf{f}$ follows a Poisson distribution, yields

$$p(\mathbf{b}|\mathbf{bl}) \propto \prod_i \frac{(\mathbf{l} \otimes \mathbf{f})_i^{\mathbf{b}_i} \exp(-(\mathbf{l} \otimes \mathbf{f})_i)}{\mathbf{b}_i!}, \tag{1.22}$$

where i indexes pixels. Its logarithmic energy is

$$\log\,(p(\mathbf{b}|\mathbf{l})) \propto \sum_i (\mathbf{b}_i \log\,(\mathbf{l} \otimes \mathbf{f})_i - (\mathbf{l} \otimes \mathbf{f})_i)\,, \tag{1.23}$$

where the constant $\mathbf{b}_i!$ term is omitted. Taking partial derivatives w.r.t. each pixel on the log energy and setting them to zero yields

$$\mathbf{f}' \otimes \left(\frac{\mathbf{b}}{\mathbf{l} \otimes \mathbf{f}} - 1\right) = 0. \tag{1.24}$$

Since $\mathbf{f}$ is a PSF, its elements sum to 1, making $\mathbf{f}' \otimes 1 = 1$. Thus Eq. (1.24) can be approximated by Richardson–Lucy deconvolution, in iterations, as

$$\mathbf{l}^{t+1} = \mathbf{l}^t \left(\mathbf{f}' \otimes \left(\frac{\mathbf{b}}{\mathbf{l}^t \otimes \mathbf{f}}\right)\right). \tag{1.25}$$

The above derivation shows that the RL method is equivalent to the Poisson maximum likelihood estimator in theory. Because there is no prior on the latent image $\mathbf{l}$, the algorithm should be stopped halfway to reduce noise and other visual artifacts. There has been research to improve RL. For example, Yuan, Sun, Quan & Shum (2008), in the multi-scale refinement scheme, applied edge-preserving bilateral filtering to the RL result. This nonlocal regularizer makes the iterative method a bit more robust against noise.

1.1.3 Recent advancements

Effective non-blind deconvolution needs to deal with noise and suppress ringing artifacts introduced by incorrect blur kernel estimates and sometimes by compression or tone management in image formation. Understanding these issues has led to better means of *regularizing* the deconvolution process in recent years, giving prior E_{prior} (denoted as $\Psi(\mathbf{l})$) a number of new forms. A general principle is that the prior should not penalize excessively estimation outliers – in order not to wrongly deviate from

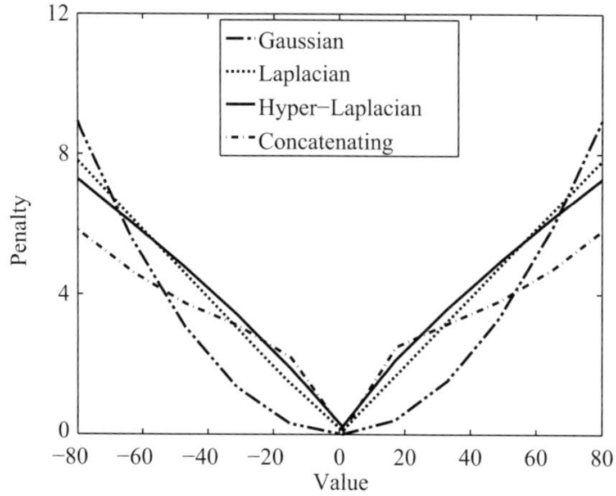

Figure 1.4 Illustration of different regularization terms. Different prior functions penalize values differently. The Gaussian prior increases energy most quickly for large absolute values.

final results. In what follows, without special mention, the overall objective function is still the one expressed in Eq. (1.6):

$$\min_{\mathbf{I}} \|\mathbf{I} \otimes \mathbf{f} - \mathbf{b}\|^2 + \lambda \Psi(\mathbf{I}). \tag{1.26}$$

Chan & Wong (1998) used a total variation regularizer, which is also known as a Laplacian prior, by setting

$$\Psi(\mathbf{I}) = \|\nabla \mathbf{I}\|^1, \tag{1.27}$$

where ∇ denotes the first-order derivative operator, i.e. $\nabla \mathbf{I} = (\partial_x \mathbf{I}, \partial_y \mathbf{I})$, a concatenation of the two gradient images. $\| \cdot \|^1$ is the L_1-norm operator for all image gradients. This prior, illustrated in Figure 1.4 by solid lines, has a stronger effect on reducing the influence of large errors than the Gaussian prior used in Eq. (1.8) (and shown as a dash–dot curve in Figure 1.4).

There are other ways to define $\Psi(\mathbf{I})$. Shan, Jia & Agarwala (2008) constructed a natural prior for the latent image by concatenating two piecewise continuous convex functions, plotted as the concatenating curve in Figure 1.4. The expression is

$$\Psi(\mathbf{I}_i) = \begin{cases} a|\nabla \mathbf{I}_i| & |\nabla \mathbf{I}_i| \leqslant \xi \\ b(\nabla \mathbf{I}_i)^2 + c & |\nabla \mathbf{I}_i| > \xi \end{cases} \tag{1.28}$$

where i indexes pixels and $\nabla \mathbf{I}_i$ represents a partial derivative for $\mathbf{I}_i$ in either the x or y direction. ξ is the value on which the linear and quadratic functions are concatenated. a, b, and c are three parameters. $\Psi(\mathbf{I})$ can actually be used to approximate natural image statistics when a is large and b is very small.

Table 1.1 Comparison of a few non-blind deconvolution methods with respect to the employed likelihood and prior.

	Likelihood	Prior
Wiener (Wiener 1949)	Gaussian	Gaussian
L-S[a] (Tikhonov *et al.* 1977)	Gaussian	Gaussian
RL (Richardson 1972, Lucy 1974)	Poisson	–
Chan & Wong 1998	Gaussian	Laplacian
Wang *et al.* 2008	Gaussian	Laplacian
Shan *et al.* 2008	Gaussian	Concatenating
Yuan *et al.* 2008	Poisson	Nonlocal (bilateral)
Krishnan & Fergus 2009	Gaussian	Hyper-Laplacian
Yang *et al.* 2009	Laplacian	Laplacian
Xu & Jia 2010	Laplacian	Laplacian

[a]L-S is an abbreviation for least square.

To make the resulting structure less smooth, Levin, Fergus, Durand & Freeman (2007) suggested a hyper-Laplacian prior, written as

$$\Psi(\mathbf{l}) = \|\nabla \mathbf{l}\|^{\alpha}, \tag{1.29}$$

where $\alpha < 1$, representing a norm corresponding to a sparser distribution.

Additionally, the methods of Yang, Zhang & Yin (2009) and Xu & Jia (2010) suppress noise via a TV-L_1 (total variation) objective, which uses the Laplacian data term, i.e. $E_{\text{data}} = \|\mathbf{l} \otimes \mathbf{f} - \mathbf{b}\|^1$, so the objective function can then be expressed as

$$\min_{\mathbf{l}} \|\mathbf{l} \otimes \mathbf{f} - \mathbf{b}\|^1 + \|\nabla \mathbf{l}\|^1. \tag{1.30}$$

This function can suppress strong Gaussian and, particularly, impulse image noise with robust constraint on the data term.

The likelihood and prior forms of different recent methods are listed in Table 1.1. Gaussian likelihood and Laplacian prior are most frequently used thanks to their simplicity in expression and reasonable ability to resist noise and error.

Albeit not quadratic, objective functions incorporating the Laplacian prior in Eq. (1.27), the concatenating term in Eq. (1.28), the hyper-Laplacian prior in Eq. (1.29), and the robust data term in Eq. (1.30), as a TV-L_1 energy can be solved efficiently through half-quadratic splitting, which decomposes the original problem into a single-variable quadratic minimization process. Details are provided in Section 1.1.4.

1.1.4 Variable splitting solver

An effective scheme to solve sparsely constrained non-blind deconvolution is variable splitting, implemented by half-quadratic penalty methods (Geman & Reynolds 1992, Geman & Yang 1995). This scheme has been used in many recent methods (Shan *et al.* 2008, Wang, Yang, Yin & Zhang 2008, Krishnan & Fergus 2009, Xu & Jia 2010). In what follows, we discuss the half-quadratic penalty solver for minimizing

$$E = \|\mathbf{l} \otimes \mathbf{f} - \mathbf{b}\|^2 + \lambda \|\nabla \mathbf{l}\|^{\alpha}, \tag{1.31}$$

with a (hyper) Laplacian prior, where $0.5 \leqslant \alpha \leqslant 1$. Objective functions with the concatenating prior expressed in Eq. (1.28) and the TV-L_1 function in Eq. (1.30) can be solved similarly.

The basic idea is to separate variables involved in convolution from those in other terms, so that they can be estimated quickly and reliably using Fourier transforms. This is realized by using a set of auxiliary variables $\psi = (\psi_x, \psi_y)$ for $\nabla \mathbf{I} = (\partial_x \mathbf{I}, \partial_y \mathbf{I})$, and adding the extra condition $\psi \approx \nabla \mathbf{I}$. Eq. (1.31) is accordingly updated to

$$E_{\mathbf{L}} = \|\mathbf{I} \otimes \mathbf{f} - \mathbf{b}\|^2 + \lambda\|\psi_x\|^\alpha + \lambda\|\psi_y\|^\alpha + \gamma\|\psi_x - \partial_x \mathbf{I}\|^2 + \gamma\|\psi_y - \partial_y \mathbf{I}\|^2, \tag{1.32}$$

given the isotropic implementation of the α-norm. γ is a weight. When its value is infinitely large, the desired conditions $\psi_x = \partial_x \mathbf{I}$ and $\psi_y = \partial_y \mathbf{I}$ can be satisfied. In this case, minimizing $E_{\mathbf{L}}$ converges to minimizing E.

Given this variable substitution, it is possible now to iterate between optimizing ψ and $\mathbf{I}$. This process is efficient and is able to converge to an optimal point, since, in each iteration, the global optimum of ψ is reached in a closed form, while a fast Fourier transform can be used to update $\mathbf{I}$.

Updating ψ

With an estimated $\mathbf{I}$ from the previous pass, Eq. (1.32) is simplified to

$$E'_\psi = \lambda\|\psi_x\|^\alpha + \lambda\|\psi_y\|^\alpha + \gamma\|\psi_x - \partial_x \mathbf{I}\|^2 + \gamma\|\psi_y - \partial_y \mathbf{I}\|^2. \tag{1.33}$$

With a few algebraic operations to decompose ψ into the set containing all elements $\psi_{i,x}$ and $\psi_{i,y}$ corresponding to all pixels i, E'_ψ can be written as a sum of sub-energy terms

$$E'_\psi = \sum_i \left(E'_{\psi_{i,x}} + E'_{\psi_{i,y}}\right), \tag{1.34}$$

where each $E'_{\psi_{i,\upsilon}}$, $\upsilon \in \{x, y\}$ only contains a single variable $\psi_{i,\upsilon} \in \psi_\upsilon$, given by

$$E'_{\psi_{i,\upsilon}} = \lambda|\psi_{i,\upsilon}|^\alpha + \gamma(\psi_{i,\upsilon} - \partial_\upsilon \mathbf{I}_i)^2, \tag{1.35}$$

where $\mathbf{I}_i$ is pixel i in $\mathbf{I}$. Each $E'_{\psi_{i,\upsilon}}$ contains only one variable $\psi_{i,\upsilon}$ so it can be optimized independently. For any α smaller than 1, minimizing Eq. (1.35) depends on two variables, i.e. joint weight γ/λ and image-dependent $\partial_\upsilon \mathbf{I}_i$. By sampling values from them, a 2D lookup table can be constructed offline, from which optimal results can be obtained efficiently. Possible errors caused by the discrepancy of actual values and nearest samples are controllable (Krishnan & Fergus 2009). For the special cases where $\alpha = 1/2$, $\alpha = 2/3$, and $\alpha = 1$, analytic solutions are available. We discuss the case $\alpha = 1$, where $\psi_{i,\upsilon}$ is expressed as

$$\psi_{i,\upsilon} = \text{sign}(\partial_\upsilon \mathbf{I}_i)\max\left\{|\partial_\upsilon \mathbf{I}_i| - \frac{\lambda}{2\gamma}, 0\right\}, \tag{1.36}$$

which is the 1D shrinkage formula (Wang *et al.* 2008). Its computation is efficient.

Updating l

With Ψ estimated above, $\mathbf{l}$ is updated by minimizing

$$E'_{\mathbf{l}} = \|\mathbf{l} \otimes \mathbf{f} - \mathbf{b}\|^2 + \gamma \|\psi_x - \partial_x \mathbf{l}\|^2 + \gamma \|\psi_y - \partial_y \mathbf{l}\|^2. \tag{1.37}$$

Since the major computation is on convolution, frequency-domain operation is applied. Denoting the Fourier transform operator and its inverse as $\mathcal{F}$ and $\mathcal{F}^{-1}$, respectively, $E'_{\mathbf{l}}$ is updated to

$$E'_{\mathcal{F}(\mathbf{l})} = \|\mathcal{F}(\mathbf{l}) \cdot \mathcal{F}(\mathbf{f}) - \mathcal{F}(\mathbf{b})\|^2 + \gamma \|\mathcal{F}(\psi_x) - \mathcal{F}(\mathbf{l}) \cdot \mathcal{F}(\partial_x)\|_2^2$$
$$+ \gamma \|\mathcal{F}(\psi_y) - \mathcal{F}(\mathbf{l}) \cdot \mathcal{F}(\partial_y)\|_2^2, \tag{1.38}$$

where $\mathcal{F}(\partial_\upsilon)$, $\upsilon \in \{x, y\}$, is the filter ∂_υ in the frequency domain. It can be obtained by the Matlab function "psf2otf".

According to Plancherel's theorem in harmonic analysis, which states that the sum of the square of a function is equal to the sum of the square of its Fourier transform, the energy equivalence $E'_{\mathbf{l}} = E'_{\mathcal{F}(\mathbf{l})}$ can be established for all possible values of $\mathbf{l}$. It further follows that the optimal $\mathbf{l}^*$ that minimizes $E'_{\mathbf{l}}$ corresponds to the counterpart $\mathcal{F}(\mathbf{l}^*)$ in the frequency domain that minimizes $E'_{\mathcal{F}(\mathbf{l})}$:

$$E'_{\mathbf{l}}|_{\mathbf{l}^*} = E'_{\mathcal{F}(\mathbf{l})}|_{\mathcal{F}(\mathbf{l}^*)}. \tag{1.39}$$

Accordingly, the optimal $\mathbf{l}^*$ is given by

$$\mathbf{l}^* = \mathcal{F}^{-1}\left(\arg\min_{\mathcal{F}(\mathbf{l})} E'_{\mathcal{F}(\mathbf{l})}\right). \tag{1.40}$$

Since $E'_{\mathcal{F}(\mathbf{l})}$ is a sum of quadratic energies of unknown $\mathcal{F}(\mathbf{l})$, it is a convex function and can be solved by simply setting the partial derivatives $\partial E'_{\mathcal{F}(\mathbf{l})}/\partial \mathcal{F}(\mathbf{l})$ to zero. The solution of $\mathbf{l}^*$ can be expressed as

$$\mathbf{l}^* = \mathcal{F}^{-1}\left(\frac{\overline{\mathcal{F}(\mathbf{f})} \cdot \mathcal{F}(\mathbf{b}) + \gamma \overline{\mathcal{F}(\partial_x)} \cdot \mathcal{F}(\psi_x) + \gamma \overline{\mathcal{F}(\partial_y)} \cdot \mathcal{F}(\psi_y)}{\overline{\mathcal{F}(\mathbf{f})} \cdot \mathcal{F}(\mathbf{f}) + \gamma \overline{\mathcal{F}(\partial_x)} \cdot \mathcal{F}(\partial_x) + \gamma \overline{\mathcal{F}(\partial_y)} \cdot \mathcal{F}(\partial_y)}\right), \tag{1.41}$$

where $\overline{(\cdot)}$ denotes the conjugate operator. The division is an element-wise one.

The above two steps, respectively, update ψ and $\mathbf{l}$ until convergence. Note that γ in Eq. (1.32) controls how strongly ψ is constrained to be similar to $\nabla \mathbf{l}$, and its value can be set with the following consideration. If γ is too large initially, the convergence is quite slow. On the other hand, if γ is overly small before convergence, the optimal solution of Eq. (1.32) must not be the same as that of Eq. (1.31). A general rule (Shan *et al.* 2008, Wang *et al.* 2008) is to adaptively adjust γ in iterations. In the early stages, γ is set small to stimulate significant gain for each step. Its value increases in every or every few iterations, making ψ gradually approach $\nabla \mathbf{l}$. γ should be sufficiently large at convergence.

1.1.5 A few results

We show a few examples to visually compare the aforementioned methods and models. In Figure 1.5, results from approaches incorporating different prior terms are shown.

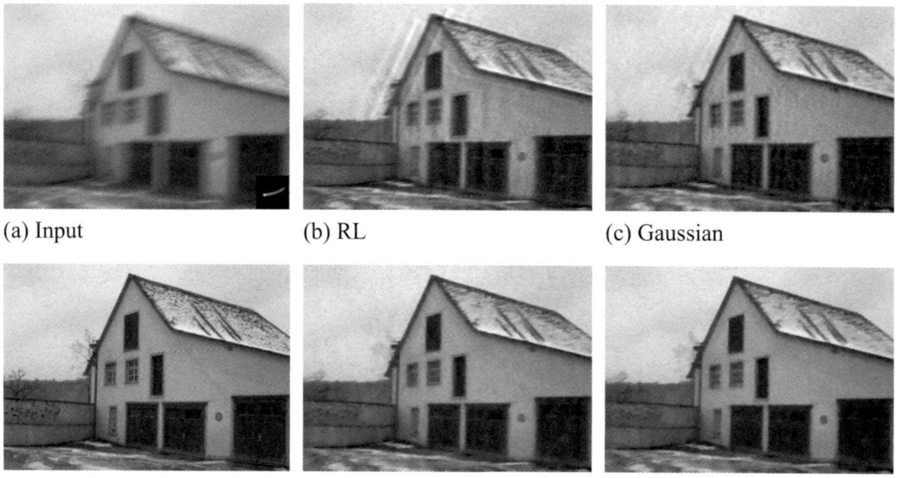

(a) Input (b) RL (c) Gaussian

(d) Ground truth (e) Hyper-Laplacian (f) Laplacian

Figure 1.5 Non-blind deconvolution results. Visual comparison of non-blind deconvolution methods with the data term defined in Eq. (1.5) and priors set as none (Richardson–Lucy), Gaussian, ground truth, hyper-Laplacian and Laplacian, respectively.

The input blurred image and PSF contain noise, making RL and the Gaussian prior method produce a level of ringing artifacts and image noise. Laplacian and hyper-Laplacian priors, in comparison, perform better in terms of robustness against these problems. Based on publicly available executables or codes, we deconvolve another input image in Figure 1.6(a), similarly containing noise and PSF errors. The results shown in (c), (e) and (f) are visually pleasing. Laplacian and hyper-Laplacian priors used to produce the results are effective in suppressing a medium level of image noise.

In terms of computation complexity, Wiener deconvolution involves the simplest operation and thus runs fastest. Methods incorporating concatenating and Laplacian priors can produce higher quality results; their corresponding algorithm is also efficient when written in optimized C. The method of Krishnan & Fergus (2009) makes use of a lookup table, which is constructed offline with respect to parameters. This table speeds up the solver considerably.

1.2 Blind deconvolution

Blind deconvolution solves shift-invariant (uniform) motion deblurring

$$\mathbf{b} = \mathbf{l} \otimes \mathbf{f} + \mathbf{n} \tag{1.42}$$

by estimating both $\mathbf{f}$ and $\mathbf{l}$. $\mathbf{n}$ represents inevitable additive noise.

There have been many blind deconvolution methods. Approaches proposed before 2005 mainly used the strategy to estimate the blur PSF and the latent image separately, which results in alternating optimization. For example, Ayers & Dainty (1988) iterated between updating the blur PSF and the latent image in a style similar to the Wiener filter;

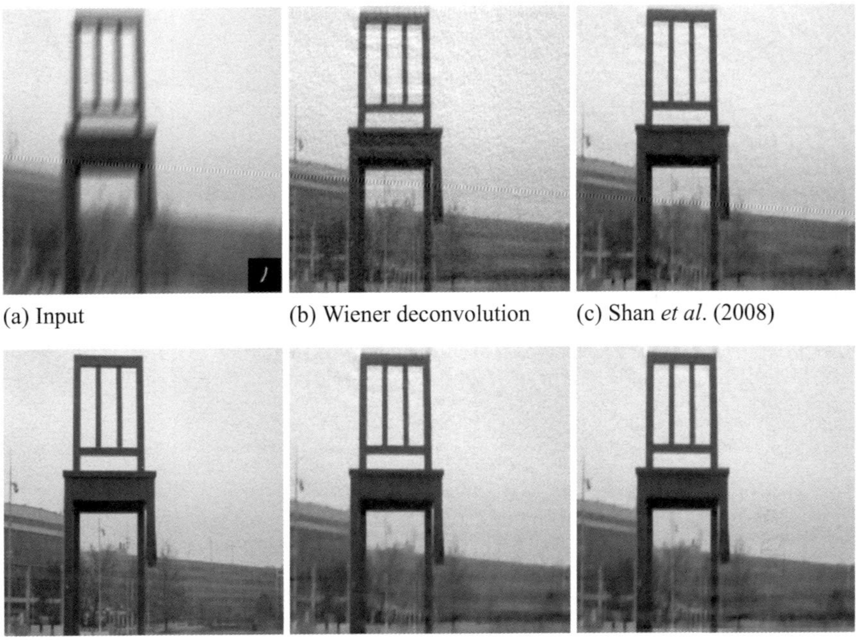

(a) Input (b) Wiener deconvolution (c) Shan *et al.* (2008)

(d) Ground truth (e) Krishnan & Fergus (2009) (f) Xu & Jia (2010)

Figure 1.6 Non-blind deconvolution results. Visual comparison of results produced by a few efficient non-blind deconvolution methods.

Fish, Brinicombe, Pike & Walker (1995) performed blind deconvolution in a maximum likelihood fashion, using the Richardson–Lucy iteration; Chan & Wong (1998) applied the total variation regularizer to both the PSF and the image. These methods are not elaborated here because they have respective limitations in handling natural image blur, especially when noise and complex-structure PSFs are present. The remainder of this chapter will focus on recent understanding and more advanced developments of models and solvers.

The major difficulty for successful natural image motion blur blind deconvolution is due to the high dimension of solution space. Any PSF $\mathbf{f}$ can be fitted into Eq. (1.42) to find a corresponding $\mathbf{l}$ and $\mathbf{n}$, making it challenging to define proper criteria for optimization. Figure 1.7 shows an example. Two solutions in the two rows on the righthand side indicate huge ambiguity for PSF and image estimation. A small change in estimation steps could significantly deviate the solution.

Modern objective functions can generally be expressed as

$$\min_{\mathbf{l},\mathbf{f}} \Phi(\mathbf{l} \otimes \mathbf{f} - \mathbf{b}) + \lambda_1 \Psi(\mathbf{l}) + \lambda_2 \Upsilon(\mathbf{f}), \tag{1.43}$$

similar to the one shown in Eq. (1.6), where λ_1 and λ_2 are two weights. Φ, Ψ, and Υ are different functions to constrain noise, latent image, and PSF, respectively. Among them, Φ and Ψ can use the same expression introduced earlier in non-blind deconvolution. Generally, $\Phi(\mathbf{l} \otimes \mathbf{f} - \mathbf{b})$ is set to $\|\mathbf{l} \otimes \mathbf{f} - \mathbf{b}\|^2$ (or $\|\nabla \mathbf{l} \otimes \mathbf{f} - \nabla \mathbf{b}\|^2$), which is a quadratic

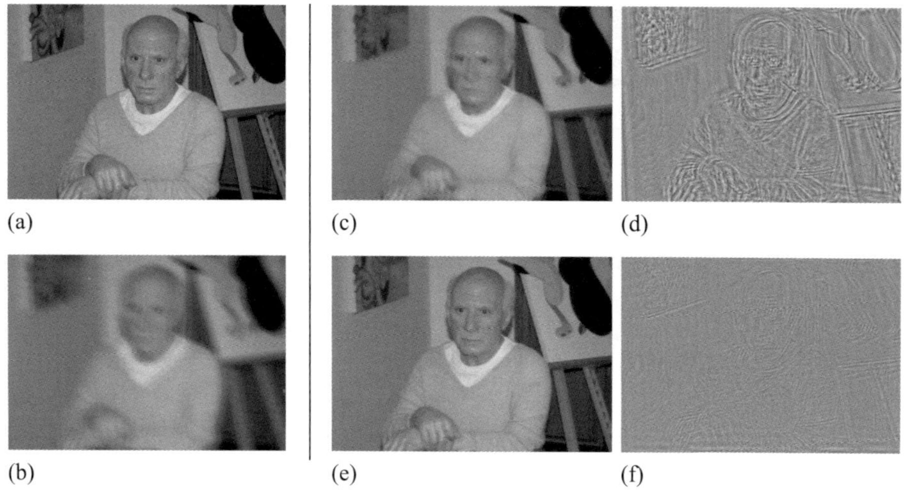

Figure 1.7 Ambiguity of solution for blind deconvolution. (a) Ground truth latent image; (b) blurred input; (c–d) one latent image estimate and the corresponding noise map $\mathbf{n}$; (e–f) another latent image result and corresponding noise map. (d) and (f) are normalized for visualization.

cost on pixel values (or their derivatives). $\Psi(\mathbf{l})$ can be set the same way as Eqs. (1.27)–(1.29) to follow sparse gradient distributions. The new Υ is ideally a sparse function since a motion PSF tends to have most elements close to zero. Its L_1-norm form is

$$\Upsilon(\mathbf{f}) = \|\mathbf{f}\|^1.$$

With these three functions, the objective function can be written as

$$\min_{\mathbf{l},\mathbf{f}} \|\mathbf{l} \otimes \mathbf{f} - \mathbf{b}\|^2 + \lambda_1 \|\nabla \mathbf{l}\|^\alpha + \lambda_2 \|\mathbf{f}\|^1, \tag{1.44}$$

where α set to 2, 1, and a value between 0 and 1 correspond to quadratic, Laplacian, and hyper-Laplacian functions, respectively. Note that different methods may alter these terms or use extra ones in the above objective, but, overall, the constraints are enough for blind deconvolution.

This objective also corresponds to a posterior probability

$$\begin{aligned} p(\mathbf{l}, \mathbf{f}|\mathbf{b}) &\propto p(\mathbf{b}|\mathbf{l}, \mathbf{f}) p(\mathbf{l}) p(\mathbf{f}), \\ &\propto \exp(-\Phi(\mathbf{l} \otimes \mathbf{f} - \mathbf{b})) \cdot \exp(-\lambda_1 \Psi(\mathbf{l})) \cdot \exp(-\lambda_2 \Upsilon(\mathbf{f})), \end{aligned} \tag{1.45}$$

in the probability framework (Shan *et al.* 2008, Fergus, Singh, Hertzmann, Roweis & Freeman 2006).

Solving Eq. (1.44) by simply estimating the PSF and the latent image iteratively cannot produce correct results. Trivial solutions or local-minima can, contrarily, be obtained. The *trivial* solution is the delta-function PSF, which contains a one in the center and zeros for all other elements, and exactly the blurred image as the latent image estimate (Levin, Weiss, Durand & Freeman 2009). Without any deblurring, the resulting

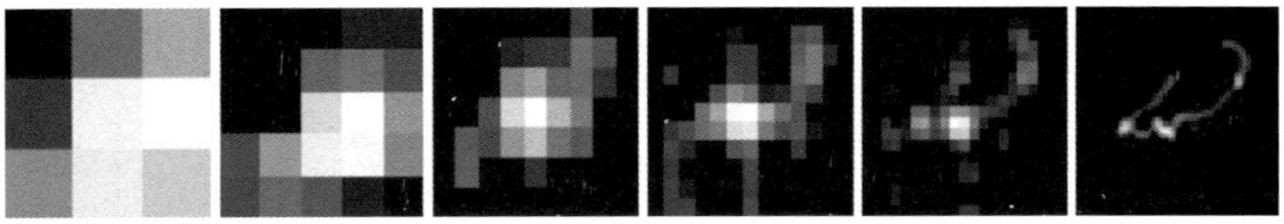

Figure 1.8 Coarse-to-fine PSF estimation in several levels.

energy in Eq. (1.44) could be even lower than that with correct deblurring for many images. In addition, simple iterative optimization is easily stuck in poor local minima.

To tackle the blind deconvolution problem, there are mainly two streams of research work for full posterior distribution approximation $p(\mathbf{l}, \mathbf{f}|\mathbf{b})$, using (i) maximum marginal probability estimation, and (ii) energy minimization directly in Eq. (1.44). Most existing methods estimate PSFs in a multi-scale framework where the PSF is first estimated on the small-resolution image in an image pyramid. The estimate is then propagated to the next level as the initialization to refine the result in a higher resolution. This process repeats for a few passes, which improves numerical stability, avoids many local minima, and even saves computation by reducing the total number of iterations. An illustration of multi-level PSFs is shown in Figure 1.8. The following discussion is based on an estimation in one image level.

1.2.1 Maximum marginal probability estimation

Theoretically, the blur PSF can be perfectly obtained by maximizing the following marginalized probability, expressed as

$$p(\mathbf{f}|\mathbf{b}) = \int p(\mathbf{l}, \mathbf{f}|\mathbf{b})\mathrm{d}\mathbf{l}, \tag{1.46}$$

where $p(\mathbf{l}, \mathbf{f}|\mathbf{b})$ is the full posterior distribution defined in Eq. (1.45). Empirically, a huge difficulty exists regarding computational tractability of integration on the latent image $\mathbf{l}$. Even if one treats the latent image as discrete, marginalization still involves summing all possible image values, which is prohibitively expensive.

To address this problem, Fergus *et al.* (2006) approximated the posterior distribution using parametric factorization, written as

$$p(\mathbf{l}, \mathbf{f}|\mathbf{b}) \approx q(\mathbf{f}, \mathbf{l}) = q(\mathbf{f})q(\mathbf{l}) \tag{1.47}$$

$$= \prod_i q(\mathbf{f}_i) \prod_j q(\mathbf{l}_j), \tag{1.48}$$

where the first equation (1.47) assumes independence between $\mathbf{f}$ and $\mathbf{l}$, making marginalization, i.e. $\int q(\mathbf{f}, \mathbf{l})\mathrm{d}\mathbf{l} = q(\mathbf{f})$, valid. The second equation (1.48) assumes pixel independence, to ease parameter estimation. Also, by assuming Gaussian distributions for both the kernel and latent images, the optimal kernel that maximizes the marginalized probability becomes the mean of the Gaussian distribution, i.e. $\mathbf{f}_i^* = \mathrm{E}_{q(\mathbf{f}_i)}(\mathbf{f}_i)$, where $\mathrm{E}_{q(\mathbf{f}_i)}$ is the expectation w.r.t. the distribution $q(\mathbf{f}_i)$. This process corresponds to the *mean field* approach.

The final approximated distributions, in this case, are Gaussian moments, obtained by minimizing a function representing the Kullback–Leibler divergence $\mathrm{KL}(q(\mathbf{l},\mathbf{f})\,\|\, p(\mathbf{l},\mathbf{f}|\mathbf{b}))$ between the approximating distribution and the true posterior, following the variational Bayesian framework (Jordan, Ghahramani, Jaakkola & Saul 1999, Miskin & MacKay 2000). More details of this approach, e.g. the use of gradients and unknown noise variance, can be found in the original papers. Note that iteratively minimizing the KL divergence cost function is very time consuming. For reference, the publicly available Matlab code takes 10 minutes to process a 255×255 image on a PC with an Intel i3 2.13 GHz CPU.

Following this line, Levin, Weiss, Durand & Freeman (2011) proposed approximating the conditional distribution $p(\mathbf{l}|\mathbf{b},\mathbf{f}) \approx q(\mathbf{l})$ instead of the joint distribution $p(\mathbf{l},\mathbf{f}|\mathbf{b})$. It leads to an expectation–maximization (EM) framework that treats $\mathbf{l}$ as a latent variable and computes expectation on it instead of integration over all possible configurations. The M-step minimizes the log-likelihood

$$\mathrm{E}_{q(\mathbf{l})}(-\ln p(\mathbf{l},\mathbf{b}|\mathbf{f})) = \mathrm{E}_{q(\mathbf{l})}(\|\mathbf{l}\otimes\mathbf{f}-\mathbf{b}\,\|^2). \tag{1.49}$$

The M-step leads to quadratic programming and can be efficiently solved using frequency-domain acceleration.

The E-step, which uses $q(\mathbf{l})$ to approximate the conditional distribution, is analogous to minimization of KL divergence in Fergus *et al.* (2006). If a Gaussian prior on the latent image is imposed, the E-step $q(\mathbf{l}) = p(\mathbf{l}|\mathbf{b},\mathbf{f})$ has a closed-form solution. Another difference, compared to Fergus *et al.* (2006), is that instead of considering the distribution of $\mathbf{f}$ (i.e. $q(\mathbf{f})$), Levin *et al.* (2011) counted in only a single $\mathbf{f}$ estimation in the M-step – which also makes it more efficient than the maximum marginal probability implementation. It reduces the running time to around 1.2 minutes for a 255×255 image based on the author-released Matlab code. Approximating $p(\mathbf{l}|\mathbf{b},\mathbf{f})$ with the general sparse image priors is still costly, especially when compared to methods employing explicit edge recovery or prediction, discussed next.

1.2.2 Alternating energy minimization

Energy minimization from Eq. (1.44) is another common way for uniform blind deconvolution. It has achieved great success based on a few milestone techniques proposed in recent years. Alternating minimization can now be applied to many natural images that are blurred with very large PSFs and/or with significant noise. The process is also efficient. For example, top performing methods (Cho & Lee 2009, Xu & Jia 2010, Xu, Zheng & Jia 2013) written in optimized C^{++}, or even Matlab, take around 5 seconds to process an 800×800 image. Additionally, this set of methods is flexibly expandable, and has been employed as key steps in many non-uniform (spatially-variant) motion deblurring approaches.

The most important empirical strategy to make the solver avoid the trivial solution is to generate an intermediate sharp-edge representation. This idea was introduced by Jia (2007), who selected object boundaries for transparency estimation and performed PSF estimation only in these regions. It is based on the observation that an opaque object

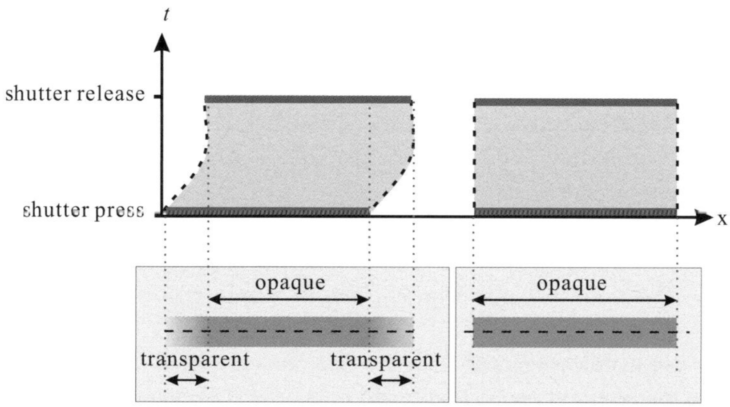

Figure 1.9 1D example of transparency related to motion blur. 1D example showing how transparency is produced on a motion-blurred object. Left: motion-blurred bar, whose two ends are blended to the background and cause semi-transparency. Right: binary transparency map without object motion blur.

with sharp edges has its boundary blended into the background after motion blur, as illustrated in Figure 1.9.

With this finding, the original energy function (1.44) can be updated to estimation of the PSF and transparency map, instead of the latent natural image. The transparency value for a blurred pixel is denoted as α^i. It ranges in [0, 1]. Its latent un-blurred value is α^o. Ideally for solid objects, the α^o map is a binary one, i.e. $\alpha^o(i) = \{0, 1\}$ for any pixel i. These variables update the original convolution model (1.42) to

$$\alpha^{\mathbf{i}} = \alpha^{\mathbf{o}} \otimes \mathbf{f} + \mathbf{n}. \tag{1.50}$$

The corresponding objective function is updated too. Note that this model does not cause the trivial solution problem. Thanks to value binarization in the latent transparency map α^o, direct optimization can lead to satisfactory results if the input transparency values α^i are accurate enough.

Later on, Joshi, Szeliski & Kriegman (2008), instead of generating the transparency map, directly detected edges and predicted step ones. These pixels are used to guide PSF estimation, also avoiding the trivial solution.

1.2.3 Implicit edge recovery

Following this line, several other methods also implicitly or explicitly predict edges from the blurred image to guide PSF estimation. A general procedure employed in Shan *et al.* (2008) iterates between PSF estimation and latent image recovery. PSF estimation is achieved by converting Eq. (1.44) to

$$\min_{\mathbf{f}} \|\mathbf{l} \otimes \mathbf{f} - \mathbf{b}\|^2 + \lambda_2 \|\mathbf{f}\|^1. \tag{1.51}$$

Latent image $\mathbf{l}$ estimation, accordingly, is obtained by a non-blind deconvolution process, expressed as

$$\min_{\mathbf{l}} \|\mathbf{l} \otimes \mathbf{f} - \mathbf{b}\|^2 + \lambda_1 \|\nabla \mathbf{l}\|^1. \tag{1.52}$$

Its solver has been presented in Section 1.1.4. Equations (1.51) and (1.52) iterate until convergence.

To solve Eq. (1.51), writing it as matrix multiplication yields

$$\min_{\nu(\mathbf{f})} \|\mathrm{A}\nu(\mathbf{f}) - \nu(\mathbf{b})\|^2 + \lambda_2 \|\nu(\mathbf{f})\|^1, \tag{1.53}$$

where A is a matrix computed from the convolution operator whose elements depend on the estimated latent image $\mathbf{l}$. $\nu(\mathbf{f})$ and $\nu(\mathbf{b})$ are the vectorized $\mathbf{f}$ and $\mathbf{b}$, respectively. Equation (1.53) is in a standard L_1-regularized minimization form, and can be solved by transforming optimization to its dual problem and computing the solution via an interior point method (Kim, Koh, Lustig & Boyd 2007) or by iterative reweighted least squares (IRLS).

This algorithm can produce reasonable kernel estimates and avoids the trivial solution because it adopts a special mechanism to set parameters λ_1 and λ_2 in Eqs. (1.51) and (1.52), which control how strong image and kernel regularization are. At the beginning of blind image deconvolution, the input kernel is not accurate; the weight λ_1 is therefore set large, encouraging the system to produce an initial latent image with mainly strong edges and few ringing artifacts, as shown in Figure 1.10(b). This also helps guide PSF estimation in the following steps to eschew the trivial delta kernel. Then, after each iteration of optimization, the λ values decrease to reduce the influence of regularization on the latent image and kernel estimate, allowing for the recovery of more details. Figure 1.10 shows intermediate results produced during this process, where the PSF is gradually shaped and image details are enhanced in iterations.

Normalized L_1 regularization

An algorithm similar to that of Shan *et al.* (2008) was later proposed by Krishnan, Tay & Fergus (2011). It incorporates a normalized L_1 regularization term on image gradients, written as $\|\nabla \mathbf{l}\|^1/\|\nabla \mathbf{l}\|^2$, where $\nabla \mathbf{l}$ denotes gradients of $\mathbf{l}$. Normalized L_1 modifies traditional L_1 regularization $\|\nabla \mathbf{l}\|^1$ by weight $1/\|\nabla \mathbf{l}\|^2$, which makes the resulting $\|\nabla \mathbf{l}\|^1/\|\nabla \mathbf{l}\|^2$ value generally smaller than that of $\|\nabla \mathbf{b}\|^1/\|\nabla \mathbf{b}\|^2$. This means the trivial blurred image solution is not favored by regularization. In this algorithm, blind deconvolution can be achieved by iteratively solving

$$\min_{\nabla \mathbf{l}} \|\nabla \mathbf{l} \otimes \mathbf{f} - \nabla \mathbf{b}\|^2 + \lambda_3 \frac{\|\nabla \mathbf{l}\|^1}{\|\nabla \mathbf{l}\|^2}, \tag{1.54}$$

and

$$\min_{\mathbf{f}} \|\nabla \mathbf{l} \otimes \mathbf{f} - \nabla \mathbf{b}\|^2 + \lambda_4 \|\mathbf{f}\|^1, \tag{1.55}$$

where λ_3 and λ_4 are two weights. Because $\frac{\lambda_3}{\|\nabla \mathbf{l}\|^2}$ in Eq. (1.54) is, in fact, a weight in each iteration, its function is similar to λ_1 in Eq. (1.52). Both weights decrease during

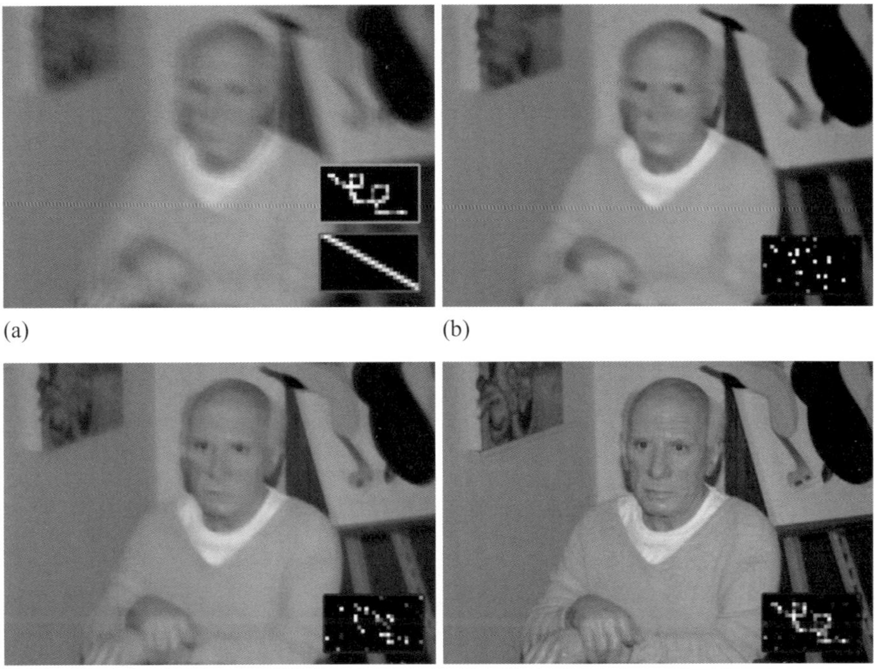

(a) (b)

(c) (d)

Figure 1.10 Illustration of optimization in iterations. (a) Blurred image. The ground truth blur kernel and simple initial kernel are shown in the two rectangles; (b–d) restored images and kernels in three iterations.

iterations to accommodate more and more details in PSF estimation, which guide blind deconvolution and avoid the trivial delta kernel solution.

1.2.4 Explicit edge prediction for very large PSF estimation

Explicit edge prediction was developed and used in Money & Kang (2008), Cho & Lee (2009), and Xu & Jia (2010), and along with shock filters in Osher & Rudin (1990). It directly restores strong edges from intermediate latent image estimates in iterations. Shock filters perform iteratively. Given an image I, in pass $t+1$, the shock filtered result $\tilde{\mathrm{I}}^{t+1}$ is expressed as

$$\tilde{\mathrm{I}}^{t+1} = \tilde{\mathrm{I}}^{t} - \mathrm{sign}(\Delta\tilde{\mathrm{I}}^{t})|\nabla\tilde{\mathrm{I}}^{t}|, \tag{1.56}$$

where Δ and ∇ are the Laplacian and gradient operators.

A shock filter can be used in iterative blind deconvolution. It produces step-like edges from intermediate latent image estimates produced in each iteration. After removing small-magnitude edges by thresholding the gradients, only a few of the strongest edges are kept, as illustrated in Figure 1.11(b). This thresholded edge map $\tilde{\mathrm{I}}$ then substitutes for $\mathbf{I}$ in Eq. (1.51) for PSF estimation in the next iteration.

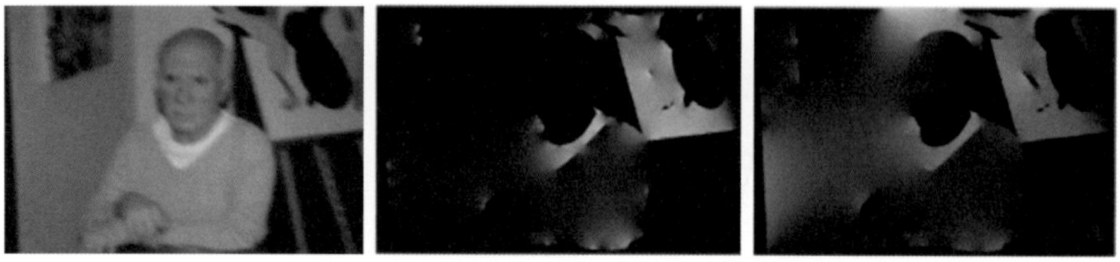

(a) Blurred image $\mathbf{b}$ (b) Thresholded $\tilde{\mathrm{I}}$ (c) Thresholded $\tilde{\mathrm{I}}$

Figure 1.11 Shock filter. Illustration of shock filter. Given the input image (a), optimization is performed in iterations. (b) and (c) are generated in two iterations.

In early iterations, the thresholded edge map $\tilde{\mathrm{I}}$ is rather coarse and is obviously different from the blur input $\mathbf{b}$. It thus effectively avoids the trivial solution. In following iterations, more details are added to the edge map, as shown in Figure 1.11(c), to further refine the PSF estimate.

This strategy, however, could suffer from the convergence problem because each shock filtering process might raise the cost resulting from Eq. (1.51) instead of reducing it in each PSF estimation iteration. Also, the shock filtered map does not guarantee to contain correct edges for large PSF estimation. To address these issues, a general and unified framework was proposed in Xu *et al.* (2013) where the edge map is predicted by a family of sparsity functions to approximate L_0-norm regularization in the new objective. It leads to consistent energy minimization and accordingly fast convergence. The L_0 scheme is mathematically established with high-sparsity-pursuit regularization. It assures only salient changes in the image are preserved and made use of.

To simplify mathematical expressions, in what follows, we describe a framework robust in optimization for large-kernel blind deconvolution, which still employs a shock filter. It is similar to that of Xu & Jia (2010). The method starts with the construction of an image pyramid with n levels. After processing the coarsest level, its result propagates to the finer one as an initialization. This procedure repeats until all levels are processed. In each level, the method takes a few iterations to select edges and initialize the PSF. The final PSF refinement is performed in the highest resolution to improve detail recovery.

Edge selection

In this phase, the PSF is estimated with salient-gradient map construction and kernel estimation. To make this process fast, coarse image restoration is adopted to obtain the $\mathbf{l}$ estimate quickly.

Initially, the blurred image is Gaussian smoothed and is then shock filtered using Eq. (1.56). Note that the output, i.e. the salient-edge map, in many cases, cannot be used directly to guide PSF estimation due to the following fact: *if the scale of an object is smaller than that of the blur kernel, the edge information of the object might adversely affect kernel estimation.*

This is explained with the example shown in Figure 1.12. Two step signals, i.e. the dashed curves in (a) and (b), are blurred with a wide Gaussian kernel, yielding signals in solid curves. Due to the small width of the latent signal, its blurred version in (a)

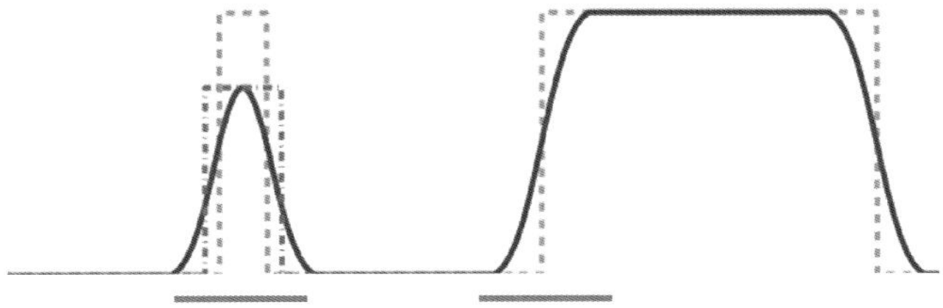

Figure 1.12 Ambiguity of motion deblurring. Two latent signals (dashed lines) in (a) and (b) are Gaussian blurred, shown as solid curves. In (a), the blurred signal is not total-variation preserving and is shorter than the input. The dot–dash curve with the same height as the blurred signal, however, is an optimal solution during deblurring. The bottom horizontal lines indicate the kernel size.

reduces in height, which can misguide PSF estimation. Specifically, the shorter dash–dot signal, compared to the taller one, has the same total variation as the blurred signal, and thus produces less energy in Laplacian regularization. It is more optimal than the ground truth signal when minimizing Eq. (1.44) with $\alpha = 1$. By contrast, the larger-scale object shown in Figure 1.12(b) has no such ambiguity because it is wider than the kernel, preserving total variation along its edges. This example indicates that if structure saliency is changed by motion blur, corresponding edges produced by a shock filter could misguide kernel estimation.

This problem can be tackled by selecting positively informative edges for PSF estimation and eliminating textured regions with fine structures. A metric to measure the usefulness of gradients is

$$r(i) = \frac{|\sum_{y \in N_{\mathrm{h}}(i)} \nabla \mathbf{b}(j)|}{\sum_{j \in N_{\mathrm{h}}(i)} |\nabla \mathbf{b}(j)| + \varepsilon}, \tag{1.57}$$

where $\mathbf{b}$ still denotes the blurred image and $N_{\mathrm{h}}(i)$ is an h × h window centered at pixel i. ε is to avoid a large r in flat regions. $\nabla \mathbf{b}(j)$ is signed. For a window containing primarily texture patterns, $\nabla \mathbf{b}$ cancels out a lot in measure $|\sum_j \nabla \mathbf{b}(j)|$. In contrast, $\sum_j |\nabla \mathbf{b}(j)|$ is the sum of absolute gradient magnitudes in $N_{\mathrm{h}}(x)$, which estimates how strong the image structure is inside the window. Their incorporation in r actually measures whether the window is a textured one or not. A large r implies that local gradients are of similar directions and are not extensively neutralized, while a small r corresponds to either a texture or a flat region. Figure 1.13(b) shows the computed r map. More explanations are provided in Xu, Yan, Xia & Jia (2012).

Pixels belonging to small r-value windows are then removed and encoded in mask

$$\mathrm{M} = H(r - \tau_r), \tag{1.58}$$

where $H(\cdot)$ is the Heaviside step function, outputting zeros for negative and zero values, and ones otherwise. τ_r is a threshold. Finally, the selected edges are formed by nonzero

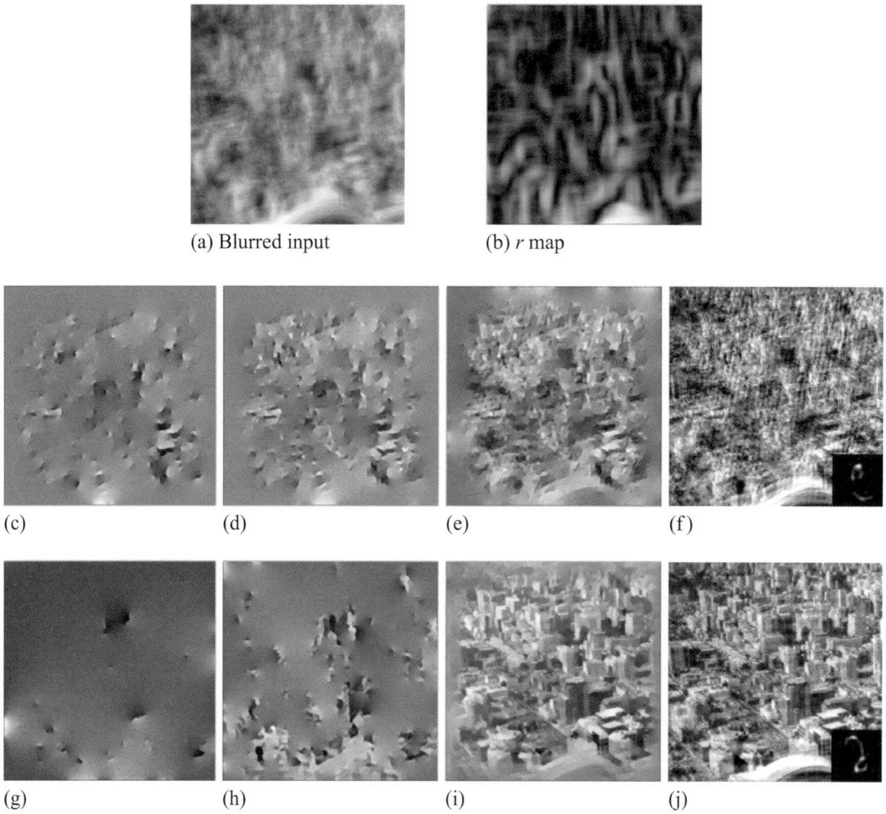

Figure 1.13 Edge selection in kernel estimation. (a) Blurred image; (b) r map (Eq. (1.57)); (c–e) I^s in the 1st, 2nd, and 7th iterations without considering r; (f) deblurring result not using the r map; (g–i) I^s maps computed according to Eq. (1.59); (j) our final result. (c–e) and (g–i) are constructed from $\nabla\mathrm{I}^s$ by solving Poisson equations.

values in $\nabla\mathrm{I}^s$, constructed as

$$\nabla\mathrm{I}^s = \nabla\tilde{\mathrm{I}} \circ H(\mathrm{M} \circ |\nabla\tilde{\mathrm{I}}| - \tau_s), \tag{1.59}$$

where $\circ$ denotes element-wise matrix multiplication, $\tilde{\mathrm{I}}$ is the shock filtered image and τ_s is a threshold of gradient magnitudes. Equation (1.59) excludes part of the gradients, whose values depend jointly on the magnitude of $|\nabla\tilde{\mathrm{I}}|$ and the prior mask M. This selection process becomes much more robust following kernel estimation.

Figure 1.13(c–e) and (g–i) illustrate the correspondingly computed I^s maps in different iterations without and with the edge selection operation, respectively. The comparison in these two rows clearly shows that including more edges does not necessarily benefit kernel estimation. Contrarily, they can confuse this process, especially in the first few iterations. So an appropriate image edge selection process is vital. To allow for inferring subtle structures eventually, one can decrease the values of τ_{r} and τ_s in iterations, to include more and more edges. The maps in (e) and (i) contain similar amounts of edges, but the quality significantly differs. The step to produce the results in (f) and (j) is detailed next.

Algorithm 1.1 Kernel Initialization

INPUT: Blur image $\mathbf{b}$ and an all-zero kernel with size h $\times$ h
Build an image pyramid with levels $\{1, 2, \cdots, n\}$.
for $l = 1$ to n **do**
  Compute gradient confidence r for all pixels using Eq. (1.57).
  for $i = 1$ to 5 **do**
    (a) Select edges $\nabla \mathrm{I}^s$ for kernel estimation (Eq. (1.59)).
    (b) Estimate kernel using Eq. (1.61).
    (c) Estimate the latent image $\mathbf{l}^l$ by fast non-blind deconvolution (Eq. (1.8)), and update $\tau_s \leftarrow \tau_s/1.1$, $\tau_r \leftarrow \tau_r/1.1$.
  end for
  Upscale image $\mathbf{l}^{l+1} \leftarrow \mathbf{l}^l \uparrow$.
end for
OUTPUT: Kernel estimate $\mathbf{f}^0$ and sharp edge gradient $\nabla \mathrm{I}^s$ for further refinement

---

Fast kernel estimation

With the selected edge maps, PSF initialization can be done quickly with a simple quadratic objective function written as

$$E_e(\mathbf{f}) = \|\nabla \mathrm{I}^s \otimes \mathbf{f} - \nabla \mathbf{b}\|^2 + \lambda \|\mathbf{f}\|^2. \tag{1.60}$$

Here, $\mathbf{f}$ is constrained in a simple quadratic term thanks to the effective gradient maps $\nabla \mathrm{I}^s$. Note that minimizing E_e makes the PSF estimate a bit more noisy compared to that constrained by the Laplacian term in Eq. (1.53). The result will be refined in the following steps.

Based similarly on Parseval's theorem and the derivation in Eq. (1.39), computing fast Fourier transforms (FFTs) on all variables and setting the derivatives w.r.t. $\mathbf{f}$ to zeros, yield the closed-form solution

$$\mathbf{f} = \mathcal{F}^{-1}\left(\frac{\overline{\mathcal{F}(\partial_x \mathrm{I}^s)}\mathcal{F}(\partial_x \mathbf{b}) + \overline{\mathcal{F}(\partial_y \mathrm{I}^s)}\mathcal{F}(\partial_y \mathbf{b})}{\mathcal{F}(\partial_x \mathrm{I}^s)^2 + \mathcal{F}(\partial_y \mathrm{I}^s)^2 + \lambda}\right), \tag{1.61}$$

where $\mathcal{F}(\cdot)$ and $\mathcal{F}^{-1}$ denote FFT and inverse FFT, respectively. $\overline{\mathcal{F}(\cdot)}$ is the complex conjugate operator.

The algorithm for multi-scale kernel initialization is sketched in Algorithm 1.1. Deconvolution starts from the coarsest level with edge selection in Eq. (1.59), kernel estimation in Eq. (1.61) and fast deconvolution using Eq. (1.8). The estimates then propagate to the next level by spatial upscaling and repeating the above steps to obtain a more accurate result. After all levels are processed, final kernel refinement is performed in the original image resolution, as described below.

Sparse kernel refinement

To remove the remaining noise from kernel $\mathbf{f}^0$ output from Algorithm 1.1, one can apply a hard or hysteresis threshold to set small values to zeros. This simple scheme however ignores the blur model, possibly making the truncated kernel less accurate. One example

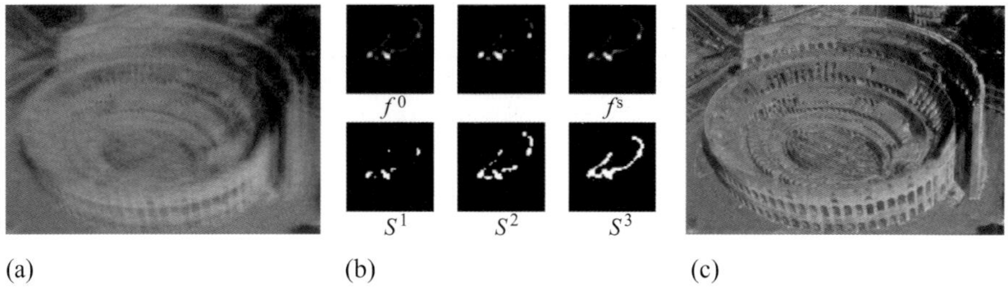

Figure 1.14 Sparse kernel refinement. (a) Blurred image; (b) top row, left to right: kernel $\mathbf{f}^0$, kernel by simply thresholding to remove small-value elements, and our kernel refinement result $\mathbf{f}^s$; bottom row, left to right: S^1–S^3, iteratively detected support regions by the ISD method; (c) restored image using $\mathbf{f}^s$.

is shown in the top, centered image of Figure 1.14(b). Keeping only the large-value elements cannot apparently correctly preserve the subtle structure in the motion PSF.

This problem is solved by iterative support detection (ISD) that ensures deblurring quality while removing noise (Wang & Yin 2009, Xu & Jia 2010). The idea is to iteratively secure PSF elements that already have large values by relaxing the regularization penalty, so that these pixels will not be significantly affected by regularization in next-round kernel refinement.

ISD is an iterative method. In each iteration i, after refining the kernel estimate $\mathbf{f}^i$, a partial support is produced to put large-value elements into set S^i and all others to set $\overline{S^i}$. This process is denoted as

$$S^i \leftarrow \{j : \mathbf{f}^i_j > \epsilon^s\}, \tag{1.62}$$

where j indexes elements in $\mathbf{f}^i$ and ϵ^s is a positive number evolving in iterations, to form the partial support. ϵ^s can be configured by applying the "first significant jump" rule. Briefly, we sort all elements in $\mathbf{f}^i$ in ascending order w.r.t. their values and compute differences $d_0, d_1, \ldots$ between each two nearby elements. Then we examine these differences sequentially starting from the head d_0 and search for the first element, d_j for example, that satisfies $d_j > \|\mathbf{f}^i\|_\infty/(2\mathrm{h} \cdot i)$, where h is the kernel width and $\|\mathbf{f}^i\|_\infty$ returns the largest value in $\mathbf{f}^i$. ϵ^s is thus assigned with the value of the jth kernel element. Readers are referred to Wang & Yin (2009) for further explanation. Examples of the detected support are shown in the bottom row of Figure 1.14(b). Elements within S are less penalized in optimization, resulting in an adaptive process.

Sparse kernel refinement in each iteration $i+1$ is achieved by minimizing

$$\mathrm{E}(\mathbf{f}) = \frac{1}{2}\left\|\nabla \mathrm{I}^s \otimes \mathbf{f} - \nabla \mathbf{b}\right\|^2 + \gamma \sum_{j \in \overline{S^i}} |\mathbf{f}_j|. \tag{1.63}$$

Soft thresholding applies to kernel refinement in regularization, which automatically maintains element sparsity faithful to the motion PSF. Algorithm 1.2 depicts the kernel refinement procedure.

Algorithm 1.2 Kernel refinement

INPUT: Initial kernel $\mathbf{f}^0$, $\nabla\mathbf{b}$, and $\nabla \mathrm{I}^s$ (output of Algorithm 1.1)
Initialize the partial support $\overline{S^0}$ with $\mathbf{f}^0$ (Eq. (1.62)) and set $i = 1$.
repeat
  Solve for $\mathbf{f}^i$ by minimizing Eq. (1.63).
  Produce $\overline{S^i}$ (Eq. (1.62)).
  $i \leftarrow i + 1$.
until $\frac{\|\mathbf{f}^{i+1} - \mathbf{f}^i\|}{\|\mathbf{f}^i\|} \leqslant \epsilon_f$ ($\epsilon_f = 1e^{-3}$ empirically)
OUTPUT: Kernel estimate $\mathbf{f}^s$

To minimize Eq. (1.63) with the partial support, iterative reweighted least squares (IRLS) can be applied. By writing convolution into the matrix multiplication form, the latent image **l**, kernel **f**, and blur input **b** are correspondingly expressed as matrix A, vector $\nu(\mathbf{f})$, and vector $\nu(\mathbf{b})$. Equation (1.63) is then minimized by iteratively solving linear equations w.r.t. $\nu(\mathbf{f})$. In the t-th pass, the corresponding linear equation is expressed as

$$[\mathrm{A}^T\mathrm{A} + \gamma \mathrm{diag}(\nu(\bar{S}))\mathrm{diag}(\varpi^{-1})]\nu(\mathbf{f})^t = \mathrm{A}^T\nu(\mathbf{b}), \tag{1.64}$$

where A^T denotes the transposed A and $\nu(\bar{S})$ is the vector form of $\bar{S}$. ϖ denotes $\max(\|\nu(\mathbf{f})^{t-1}\|^1, 1e^{-5})$, which is the weight related to the kernel estimate from the previous iteration. diag$(\cdot)$ produces a diagonal matrix from the input vector. Equation (1.64) can be solved by a conjugate gradient method in each pass.

The finally refined kernel $\mathbf{f}^s$ is shown in Figure 1.14(b). It maintains small-value elements, which exist in almost all motion kernels. In the meantime, it is reasonably sparse. Optimization in this phase converges in less than 3 iterations referring to the loop in Algorithm 1.2.

Finally, given the PSF estimate **f** output from this algorithm, high quality latent image reconstruction can be applied by non-blind deconvolution using Eq. (1.31). Figure 1.14(c) shows the restored image that contains correctly reconstructed texture and structure, manifesting the effectiveness of this blind deconvolution framework.

1.2.5 Results and running time

Two blind deconvolution examples and their results are presented in Figures 1.15 and 1.16. The input blurred natural images are with motion kernels with resolutions 50×50 and 85×85, respectively, in a spatial-invariant manner. All methods that are compared in this section can remove part or all of the blur. Differences can be observed by comparing the motion kernel estimates shown in the bottom right hand corner of each result and the finally deblurred images, which depend on the quality of kernel estimates and the different non-blind deconvolution strategies employed during kernel estimation or after it. Note that these uniform blind deconvolution methods provide basic and vital tools, which avail research in recent years, and in future will remove spatially-variant blur from natural images caused by camera rotation and complex object motion.

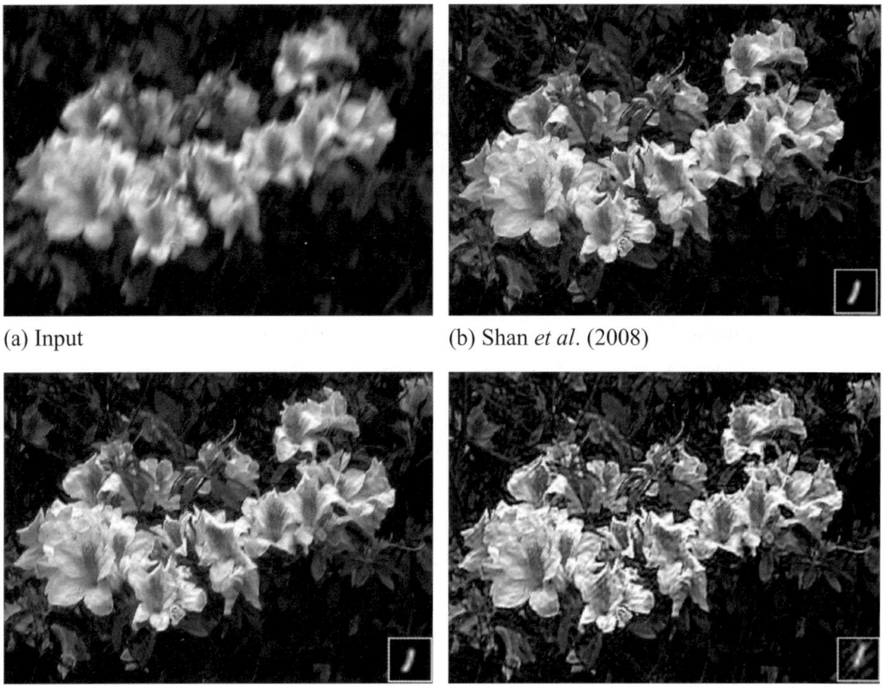

(a) Input (b) Shan *et al.* (2008)

(c) Cho & Lee (2009) (d) Levin *et al.* (2011)

Figure 1.15 Visual comparison of different blind deconvolution results. The input is a camera captured blurred natural image.

Figure 1.17 shows a very challenging example where the input image is blurred with a kernel as large as 95×95 pixels. The method of Xu & Jia (2010) can deblur the input image and produce a high quality deconvolution result. The running time of different methods on CPU is given in Table 1.2, tested using a PC with an Intel i7 2.93GHz CPU and 8 GB memory. There is much room to speed up the methods written in Matlab by programming them in C^{++}. In general, as the alternating energy minimization strategy does not need to estimate marginal probability or sample distributions, it is generally more efficient and flexible.

Finally, code or executables for several representative natural image uniform blind deblurring methods is publicly available at the links listed below.

- Fergus *et al.* (2006) (website and code)
  `http://cs.nyu.edu/%7efergus/research/deblur.html`
- Shan *et al.* (2008) (website and executable)
  `http://www.cse.cuhk.edu.hk/leojia/projects/motion%5fdeblurring`
- Cho & Lee (2009) (website and executable)
  `http://cg.postech.ac.kr/research/fast%5fmotion%5fdeblurring/`
- Xu & Jia (2010) (website and software)
  `http://www.cse.cuhk.edu.hk/leojia/projects/robust%5fdeblur`

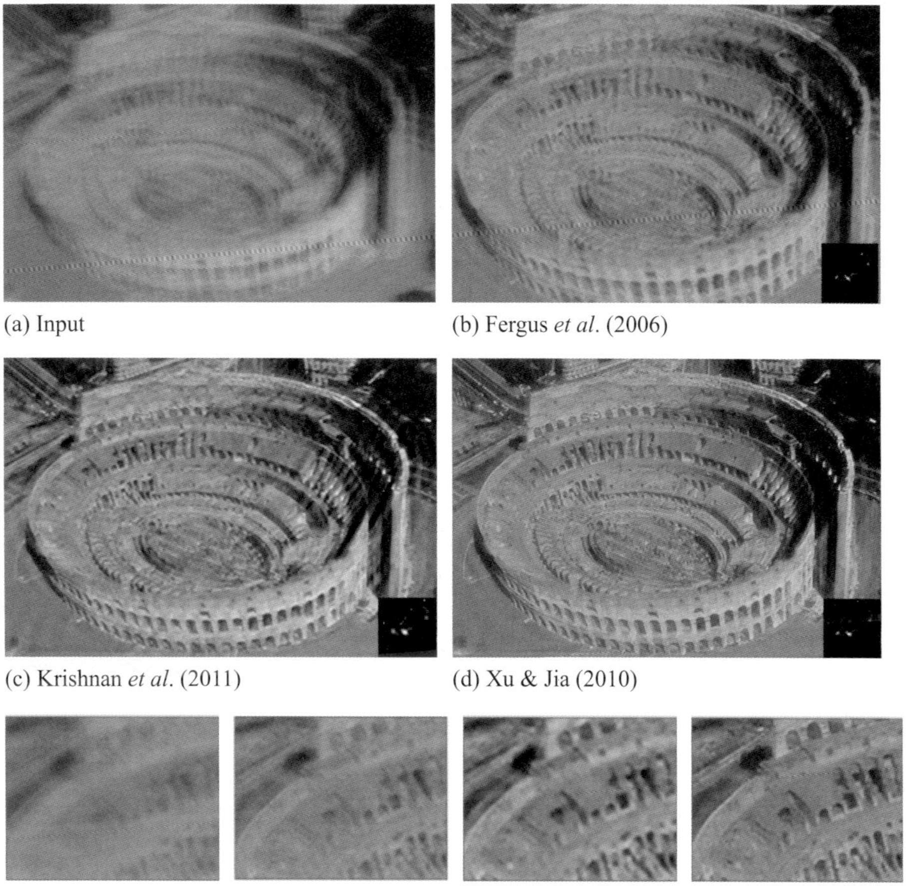

(a) Input (b) Fergus *et al.* (2006)

(c) Krishnan *et al.* (2011) (d) Xu & Jia (2010)

(e) Input (f) Close-up of (b) (g) Close-up of (c) (h) Close-up of (d)

Figure 1.16 Visual comparison of different blind deconvolution results. The input is a camera captured blurred natural image. Close-ups are shown from (e–h).

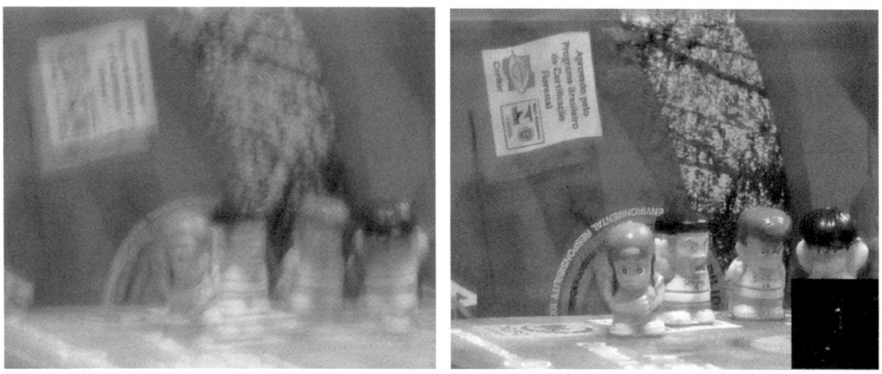

(a) Input (b) Xu & Jia (2010)

Figure 1.17 Deconvolution of a significantly blurred natural image. The blur kernel is as large as 95×95.

Table 1.2 Blind deconvolution running time. Running time (in seconds) of different methods on CPU with respect to three image resolutions. The kernel sizes are, respectively, $31 \times 31, 41 \times 41$, and 51×51 for the three sets of data. Note that Matlab implementation can generally be sped up when reprogrammed in C or C++ .

Methods (Implementation) version	255×255	800×800	1024×768
Fergus *et al.* (2006) (Matlab) v1.2	649.11	4343.24	6916.44
Shan *et al.* (2008) (C++) v1.0	73.28	417.22	700.23
Cho & Lee (2009) (C++)	0.79	5.78	11.60
Xu & Jia (2010) (C++) v3.0	0.80	5.75	13.26
Levin *et al.* (2011) (Matlab)	76.21	1084.19	1737.27
Krishnan *et al.* (2011) (Matlab)	25.60	215.22	273.85
Xu *et al.* (2013) (Matlab)	1.05	5.77	12.23

- Krishnan *et al.* (2011) (website and code)
  `http://cs.nyu.edu/%7edilip/research/blind-deconvolution/`
- Levin *et al.* (2011) (code)
  `http://www.wisdom.weizmann.ac.il/%7elevina/papers/LevinEtalCVPR2011Code.zip`
- Xu *et al.* (2013) (website and software)
  `http://www.cse.cuhk.edu.hk/leojia/projects/l0deblur`

Acknowledgements

Portions of text and figures are based on the work that appeared in Xu & Jia (2010), with kind permission of Springer Science+Business Media.

References

Andrews, H. & Hunt, B. (1977). *Digital Image Restoration*, Prentice-Hall Signal Processing Series. Englewood Cliffs: Prentice-Hall.

Ayers, G. & Dainty, J. (1988). Iterative blind deconvolution method and its applications. *Optics Letters*, **13**(7), 547–9.

Chan, T. & Wong, C. (1998). Total variation blind deconvolution. *IEEE Transactions on Image Processing*, **7**(3), 370–5.

Cho, S. & Lee, S. (2009). Fast motion deblurring. *ACM Transactions on Graphics*, **28**(5), 145:1–8.

Fergus, R., Singh, B., Hertzmann, A., Roweis, S. T. & Freeman, W. T. (2006). Removing camera shake from a single photograph. *ACM Transactions on Graphics*, **25**(3), 787–94.

Fish, D., Brinicombe, A., Pike, E. & Walker, J. (1995). Blind deconvolution by means of the Richardson–Lucy algorithm. *Journal of the Optical Society of America*, **12**(1), 58–65.

Geman, D. & Reynolds, G. (1992). Constrained restoration and the recovery of discontinuities. *IEEE Transactions on Pattern Analysis and Machine Intelligence*, **14**(3), 367–83.

Geman, D. & Yang, C. (1995). Nonlinear image recovery with half-quadratic regularization. *IEEE Transactions on Image Processing*, **4**(7), 932–46.

Hunt, B. (1973). The application of constrained least squares estimation to image restoration by digital computer. *IEEE Transactions on Computers*, **100**(9), 805–12.

Jia, J. (2007). Single image motion deblurring using transparency. In *IEEE Conference on Computer Vision and Pattern Recognition*, pp. 1–8.

Jordan, M., Ghahramani, Z., Jaakkola, T. & Saul, L. (1999). An introduction to variational methods for graphical models. *Machine Learning*, **37**(2), 183–233.

Joshi, N., Szeliski, R. & Kriegman, D. J. (2008). PSF estimation using sharp edge prediction. In *IEEE Conference on Computer Vision and Pattern Recognition*, pp. 1–8.

Kim, S.-J., Koh, K., Lustig, M. & Boyd, S. P. (2007). An efficient method for compressed sensing. In *IEEE International Conference on Image Processing*, pp. 117–20.

Krishnan, D. & Fergus, R. (2009). Fast image deconvolution using hyper-Laplacian priors. In *Neural Information Processing Systems Conference*, pp. 1033–41.

Krishnan, D., Tay, T. & Fergus, R. (2011). Blind deconvolution using a normalized sparsity measure. In *IEEE Conference on Computer Vision and Pattern Recognition*, pp. 233–40.

Levin, A., Fergus, R., Durand, F. & Freeman, W. T. (2007). Image and depth from a conventional camera with a coded aperture. *ACM Transactions on Graphics*, **26**(3), 70:1–10.

Levin, A., Weiss, Y., Durand, F. & Freeman, W. T. (2009). Understanding and evaluating blind deconvolution algorithms. In *IEEE Conference on Computer Vision and Pattern Recognition*, pp. 1964–71.

Levin, A., Weiss, Y., Durand, F. & Freeman, W. T. (2011). Efficient marginal likelihood optimization in blind deconvolution. In *IEEE Conference on Computer Vision and Pattern Recognition*, pp. 2657–64.

Lucy, L. (1974). An iterative technique for the rectification of observed distributions. *Journal of Astronomy*, **79**, 745.

Miller, K. (1970). Least squares methods for ill-posed problems with a prescribed bound. *SIAM Journal on Mathematical Analysis*, **1**, 52.

Miskin, J. & MacKay, D. (2000). Ensemble learning for blind image separation and deconvolution. *Advances in Independent Component Analysis*, pp. 123–41.

Money, J. & Kang, S. (2008). Total variation minimizing blind deconvolution with shock filter reference. *Image and Vision Computing*, **26**(2), 302–14.

Osher, S. & Rudin, L. I. (1990). Feature-oriented image enhancement using shock filters. *SIAM Journal on Numerical Analysis*, **27**(4), 919–40.

Richardson, W. (1972). Bayesian-based iterative method of image restoration. *Journal of the Optical Society of America*, **62**(1), 55–9.

Shan, Q., Jia, J. & Agarwala, A. (2008). High-quality motion deblurring from a single image. *ACM Transactions on Graphics*, **27**(3), 73:1–10.

Tikhonov, A. N. & Arsenin, V. Y. (1977). *Solutions of Ill Posed Problems*. V. H. Winston and Sons.

van Cittert, P. (1931). Zum einfluß der spaltbreite auf die intensitätsverteilung in spektrallinien. ii. *Zeitschrift für Physik A Hadrons and Nuclei*, **69**(5), 298–308.

Wang, Y., Yang, J., Yin, W. & Zhang, Y. (2008). A new alternating minimization algorithm for total variation image reconstruction. *SIAM Journal on Imaging Sciences*, **1**(3), 248–72.

Wang, Y. & Yin, W. (2009). *Compressed Sensing via Iterative Support Detection*, CAAM Technical Report TR09-30.

Wiener, N. (1949). Extrapolation, interpolation, and smoothing of stationary time series: with engineering applications. *Journal of the American Statistical Association*, **47**(258).

Woods, J. & Ingle, V. (1981). Kalman filtering in two dimensions: Further results. *IEEE Transactions on Acoustics, Speech and Signal Processing*, **29**(2), 188–97.

Xu, L. & Jia, J. (2010). Two-phase kernel estimation for robust motion deblurring. In *European Conference on Computer Vision*, pp. 157–70.

Xu, L., Yan, Q., Xia, Y. & Jia, J. (2012). Structure extraction from texture via relative total variation. *ACM Transactions on Graphics*, **31**(6), 139:1–10.

Xu, L., Zheng, S. & Jia, J. (2013). Unnatural l0 sparse representation for natural image deblurring. In *IEEE Conference on Computer Vision and Pattern Recognition*, pp. 1107–14.

Yang, J., Zhang, Y. & Yin, W. (2009). An efficient TVL1 algorithm for deblurring multichannel images corrupted by impulsive noise. *SIAM Journal on Scientific Computing*, **31**(4), 2842–65.

Yuan, L., Sun, J., Quan, L. & Shum, H.-Y. (2008). Progressive inter-scale and intra-scale non-blind image deconvolution. *ACM Transactions on Graphics*, **27**(3), 74:1–10.

2 Spatially-varying image deblurring

Neel Joshi, Sing Bing Kang, and Richard Szeliski

Camera shake is one of the most common causes of image blur. This type of blur arises when a long exposure is required (due to low light levels, for example), and the camera is not held still.

Removing blur due to camera shake is a very active area of research. With just a single photograph as input, this blur removal is known as the blind deconvolution problem, i.e. simultaneously recovering both the blurring function (point spread function or PSF) and the deblurred, latent image. Unfortunately, the blind deconvolution problem is inherently ill-posed, as the observed blurred image provides only a partial constraint on the solution. Therefore, the problem cannot be solved without assuming constraints or priors on the blur or the deblurred image. The most common assumption is that the blur is spatially invariant, but this can handle only a limited set of camera motions. The deblurred image is typically assumed to have natural image statistics (Fergus, Singh, Hertzmann, Roweis & Freeman 2006). A key open problem is to model general camera motions, which are quite common and can cause the blur kernels to vary spatially.

2.1 Review of image deblurring methods

Image deblurring has received a lot of attention in the computer vision community. Deblurring is the combination of two tightly coupled sub-problems: PSF estimation and non-blind image deconvolution. These problems have been addressed both independently and jointly (Richardson 1972).

Single-image methods use image priors and kernel priors to constrain an optimization for the blur kernel and the latent image (Shan, Jia & Agarwala 2008, Likas & Galatsanos 2004, Fergus *et al.* 2006, Yuan, Sun, Quan & Shum 2007, Joshi, Zitnick, Szeliski & Kriegman 2009, Levin 2007, Joshi, Szeliski & Kriegman 2008, Jia 2007).

Jin *et al.* (Jin, Favaro & Cipolla 2005) address the problem of simultaneous tracking and deblurring in motion-blurred image sequences. Cho *et al.*'s work (Cho, Matsushita & Lee 2007) uses segmentation masks to simultaneously estimate multiple (parametric) motion blurs over the whole image as well as handle occlusion. It can be regarded as an extension of the previous technique (Jin *et al.* 2005), since it uses a similar blur constraint and also assumes that the blur kernel is a 1D Gaussian. Bascle *et al.* (Bascle, Blake & Zisserman 1996) jointly perform deblurring and super-resolution on image sequences using a generative back-projection technique.

The technique assumes that the 1D motion that produces blur is completely determined by affine motion between images. Onogi and Saito (Onogi & Saito 2005) deblur video frames for panorama generation. In their case, they perform feature tracking with the assumption that there are only foreground and background global motions. Like Bascle *et al.*, they derive the blur directly from the estimated motions, implicitly assuming a known duty cycle, and invert each frame using a Wiener filter.

The most common commercial approach for reducing image blur is image stabilization (IS). These methods, used in high-end lenses and now appearing in point and shoot cameras and cell phones, use mechanical means to dampen camera motion by offsetting lens elements or translating the sensor. There are also techniques for deblurring that either use specialized hardware or assume a highly controlled capture process. For instance, Ben-Ezra and Nayar (Ben-Ezra & Nayar 2004) use a hybrid camera system to estimate camera motion for deblurring purposes. Agrawal *et al.* (Agrawal, Xu & Raskar 2009) use multi-image deconvolution to address the problem of zeros in the frequency domain of blur kernels.

There is relatively little work on handling spatially-varying blur. Tai *et al.* (Tai, Du, Brown & Lin 2008) developed a hybrid camera that captured a high frame rate video and a blurred image. Optical flow vectors from the video are used to guide the computation of spatially-varying blur kernels, which are in turn used for deblurring. Tai *et al.* (Tai, Kong, Lin & Shin 2010) use a coded exposure to produce a stroboscopic motion image and estimate motion homographies for the discrete motion steps with some user interaction. Dai and Wu (Dai & Wu 2008) propose a method to estimate spatially-varying blur kernels based on values of the alpha map. The technique of Shan *et al.* (Shan, Xiong & Jia 2007) handles rotational motion blur, but it requires user interaction for rotation cues and also relies on constraints from the alpha matte.

One approach to model the spatial variation of blur kernels is to run a blind deconvolution method at each pixel. Joshi *et al.* (Joshi *et al.* 2008) partly accomplished this; they run their method for non-overlapping windows in an image and use this to remove spatially-varying defocus blur and chromatic aberration. However, they do not address camera motion blur, nor do they try to recover a global model of the blur. Levin *et al.* (Levin, Fergus, Durand & Freeman 2007) take a similar approach for object motion blur, where an image is segmented into several areas of different motion blur and then each area is deblurred independently. Hirsch *et al.* (Hirsch, Sra, Schölkopf & Harmeling 2010) also propose a multi-frame patch-based deblurring approach but do not impose any global camera motion constraints on the spatially-varying blur. Whyte *et al.* (Whyte, Sivic, Zisserman & Ponce 2010) describe a framework where they recover the integrated 3-dimensional rotational camera motion (roll, pitch and yaw) to explain the spatially-varying blur.

2.2 A unified camera-shake blur model

In this section, we develop a unified model relating the camera motion, the latent image, and the blurred image for a planar scene.

Let i be the latent image of a planar scene and b be the recorded blurred image. The blurred image can be written as a convolution of the latent image with a kernel k and the addition of some noise n

$$b = k \otimes i + n, \tag{2.1}$$

where $n \sim \mathcal{N}(0, \sigma^2)$. This convolution model does not account for variations in depth and view-dependent illumination changes; we do not handle them here.

This convolution model can also be written as a matrix-vector product

$$B = \mathcal{K}I + N, \tag{2.2}$$

where I, B, and N denote the column-vector forms of i, b, and n respectively. $\mathcal{K}$ is an image resampling matrix (or blur matrix) that applies the convolution, with each row of $\mathcal{K}$ being the blur kernel placed at each pixel location and unraveled into a row vector. If the blur is spatially invariant, each row will just be a shifted version of another. This matrix-vector form becomes particularly useful for formulating spatially-varying blur; each row can be set with a different blur kernel for each pixel (Seitz & Baker 2009), as we will discuss in the next section.

2.2.1 Blur matrices

We assume the camera initially lies at the world origin with its axes aligned with the world axes. A camera motion is a sequence of camera poses where each pose can be characterized by six parameters: three rotations and three translations. Any camera motion can be represented as a 1D continuous path through this six-dimensional space, which we call *camera pose space*. In a discretized version of this space, the camera spends a fraction of the exposure time at each pose; we call this proportion the *density* at that pose. Taken together, these densities form a *motion density function* (MDF) from which a blur kernel can be directly determined for any point on the image. The MDF for all the camera poses forms a column vector over the discrete positions in the camera pose space. We denote the MDF by a vector A with elements a_j that denote the density at the camera pose j.

The observed blurred image B is an integration over the images seen by the camera in all the poses in its path. In the discrete motion space, B is a summation over the images seen by the camera in all possible poses, each weighted by the proportion of time spent by the camera in that pose, which in our notation is the pose's *density*. We write this mathematically as

$$B = \sum_j a_j(K_j I) + N, \tag{2.3}$$

where K_j is a warping transformation from I (the latent image or the un-blurred image seen by the camera in the original pose) to the image seen in pose j. N is the Gaussian noise. Given a particular 6D pose (indexed by j) of a camera, the corresponding homography that warps a scene at depth d is P_j

$$P_j = C\left(R_j + \frac{1}{d}t_j[0\ 0\ 1]\right)C^{-1}, \tag{2.4}$$

where R_j and t_j are the rotation matrix and translation vector for pose j, and C is the matrix of camera intrinsics. For now, we assume the depth d is known. K_j is a resampling matrix where each row contains the weights used to compute the values of pixels in the warped image by applying the inverse homography. We use bilinear interpolation for the warps and thus there are at most four nonzero values per row of K_j. For clarity, we note that K_j is a sparse square matrix where each dimension is the width times the height of the image i.

From Eqs. (2.2) and (2.3), we can write the blur matrix $\mathcal{K}$ as

$$\mathcal{K} = \sum_j a_j K_j. \tag{2.5}$$

This shows that the K_j's form a basis set whose elements can be linearly combined using the MDF to get the corresponding blur matrix for any camera path. By definition, the blur matrix also gives us the blur kernels for each pixel location in the image.

2.2.2 Spatially-varying deconvolution

If these values are known, the image can be deblurred using non-blind deconvolution. We modify the formulation of Levin *et al.* (2007) to use our spatially-varying blur model.

For single image deconvolution, we find the most likely estimate of the sharp image I, given the observed blurred image B, the blur matrix A, and noise level σ^2 using a *maximum a posteriori* (MAP) technique.

We express this as a maximization over the probability distribution of the posterior using Bayes' rule. The result is a minimization of a sum of negative log likelihoods

$$P(I|B,A) = P(B|I)\frac{P(I)}{P(B)} \tag{2.6}$$

$$\underset{I}{\operatorname{argmax}}\, P(I|B) = \underset{I}{\operatorname{argmin}}[L(B|I) + L(I)]. \tag{2.7}$$

The problem of deconvolution is now reduced to minimizing the negative log likelihood terms. Given the blur formation model Eq. (2.3), the "data" negative log likelihood is

$$L(B|I) = \frac{\|B - \mathcal{K}I\|^2}{\sigma^2}. \tag{2.8}$$

The contribution of our deconvolution approach is this new data term that uses the spatially-varying model derived in the previous section.

The "image" negative log likelihood is derived from a natural image prior that models images as having a heavy tailed gradient distribution (Weiss & Freeman 2007, Levin *et al.* 2007, Shan *et al.* 2008). The minimization is performed using iteratively re-weighted least squares (Stewart 1999).

2.3 Single image deblurring using motion density functions

In the previous section, we described the blur kernel formation model based on a known MDF and how we use it to deblur an image. In this section, we show how to estimate the MDF from just an image. We discuss an approach for estimating the full 6D space for camera motion blur using a 3D space of 2D translation and in-plane rotation. Starting from a single image, we seek to recover the latent image and the spatially-varying blur kernels that arise from this 3D camera motion model.

Although a full model of motion would require six degrees of freedom (three translations plus three rotations), 6D general motion can be reasonably approximated by modeling only in-plane rotation and translation, i.e. 3D camera motion. Our approach is similar to parallel work by Whyte *et al.* (2010) who use a rotational camera motion (*roll*, *pitch* and *yaw*) instead. Whyte *et al.*'s model will be approximately the same as ours for the longer focal lengths. For the shorter focal lengths, their system will have more artifacts if the underlying camera motion is mostly translations instead of out-of-plane rotations. Our system will have more artifacts if the underlying camera motion is mostly out-of-plane rotations instead of translations.

2.3.1 Optimization formulation

Equation (2.3) relates the MDF to the latent image and the blurred image. In the process of deblurring, the basis (given by K_j's) is pre-computed, and we solve for the variables I and A. To do this, we pose the problem in a Bayesian framework and seek to recover the latent image and MDF that is most likely given the observation and priors on the image and MDF.

We compute the MAP estimate, which we formulate as the minimization of the following energy function

$$E = \left\| \left[\sum_j a_j K_j \right] I - B \right\|^2 + prior(A) + prior(I) \tag{2.9}$$

$$prior(A) = \lambda_1 ||A||^{\gamma} + \lambda_2 ||\nabla A||^2 \tag{2.10}$$

$$prior(I) = \phi(|\partial_x I|) + \phi(|\partial_y I|). \tag{2.11}$$

The $prior(I)$ is the image prior with the same parameter settings as used by Shan *et al.* (2008).

We model priors that are important to recovering an accurate MDF. The first is a sparsity prior on the MDF values. We note that while blur kernels in the 2D image space may seem quite dense, in the higher dimensional MDF space, a 1D path represents an extremely sparse population of the space. The second prior is a smoothness prior on the MDF, which also incorporates the concept of the MDF representing a path, as it enforces continuity in the space and captures the conditional probability that a particular pose is more likely if a nearby pose is likely. We also note that we can choose to use the whole blurred image for the optimization or some selected parts by masking out rows in *I*, *B*, and the corresponding matrices.

The number of matrices K_j is the number of motion poses that we sample. The number of samples along each depth of field (DOF) affects the size of the data stored and hence we want to keep the sampling as coarse as possible. We hypothesize that the sampling need only be dense enough that the neighboring voxels in the discretized motion space project to within a pixel width at any image location. The range of the motion can be chosen to cover the estimate of the kernel size that is initially specified by the user. We automatically choose the sampling range and resolution along the three degrees of freedom and pre-compute the K_j's.

2.3.2 Our system for MDF-based deblurring

The proposed optimization in Eq. (2.9) is nonlinear in the variables *I* and *A*. We solve this using an alternating, iterative expectation–maximization (EM)-style procedure, which takes an initialization for *I* and then optimizes for *A* and *I* successively.

For the initial estimate, we first select uniformly distributed patches on the blurred image, which we independently deblur using the blind deconvolution approach proposed by Shan *et al.* (2008). The patch sizes are proportional to the estimated maximum blur kernel size in the image, which is an input parameter for our system. Since kernel estimation techniques require a good distribution of edge orientations (Joshi *et al.* 2008), we filter out the patches having a low average value of the Harris corner metric. The Harris corner metric measures the presence of gradients in orthogonal directions in an image region and hence is a good estimate for the success of kernel estimation. We empirically choose this threshold value to be 0.1. These deblurred patches are the initial estimates for the latent image in corresponding regions.

2.3.3 Experiments and results

Figure 2.1 shows results for real-world blurred images of scenes captured using a Canon EOS-1D camera. We perform all our comparisons with the recent blind deconvolution method of Shan *et al.* (2008), who have code available online. We show the original blurred image (A), the deblurred result using spatially-invariant deconvolution (B), our deblurred result (C), and the inset comparisons between (B) and (C). Our approach shows a significant improvement over the spatially-invariant approach in all the cases. Our current implementation does not handle depth variance in the scene. See Gupta, Joshi, Zitnick, Cohen & Curless (2010) for additional comparison sets with real and synthetic data.

2.4 Image deblurring using inertial measurement sensors

In this section, we describe a combined hardware and software-based approach for estimating spatially-varying blur that uses a novel hardware attachment that can be affixed

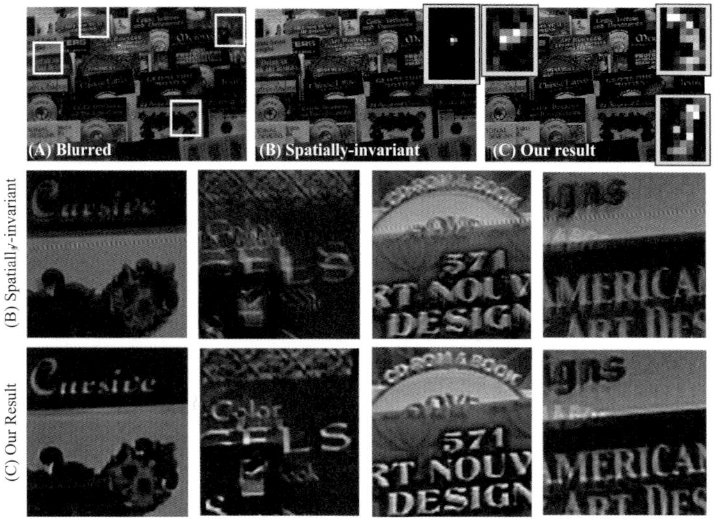

Figure 2.1 Deblurring results. (A) Original blurred image; (B) deblurred image using spatially-invariant approach; (C) deblurred result using our system. Recovered blur kernels at a few locations are shown in boxes in the corner of the images. [Reproduced from Gupta *et al.* (2010) with permission from Springer.]

to any consumer camera. The device uses inexpensive gyroscopes and accelerometers to measure a camera's acceleration and angular velocity during an exposure. (We note that sensors are already in many cameras and smartphones.) These data are used as input to a novel "aided blind deconvolution" algorithm that computes the spatially-varying image blur and latent deblurred image. We derive a model that handles spatially-varying blur due to full 6-DOF camera motion and spatially-varying scene depth; however, our system assumes spatially-invariant depth.

By instrumenting a camera with inertial measurement sensors, we can obtain relevant information about the camera motion and thus the camera-shake blur; however, there are many challenges in using this information effectively. Motion tracking using inertial sensors is prone to significant error when tracking over an extended period of time. This error, known as "drift", occurs due to the integration of the noisy measurements, which leads to increasing inaccuracy in the tracked position over time. As we will show in our experiments, using inertial sensors directly is not sufficient for camera tracking and deblurring.

Instead, we use the inertial data and the recorded blurry image together with an image prior in a novel "aided blind deconvolution" method that computes the camera-induced motion blur and the latent deblurred image using an energy minimization framework. We consider the algorithm to be "aided blind deconvolution", since it is only given an estimate of the PSF from the sensors. Our method is completely automatic, handles per-pixel, spatially-varying blur, out-performs current leading image-based methods, and our experiments show that it can handle large kernels – up to 100 pixels, with a typical size of 30 pixels.

2.4.1 Deblurring using inertial sensors

In this section, we describe the design challenges and decisions for building our sensing platform for image deblurring and how it is used to estimate image blur.

Estimating camera motion using inertial sensors

Given the spatially-varying blur model introduced in Section 2.2, our goal is to estimate the blur matrix $\mathcal{K}$ by directly solving for the time-varying extrinsics, R_j and t_j, and depth d in Eq. (2.4). In this section, we discuss how to recover the camera rotations and translations, and in Section 2.4.2 we address recovering camera intrinsics.

Any motion of a rigid body and any point on that body can be parameterized as a function of six unknowns, three for rotation and three for translation. We now describe how to recover these quantities given inertial measurements from accelerometers and gyroscopes.

Accelerometers measure the total acceleration at a given point along an axis, while gyroscopes measure the angular velocity at a given point around an axis. Note that for a moving rigid body, the pure rotation at all points is the same, while the translations for all points are not the same when the body is rotating.

Before deriving how to compute camera motion from inertial measurements, we first present our notation, as summarized in Table 2.1.

A rigid body, such as a camera, with a three-axis accelerometer and three-axis gyroscope (three accelerometers and gyroscopes mounted along x, y, and z in a single chip, respectively) measures the following accelerations and angular velocities

$$\omega_t^t = R_t \omega_t, \tag{2.12}$$

$$b_t^t = R_t \left(a_t + g + (\omega_t \times (\omega_t \times r)) + (\alpha_t \times r) \right). \tag{2.13}$$

The measured acceleration is the sum of the acceleration due to translation of the camera, centripetal acceleration due to rotation, the tangential component of angular acceleration, and gravity, all rotated into the current frame of the camera. The measured

Table 2.1 Quantities in **bold** indicate measured or calibrated values. Note that these are vectors, i.e. three axis quantities. The "superscript" character indicates the coordinate frame of the value and the "subscript" indicates the time the value was measured, i.e. the current frame is t. The initial frame is where $t = 0$. For readability, we will generally omit the superscript or subscript when the frame or time is the initial frame.

Symbol	Description
R_t	Rotation from the initial coordinate frame to the frame at time t
R^t	Rotation from frame at time t to the initial coordinate frame
θ_t, ω_t, α_t	Angular position, velocity, and acceleration at time t in the initial frame
ω_t^t	**Angular velocity at time t in the frame at time t**
x_t,v_t,a_t	Position, velocity, and acceleration at time t in the initial coordinate frame
b_t^t	**Accelerometer reading at time t in the coordinate frame at time t**
y_t	**Position of the accelerometer in the initial coordinate frame**
r	Vector from the accelerometer position to center of rotation
g	Gravity in the camera's initial coordinate frame

angular velocity is the camera's angular velocity also rotated in the current frame of the camera. To recover the relative camera rotation, it is necessary to recover the angular velocity for each time-step t in the coordinate system of the initial frame ω_t, which can be integrated to get the angular position. To recover relative camera translation, we need to first compute the accelerometer position for each time-step relative to the initial frame. From this, we can recover the camera translation.

The camera rotation can be recovered by sequentially integrating and rotating the measured angular velocity into the initial camera frame

$$\theta_t = (R^{t-1}\omega_{t-1}^{t-1})\Delta t + \theta_{t-1} \tag{2.14}$$

$$R_t = angleAxisToMat(\theta_t), \tag{2.15}$$

where "*angleAxisToMat*" converts the angular position vector to a rotation matrix. Since we are only concerned with relative rotation, the initial rotation is zero

$$\theta_{t=0} = 0, \quad R_{t=0} = Identity. \tag{2.16}$$

Once the rotations are computed for each time-step, we can compute the acceleration in the initial frame's coordinate system

$$b_t = R^t b_t^t, \tag{2.17}$$

and integrate the acceleration, minus the constant acceleration of gravity, to get the accelerometer's relative position at each time-step

$$w_t = (b_{t-1} - g)\Delta t + w_{t-1}, \tag{2.18}$$

$$y_t = 0.5(b_{t-1} - g)\Delta t^2 + w_{t-1}\Delta t + y_{t-1}. \tag{2.19}$$

As we are concerned with relative position, we set the initial position to zero, and we also assume that the initial velocity is zero

$$y_{t=0} = w_{t=0} = [0, 0, 0]. \tag{2.20}$$

The accelerometer's translation (its position relative to the initial frame) in terms of the rigid body rotation and translation is

$$y_t = R_t y_{t=0} + x_t. \tag{2.21}$$

Given this, we can compute the camera position at time t

$$x_t = R_t y_{t=0} - y_t. \tag{2.22}$$

In Eq. (2.18), it is necessary to subtract the value of gravity in the initial frame of the camera. We note, however, that the initial rotation of the camera relative to the world is unknown, as the gyroscopes only measure velocity. The accelerometers can be used to estimate the initial orientation if the camera initially has no external forces on it other than gravity. We have found this assumption unreliable as the camera is often not at rest, so we instead make the assumption that the measured acceleration is normally distributed about the constant force of gravity. We have found this reliable when the camera motion is due to high-frequency camera shake. Thus we set the direction of mean acceleration vector as the direction of gravity

$$g = mean(b_t, [0 \ldots T]). \tag{2.23}$$

To summarize, the camera rotation and translation are recovered by integrating the measured acceleration and angular velocities that are rotated into the camera's initial coordinate frame. This gives us the relative rotation and translation over time, which is used to compute the spatially-varying PSF matrix. If the measurements are noise-free, this rotation and motion information is sufficient for deblurring; however, in practice, the sensor noise introduces significant errors. Furthermore, even if the camera motion is known perfectly, one still needs to know the scene depth, as discussed in Section 2.4.1. Thus additional steps are needed to deblur an image using the inertial data.

Drift compensation and deconvolution

It is well known that computing motion by integrating differential sensors can lead to drift in the computed result. This drift is due to the noise present in the sensor readings. The integration of a noisy signal leads to a temporally increasing deviation of the computed motion from the true motion.

We have measured the standard deviation of our gyroscope's noise to be 0.5°/s and the accelerometer noise is $0.006\,\text{m/s}^2$, using samples from when the gyroscopes and accelerometers are held stationary (at zero angular velocity and constant acceleration, respectively). In our experiments, there is significantly less drift in rotation, due to the need to perform only a single integration step on the gyroscope data. The necessity to integrate twice to get positional data from the accelerometers causes more drift.

To get high-quality deblurring results, we must overcome the drift. We propose a novel aided blind deconvolution algorithm that computes the camera motion, and in turn the image blur function, that best matches the measured acceleration and angular velocity while maximizing the likelihood of the deblurred latent image according to a natural image prior.

Our deconvolution algorithm compensates for positional drift by assuming it is linear in time, which can be estimated if one knows the final end position of the camera. We assume the rotational drift is minimal. The final camera position is, of course, unknown; however, we know the drift is bounded and thus the correct final position should lie close to our estimate from the sensor data. Thus in contrast to a traditional blind deconvolution algorithm that solves for each value of a kernel or PSF, our algorithm only has to solve for a few unknowns. We solve for these using an energy minimization framework that performs a search in a small local neighborhood around the initially computed end point.

In our experiments, we have found that the camera travels in the order of a couple of millimeters during a long exposure (the longest we have tried is half a second). We note that a few millimeter translation in depth (z) has little effect on the image for lenses of common focal lengths, thus the drifts in x and y are the only significant sources of error. We set our optimization parameters to search for the optimal end point within a 1 mm radius of the initially computed end point, subject to the constraints that the acceleration along that recovered path matches the measured accelerations best in the least-squares sense. The optimal end point is the one that results in a deconvolved image with the highest log-likelihood as measured by the hyper-Laplacian image prior (discussed in Section 2.4.1).

Specifically, let us define a function g that, given a potential end point (u, v), computes the camera's translational path that best matches, in the least squares sense, the observed acceleration and terminates at (u, v)

$$g(a, u, v) = \underset{x}{\operatorname{argmin}} \sum_{t=0}^{T} \left(\frac{d^2 x_t}{dt^2} - a_t \right)^2 + (\theta_{x,T} - u)^2 + (\theta_{y,T} - v)^2. \tag{2.24}$$

For notational convenience, let ρ define a function that forms the blur sampling matrix from the camera intrinsics, extrinsics, and scene depth using the rigid-body dynamics and temporal integration processes discussed in Section 2.4.1

$$A(d) = \rho(\theta, x, d, K). \tag{2.25}$$

The drift-compensated blur matrix and deconvolution equations are

$$A(d, u, v) = \rho(\omega, g(a, u, v), d, K), \tag{2.26}$$

$$I = \underset{I,d,u,v}{\operatorname{argmin}} \left[\|B - A(d, u, v)I\|^2/\sigma^2 + \lambda \|\nabla I\|^{0.8} \right]. \tag{2.27}$$

We then search over the space of (u, v) to find the (u, v) that results in the image I that has the highest likelihood given the observation and image prior. We perform this energy minimization using the Nelder–Mead simplex method. The spatially-varying deconvolution method discussed in Section 2.2.2 is used as the error function in the inner loop of the optimization. We perform the optimization on one-tenth down-sampled versions (of our 21 MP images).

The results from our search process are shown as plots of the camera motion in Figure 2.2 and visually in Figure 2.5, when used to deblur images. The running time for this search method is about five minutes on a 0.75 MP image. The search only needs to be run once either for the entire image, or it could be run on a subsection of the image if that is preferable. Once the drift is corrected for, the PSF is more accurate for the entire image, and the image can be deblurred using the method described in Section 2.2.2.

Computing scene depth

Note that a spatially-invariant scene depth is implicitly computed during the drift compensation process, as scaling the end point equally in the x and y dimensions is equivalent to scaling the depth value. Specifically, the optimization is over $u' = u/d$ and $v' = v/d$ and thus solves for a single depth value for the entire scene.

2.4.2 Deblurring system

In this section, we describe our hardware for recording the accelerometer and gyroscope data and implementation related concerns and challenges.

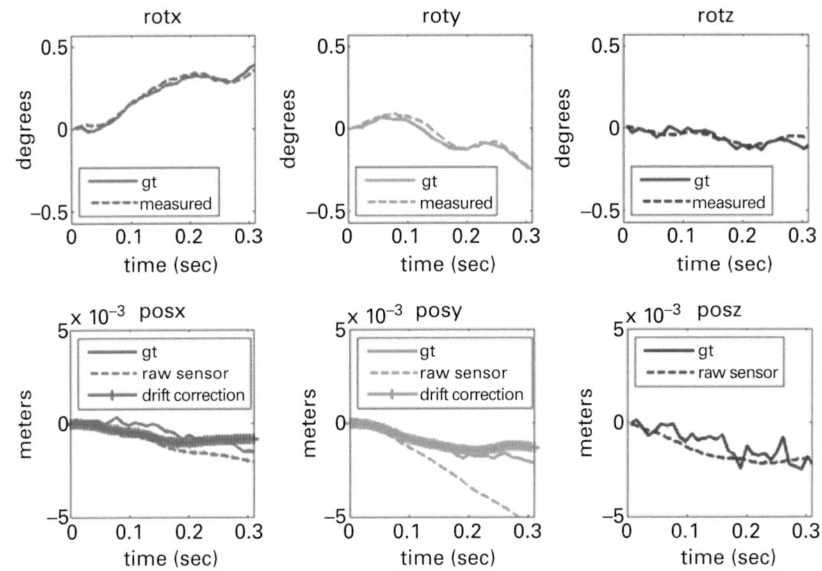

Figure 2.2 Drift compensation: angular and translational position of the camera vs time is shown for the image in the horizontal box in Figure 2.5. The "ground truth" motion, computed using structure from motion, is shown as solid lines. The dashed lines are the motions from using the raw sensor data and the "+" marked lines are the result after drift compensation. The drift compensated results are much closer to the ground truth result. Note the bottom right plot does not show a drift corrected line as we only compensate for x and y drift. [Reproduced from (Joshi *et al.* 2010) with permission from ACM.]

Hardware design

Since there are six unknowns per time-step, the minimal configuration of sensors is six. It is possible to recover rotation and translation using six accelerometers alone, by sensing each axis in pairs at three different points on a rigid-body; however, after experimenting with this method we found accelerometers alone to be too noisy for reliable computation of rotation. Thus our prototype hardware system, shown in Figure 2.3, is a minimal configuration consisting of a three-axis ± 1.5 g MEMS accelerometer package and three single axis $\pm 150°$/s MEMS gyroscopes wired to an Arduino controller board with a Bluetooth radio. All parts are off-the-shelf components. Additionally, the SLR's hot-shoe, i.e. flash trigger signal, is wired to the Arduino board. The Arduino board is interrupt-driven such that when the trigger signal from the SLR fires, the accelerometers and gyroscopes are polled at 200 Hz during the exposure window. Each time the sensors are read, the values are sent over the Bluetooth serial port interface. The sensors and Arduino board are mounted to a laser-cut acrylic base that secures the board, the sensors, and a battery pack.

Our hardware attachment has an optional feature used for calibration and validation experiments: mounting holes for a Point Grey high-speed camera. When the Arduino board is sampling the inertial sensors, it can also send a 100 Hz trigger to the high-speed

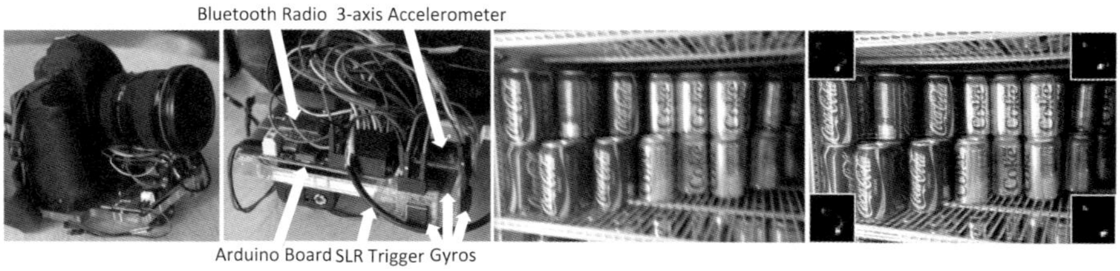

Figure 2.3 First two images: an SLR camera instrumented with our image deblurring attachment that uses inertial measurement sensors and the input image in an "aided blind deconvolution" algorithm to automatically deblur images with spatially-varying blurs; third image: a blurry input image; fourth image: the result of our method. The blur kernel at each corner of the image is shown at 2× size. [Reproduced from Joshi *et al.* (2010) with permission from ACM.]

camera. We use the high-speed data to help calibrate our sensors and to acquire data to get ground truth measurements for motion blur.

Calibration and ground truth measurements

It is necessary to calibrate several aspects of our system: the sensor ranges, the position of the accelerometer relative to the camera's optical center, and the camera intrinsics.

We calibrate these values in two stages. The first stage is to calibrate the sensors' responses. We do this by rotating the gyroscopes at a known constant angular velocity to recover the mapping from the 10-bit A/D output to degrees per second. We performed this measurement at several known angular velocities to confirm that the gyroscopes have a linear response. To calibrate the accelerometers, we held them stationary in six orientations with respect to gravity, $(\pm x, \pm y, \pm z)$, which allowed us to map the A/D output to units of m/s^2.

To calibrate our setup and to take ground truth measurements for camera shake, we developed a method to accurately recover a camera's position during an exposure using a method inspired by Ben-Ezra and Nayar (Ben-Ezra & Nayar 2004). However, instead of tracking 2D motion, we track 6D motion. We attached a high-speed camera (200 FPS Point Grey Dragonfly Express) to our sensor platform, and our Arduino microcontroller code is set to trigger the high-speed camera at 100 FPS during the SLR's exposure window. In a laboratory setting, we created a scene with a significant amount of texture and took about 10 images with exposures ranging from one-tenth to one-half of a second. For each of these images, accelerometer and gyro data were recorded in addition to high-speed frames. We took the high-speed frames from these shots and acquired additional wide-baseline shots with the SLR and high-speed camera. Using all of these data, we created a 3D reconstruction of the scene using 3D structure from motion.

This reconstruction process gives us a collection of sparse 3D points, camera rotations, and camera translations, with an unknown global transformation and a scale ambiguity between camera depth and scene depth. We resolve the scale ambiguity using a calibration grid of known size.

2.4.3 Results

We will now describe the results of our ground truth camera-shake measurements and compare these results, using our deblurring method, to the ground truth measurements. We also compare our results to those of Shan *et al.* (2008) and Fergus *et al.* (2006), using the implementations the authors have available online. We also show results of our methods running on natural images acquired outside of a laboratory setup and in addition compare these to results using previous work.

Laboratory experiments and camera-shake study

In Figure 2.4, we show visualizations of the ground truth spatially-varying PSFs for an image from our laboratory setup. This image shows some interesting properties. There is a significant variation across the image plane. Also, the kernels displayed are for a fixed depth, thus all the spatial variance is due to rotation. To demonstrate the importance of accounting for spatial variance, on the bottom row of Figure 2.4, we show a result where we have deconvolved using the PSF for the correct part of the image and the PSF for a different, non-corresponding area. These results are quite interesting as they show that some of the common assumptions made in image deconvolution do not always hold. Most deconvolution work assumes spatially-invariant kernels, which really only applies for camera motion under an orthographic model; however, with a typical imaging setup (we use a 40 mm lens), the perspective effects are strong enough to induce a spatially-varying blur. We also note that there is often a roll component to the blur, something that is also not modeled by spatially-invariant kernels. Lastly, we observe that translation,

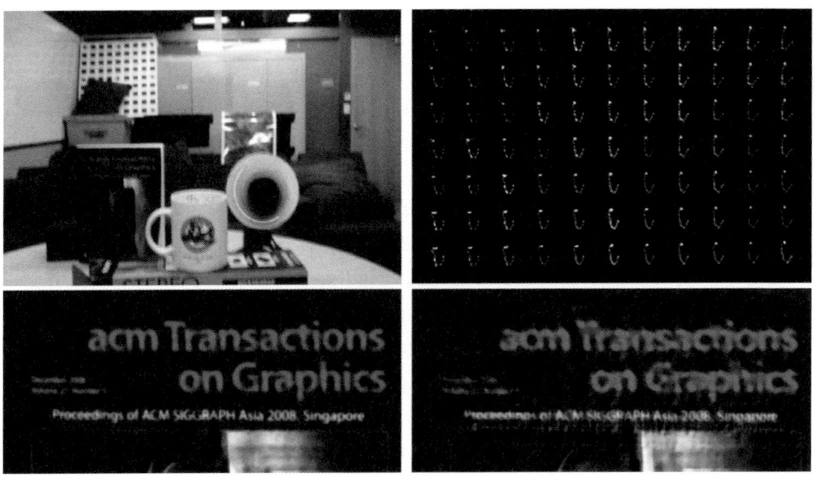

Figure 2.4 Visualization of ground truth for spatially-varying PSFs. For the blurry image (top left), we show a sparsely sampled visualization of the blur kernels across the image plane (top right). There is quite a significant variation across the image plane. To demonstrate the importance of accounting for spatial variance, in the bottom row we show a result where we have deconvolved using the PSF for the correct part of the image and a non-corresponding area – the center of the image. [Reproduced from Joshi *et al.* (2010) with permission from ACM.]

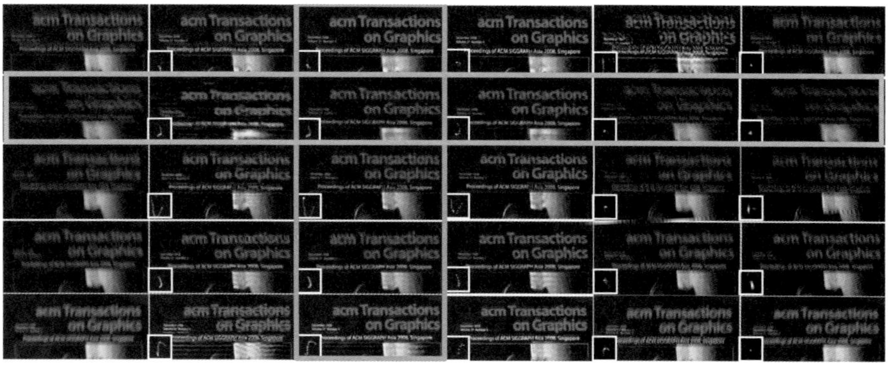

Figure 2.5 Deblurring results and comparisons. Here we show deblurring results for a cropped portion of the scene shown in Figure 2.4. Each row is a different capture. These sections are cropped from an 11 MP image. The results in the column with the vertical box are the final output of our method, where the sensor data plus our drift compensation method are used to compute the camera motion blur. Subtle differences in the PSF before and after drift compensation can have a big result on the quality of the deconvolution. [Reproduced from Joshi *et al.* (2010) with permission from ACM.]

and thus depth-dependent effects, can be significant, which is interesting as it is often thought that most camera-shake blur is due to rotation.

In Figure 2.5, using the same scene as above, we show comparisons of deconvolution results using our method, ground truth, and two others. The images shown of the laboratory calibration scene were taken at exposures from one-third to one-tenth of a second. For each image we show the input, the result of deconvolving with PSFs from the initial motion estimate and after performing drift correction, and compare these to a deconvolution using PSFs from the recovered ground truth motions, and PSFs recovered using the methods of Shan *et al.* (2008) and Fergus *et al.* (2006). For these latter two comparisons, we made a best effort to adjust the parameters to recover the best blur kernel possible. To make the comparison fair, all results were deblurred using exactly the same deconvolution method, that of Levin *et al.* (2007). The results in Figure 2.5 show a wide variety of blurs, yet our method recovers an accurate kernel and provides deconvolution results that are very close to that of the ground truth. In all cases, our results are better than those using Shan *et al.*'s and Fergus *et al.*'s methods.

Real images

After calibrating our hardware system using the method discussed in Section 2.4.2, we took the camera outside of the laboratory and used a laptop with a Bluetooth adapter to capture the inertial sensor data. In Figure 2.6, we show several results where we have deblurred images using our method, Shan *et al.*'s method, and Fergus *et al.*'s method. The Shan *et al.* and Fergus *et al.* results were deconvolved using Levin *et al.*'s method (Levin *et al.* 2007), and our results are deconvolved using our spatially-varying deconvolution method discussed in Section 2.4.1.

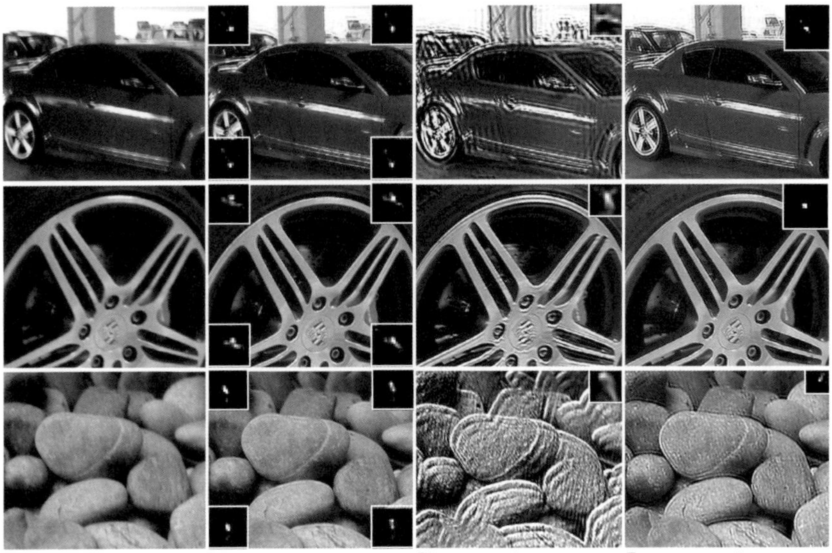

Input IMU deblurring (our final output) Shan *et al.* Fergus *et al.*

Figure 2.6 Natural images deblurred with our setup. For all the images our results show a clear improvement over the input blurry image. The blur kernel at each corner of the image is shown at $2\times$ size. The Shan *et al.* results are of varying quality, and many show large ringing artifacts that are due to kernel misestimation and over-sharpening; the Fergus *et al.* results are generally blurrier than our results. With the stones image, the intentionally defocused background stones are not sharpened in our result. [Reproduced from Joshi *et al.* (2010) with permission from ACM.]

All the images shown here were shot with one-half to one-tenth of a second exposures with a 40 mm lens on a Canon 1Ds Mark III. For additional results, see Joshi *et al.* (2010).

2.5 Generating sharp panoramas from motion-blurred videos

A convenient way to generate a panorama is to take a video while panning, then stitch the frames using a commercial tool such as AutoStitch, Hugin, Autodesk Stitcher, Microsoft Image Composite Editor, or Microsoft Photosynth. However, if there is significant camera motion, the frames in the video can be very blurry. Stitching these frames will result in a blurry panorama, as shown in Figure 2.7. In this section, we describe a new technique that is capable of generating sharp panoramas such as that shown in the third row.

Our framework assumes that the scene is static and adequately far away from the camera. Hence the apparent motion and motion blur in the video are mainly due to camera rotation. This allows us to parameterize the image motion as a homography (Tai, Tan, Gao & Brown 2009). Moreover, we assume that the camera motion is piecewise linear (i.e. the velocity is constant between successive frames).

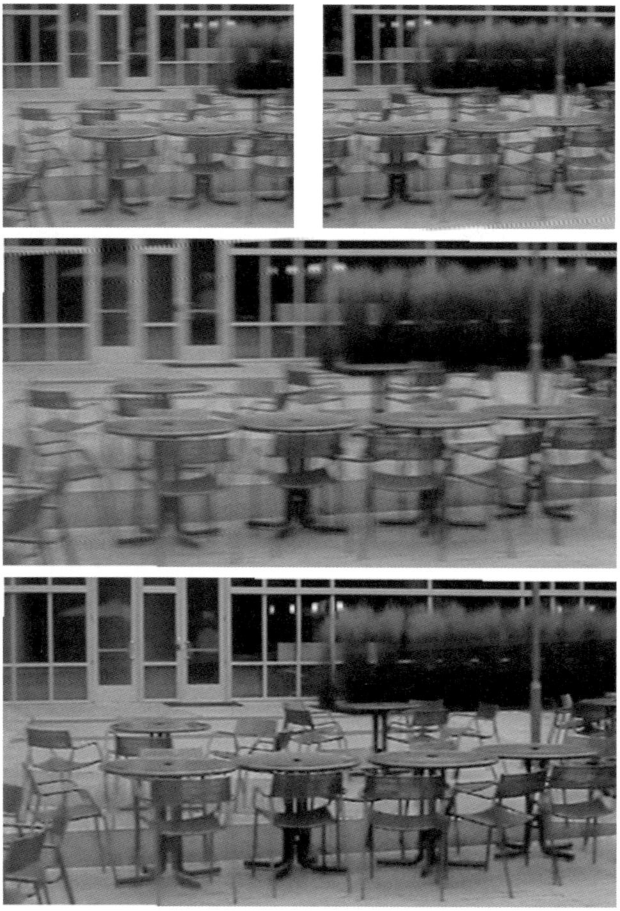

Figure 2.7 Stitching example. First row: input frames (only first and last frames shown); second row: result of directly stitching the input frames; third row: result of our technique. [Reproduced from Li *et al.* (2010) with permission from IEEE.]

We pose the problem of generating a sharp panorama from a sequence of blurry input photos as that of estimating the camera motion, its duty cycle (i.e. the amount of time the sensor is exposed relative to maximum exposure given the video frame rate), and the sharpened images, where the motion and the duty cycle give us the blur kernel for sharpening. In our approach, all these are estimated jointly by minimizing an energy function in a multi-image deconvolution framework.

The energy minimization is carried out using gradient descent. This is viable under our model since we can compute the derivative of the energy function with respect to motion, duty cycles, as well as latent images. We initialize motion by computing global parametric optical flow between successive video frames using the Lucas–Kanade method (Lucas & Kanade 1981). Duty cycles are set to an initial guess, which does not need to be accurate. Subsequently, we alternate between updating latent images while holding motion and duty cycles constant (i.e. performing deconvolution) and updating motion and duty cycles while holding the latent images constant. The updates are described in Sections 2.5 and 2.5.1, respectively. Although in theory all three sets of variables can be optimized simultaneously, the alternating scheme has a much smaller and hence manageable memory footprint and is effective in practice.

Multi-image deconvolution

Multi-image deconvolution can be formulated as an extension of the single image model discussed in Section 2.2.2. Let $B_1, \ldots, B_n$ be the sequence of observed images with motion blur and $I_1, \ldots, I_n$ be the corresponding underlying latent images without motion blur. If we regard each image as an m-dimensional vector, where m is the number of pixels, then the spatially-varying motion blur can be represented as sparse $m \times m$ matrices $\mathcal{K}_1, \ldots, \mathcal{K}_n$, namely

$$B_i = \mathcal{K}_i I_i + N_i \tag{2.28}$$

for each image $i \in \{1, \ldots, n\}$, where N_i is the noise. Recall that $\mathcal{K}_i$ is parameterized by camera motion, as discussed in Section 2.2. Similarly, let $A_{i,j}$ denote the warping to frame i from frame j, which is determined also by the relative motion, i.e.

$$I_i = A_{i,j} I_j. \tag{2.29}$$

Hence

$$B_i = \mathcal{K}_i A_{i,j} I_j + N_i. \tag{2.30}$$

Assuming Gaussian noise, the maximum-likelihood estimate for frame j is then obtained by minimizing the energy function

$$E_{\mathrm{ML}}(I_j) = \sum_{i=j_-}^{j_+} \left\| D_i^{-1} \left(\mathcal{K}_i A_{i,j} I_j - B_i \right) \right\|^2, \tag{2.31}$$

where $j_- = \max(j - r, 1), j_+ = \min(j + r, n)$, D_i is a diagonal matrix whose entries are the standard deviations of noise at each pixel in the ith image, and r is the number of nearby observations to include in each temporal direction. In our work, r is typically in the range of 1 to 3. Note that if $r = 0$, the problem reduces to single image deconvolution. Because of noise in the observation B_i as well as empirical errors in the recovered warping $A_{i,j}$ and blur $\mathcal{K}_i$, a common approach is to introduce an image prior on I that typically regularizes its gradients (e.g. Levin *et al.* (2007)). Hence the maximum a posteriori estimate corresponds to the minimum of the energy function

$$E_{\mathrm{MAP}}(I_j) = \sum_{i=j_-}^{j_+} \left\| D_i^{-1} \left(\mathcal{K}_i A_{i,j} I_j - B_i \right) \right\|^2 + \phi(I_j), \tag{2.32}$$

where $\phi(\cdot)$ is the functional form of the prior. The overall energy function can be minimized with respect to the latent images I_j using gradient-based MAP-deconvolution techniques (e.g. Levin *et al.* (2007), Nocedal (1980)).

2.5.1 Motion and duty cycle estimation

In this section, we describe how to refine motion and duty cycles given the latent images. Again, let $\mathbf{B} = (B_1, \ldots, B_n)$ be the blurred video frames and $\mathbf{I} = (I_1, \ldots, I_n)$ be the underlying sharp frames that we want to recover. Let $\mathbf{H} = (H_1, \ldots, H_n)$ be the warps to each frame from some reference frame. Let $\boldsymbol{\tau} = (\tau_1, \ldots, \tau_n)$ denote the duty cycles of each frame. We denote $\boldsymbol{\psi} = (\mathbf{H}, \boldsymbol{\tau})$ for notational convenience. Hence both $A_{i,j}$ and $\mathcal{K}_i$ (defined in previous sections) are functions of $\mathbf{H}$ and $\boldsymbol{\tau}$, and we will subsequently write them as $A^{\boldsymbol{\psi}}_{i,j}$ and $\mathcal{K}^{\boldsymbol{\psi}}_i$ to reflect this. Since the correct warps and duty cycles should result in a deblurred output with lower energy than incorrect ones, it is desirable to minimize Eq. (2.32) over the whole sequence with respect to these variables as well. Hence we aim to minimize the following energy function

$$E(\mathbf{I}, \boldsymbol{\psi}) = \sum_{j=1}^{n} \sum_{i=j_-}^{j_+} \left\| D_i^{-1} \left(\mathcal{K}^{\boldsymbol{\psi}}_i A^{\boldsymbol{\psi}}_{i,j} I_j - B_i \right) \right\|^2 + \phi\left(I_j\right). \tag{2.33}$$

The minimization of Eq. (2.33) with respect to $\mathbf{I}$ amounts to MAP-deconvolution, which we already addressed in the previous section; therefore the rest of this section will describe how to minimize it with respect to $\boldsymbol{\psi}$.

Pure translation

We start with pure translation for simplicity of presentation. In this case, the warps $\mathbf{H}$ can be represented by the corresponding 2-component vectors $\mathbf{h} = (\mathbf{h_1}, \ldots, \mathbf{h_n})$. Thus we let $\boldsymbol{\psi} = (\mathbf{h}, \boldsymbol{\tau})$ for pure translation. Since the image prior $\phi(I_j)$ does not depend on $\boldsymbol{\psi}$, it can be ignored as far as the optimization of $\boldsymbol{\psi}$ is concerned. Also for conciseness we omit the noise matrices D^{-1} from the subsequent notation, since it is simply a weighting factor on each pixel. Furthermore, we will write $E(\mathbf{I}, \boldsymbol{\psi})$ as simply $E(\boldsymbol{\psi})$ since minimization is with respect to $\boldsymbol{\psi}$. Hence

$$E(\boldsymbol{\psi}) = \sum_{j=1}^{n} \sum_{i=j_-}^{j_+} \left\| \mathcal{K}^{\boldsymbol{\psi}}_i A^{\boldsymbol{\psi}}_{i,j} I_j - B_i \right\|^2. \tag{2.34}$$

Let $I^{\boldsymbol{\psi}}_{i,j} = A^{\boldsymbol{\psi}}_{i,j} I_j$ and $B^{\boldsymbol{\psi}}_{i,j} = \mathcal{K}^{\boldsymbol{\psi}}_i I^{\boldsymbol{\psi}}_{i,j} = \mathcal{K}^{\boldsymbol{\psi}}_i A^{\boldsymbol{\psi}}_{i,j} I_j$, i.e. $I^{\boldsymbol{\psi}}_{i,j}$ is the sharp frame i obtained by warping I_j, and $B^{\boldsymbol{\psi}}_{i,j}$ is the blurred version of $I^{\boldsymbol{\psi}}_{i,j}$. Therefore,

$$E(\boldsymbol{\psi}) = \sum_{j=1}^{n} \sum_{i=j_-}^{j_+} \sum_{\mathbf{p} \in \text{pixels of } B_i} \delta B_{i,j,\mathbf{p}}(\boldsymbol{\psi}), \tag{2.35}$$

where

$$\delta B_{i,j,\mathbf{p}}(\boldsymbol{\psi}) = \left(B^{\boldsymbol{\psi}}_{i,j}(\mathbf{p}) - B_i(\mathbf{p}) \right)^2. \tag{2.36}$$

Here we use $\mathbf{p}$ to denote the 2-component vector representing a pixel's location $(x_p, y_p)^T$ in the image. Thus it suffices to find the derivative of $\delta B_{i,j,\mathbf{p}}(\boldsymbol{\psi})$, which in turn depends on the derivative of $B^{\boldsymbol{\psi}}_{i,j}(\mathbf{p})$ (with respect to $\boldsymbol{\psi}$). Recall that in our model the blur kernel is determined by the relative warps to the two adjacent frames due to the assumption of piecewise linear motion, i.e.

$$B_{i,j}^{\psi}(\mathbf{p}) = \int_{t=-\tau_i}^{\tau_i} I_{i,j}^{\psi}\left(\mathbf{p} + t\left(\mathbf{h}_{\mathbf{i}+\mathrm{sign}(\mathbf{t})} - \mathbf{h}_{\mathbf{i}}\right)\right) dt, \tag{2.37}$$

where sign(t) is 1 if t is positive and -1 otherwise. Thus it can be approximated by a sequence of sampled points on the motion path for each point,

$$\begin{aligned} B_{i,j}^{\psi}(\mathbf{p}) &= \frac{1}{2s+1} \sum_{k=-s}^{s} I_{i,j}^{\psi}\left(\mathbf{p} + \frac{|k|}{2s+1} \tau_i \left(\mathbf{h}_{\mathbf{i}+\mathrm{sign}(\mathbf{k})} - \mathbf{h}_{\mathbf{i}}\right)\right) \\ &= \frac{1}{2s+1} \sum_{k=-s}^{s} I_j\left(\mathbf{p} + \frac{|k|}{2s+1} \tau_i \left(\mathbf{h}_{\mathbf{i}+\mathrm{sign}(\mathbf{k})} - \mathbf{h}_{\mathbf{i}}\right) + \left(\mathbf{h}_{\mathbf{i}} - \mathbf{h}_{\mathbf{j}}\right)\right), \end{aligned} \tag{2.38}$$

where constant s is the number of samples in each temporal direction (independent of the duty cycles) used to approximate the blur kernel. In our case $s = 50$. The derivative of $B_{i,j}^{\psi}(\mathbf{p})$ is

$$\frac{d}{d\psi} B_{i,j}^{\psi}(\mathbf{p}) = \nabla I_j\left(\mathbf{p}_{\mathbf{i},\mathbf{j},\mathbf{k}}^{\psi}\right) \cdot \frac{1}{2s+1} \sum_{k=-s}^{s} \frac{d}{d\psi} \mathbf{p}_{\mathbf{i},\mathbf{j},\mathbf{k}}^{\psi}, \tag{2.39}$$

where

$$\mathbf{p}_{\mathbf{i},\mathbf{j},\mathbf{k}}^{\psi} = \mathbf{p} + \frac{|k|}{2s+1} \tau_i \left(\mathbf{h}_{\mathbf{i}+\mathrm{sign}(\mathbf{k})} - \mathbf{h}_{\mathbf{i}}\right) + \left(\mathbf{h}_{\mathbf{i}} - \mathbf{h}_{\mathbf{j}}\right) \tag{2.40}$$

and ∇I_j is the image gradient of I_j. In the case of $i = n$ and $k > 0$, $\mathbf{h}_{\mathbf{i}} - \mathbf{h}_{\mathbf{i}+\mathbf{1}}$ is replaced with $\mathbf{h}_{\mathbf{i}-\mathbf{1}} - \mathbf{h}_{\mathbf{i}}$ as an approximation (since $\mathbf{h}_{\mathbf{n}+\mathbf{1}}$ does not exist). The case of $i = 1$ and $k < 0$ is handled similarly.

Full homography

The case for a homography is analogous. Recall that $\mathbf{H} = (H_1, \ldots, H_n)$ defines the warps to each frame from some reference frame, and the relative warp to frame i from frame j is thus $H_{i,j} = H_j^{-1} H_i$. Let $\mathbf{p}$ now denote the homogeneous coordinates of a pixel, i.e. $\mathbf{p} = (x_p, y_p, 1)^T$. Thus to extend the relationship between the energy function and ψ from translation to general homography, we only need to rewrite Eqs. (2.38) and (2.40), so that

$$B_{i,j}^{\psi}(\mathbf{p}) = \frac{1}{2s+1} \sum_{k=-s}^{s} I_{i,j}^{\psi}\left(\mathbf{p}_{\mathbf{i},\mathbf{j},\mathbf{k}}^{\psi}\right) \tag{2.41}$$

with

$$\mathbf{p}_{\mathbf{i},\mathbf{j},\mathbf{k}}^{\psi} = \Pi\left(H_{i,j}\left[\mathbf{p} + \frac{|k|}{2s+1} \tau_i \left(\Pi\left(H_{i+\mathrm{sign}(k),i} \cdot \mathbf{p}\right) - \mathbf{p}\right)\right]\right) \tag{2.42}$$

where $\Pi(\cdot)$ is the projection of points in homogeneous coordinates onto the image plane $z = 1$, i.e.

$$\Pi\left((x, y, z)^T\right) = \left(\frac{x}{z}, \frac{y}{z}, 1\right)^T. \tag{2.43}$$

Equation (2.42) uses the approximation that every pixel is moving at constant velocity in between successive frames. The approximation is reasonable for videos since one can expect the perspective change between successive frames to be small. We make the further approximation that $\Pi(H_{i+\text{sign}(k),i} \cdot \mathbf{p}) \approx H_{i+\text{sign}(k),i} \cdot \mathbf{p}$, which is valid since the perspective change in $H_{i+\text{sign}(k),i}$ is small. Hence Eq. (2.42) simplifies to

$$\mathbf{p}^{\psi}_{\mathbf{i},\mathbf{j},\mathbf{k}} = \Pi\left(H_j^{-1}\left[H_i + \frac{|k|}{2s+1}\tau_i\left(H_{i+\text{sign}(k)} - H_i\right)\right]\mathbf{p}\right). \tag{2.44}$$

The derivatives with respect to warps and duty cycles can be obtained using standard matrix calculus with the chain rule. Hence, the energy function can be minimized via gradient-based optimization methods (L-BFGS (Nocedal 1980) in our case).

2.5.2 Experiments

We evaluate our model on both synthetic and real videos. For energy minimization, we use limited-memory BFGS (L-BFGS) of Nocedal (1980). Compared with iterative reweighted least squares (IRLS), L-BFGS does not have the reweighting step and requires only the value and the derivative of the energy function. In our experiments we found it to converge slightly faster than IRLS and reach comparable energy levels. As outlined in Section 2.5, an image prior is used to regularize the output. The noise levels of the input videos are automatically estimated using the method of Liu *et al.* (Liu, Szeliski, Kang, Zitnick & Freeman 2008).

Synthetic videos

For quantitative evaluation, we generated synthetic blurred videos with ground truth information. This is done by moving a virtual video camera in front of a high resolution image, which serves as the scene to be recorded. Motion blur is generated by temporally averaging consecutive frames. Since the virtual camera can have an arbitrarily high frame rate, we have dense enough samples to accurately approximate the motion blur kernel. We output the middle frame of the samples (of successive frames) that are used to produce the blurred frame as the corresponding "true" sharp frame.

Figure 2.8 shows a sample frame from deblurring a sequence with synthesized motion blur. The goal is to compare multi-image deconvolution against single-image deconvolution. In this particular experiment, motion from an initial motion estimation and the known duty cycle value were used. Hence no iterative refinement was performed. We also did not use any image prior in this case since the purpose here is to compare multi-image deconvolution against the single-image counterpart, the two of which differ only in the data term (i.e. the part of the energy function excluding the image prior). For this

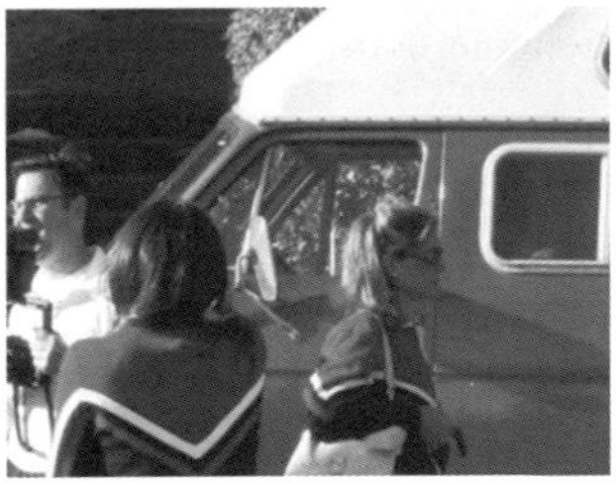

Ground truth

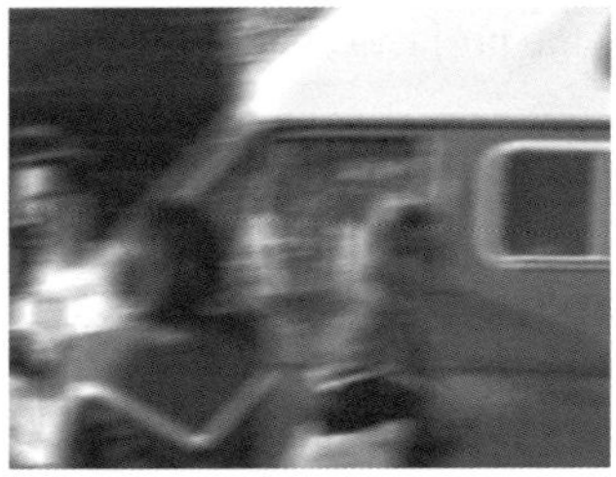

Blurred

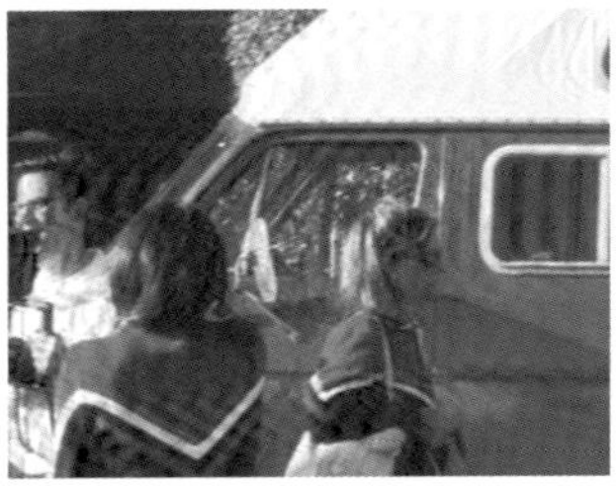

Single-image deconvolution

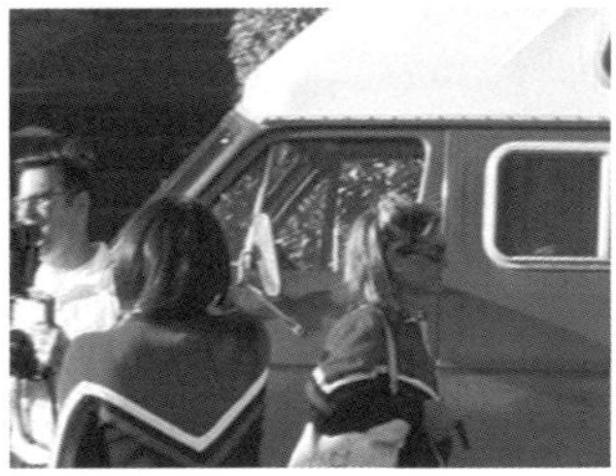

Multi-image deconvolution

Figure 2.8 Sample output frame from a synthetic video sequence. The camera undergoes translational motion. The result of multi-image deconvolution is sharper and has fewer ringing artifacts than that of single-image deconvolution. [Reproduced from Li *et al.* (2010) with permission from IEEE.]

experiment, seven blurred frames are used for reconstructing each deblurred frame for multi-image deconvolution (the temporal radius $r = 3$). The results show that the image restored using multi-image deconvolution exhibits noticeably fewer ringing artifacts, demonstrating the advantage of using multiple observations.

Figure 2.9 shows the errors of motion, duty cycles, and restored frames, over the whole 11-frame video sequence, plotted against the number of iterations completed. The plots correspond to four different sequences with combinations of two types of camera motions (translation and perspective rotation) and two duty cycle values (approximately half and full). A temporal radius of 1 is used in these experiments. Motion errors are expressed as average end-point errors (Baker, Scharstein, Lewis, Roth, Black & Szeliski 2007) of warps between successive frames; restoration errors are in the range 0–255. In all these experiments, the duty cycles are initialized to 1 for the entire sequence. As can be observed from the figures, the duty cycles converge to the correct values even when the initial value is quite far away from the true values. In the case where the initialization is close to the ground truth, the estimated values remain close to the true values over the iterations. The experiments show that the estimation of duty cycles is quite robust. Finding the correct duty cycles is essential for determining the blur kernel size and subsequently for achieving good deblur performance. This is reflected in the plots, where the restoration error is highly correlated with the duty cycle error. Figure 2.10 shows a sample frame of the deblurred output using the initial incorrect duty cycles (left) and using the refined duty cycles (right) (after the last iterations). The difference is dramatic.

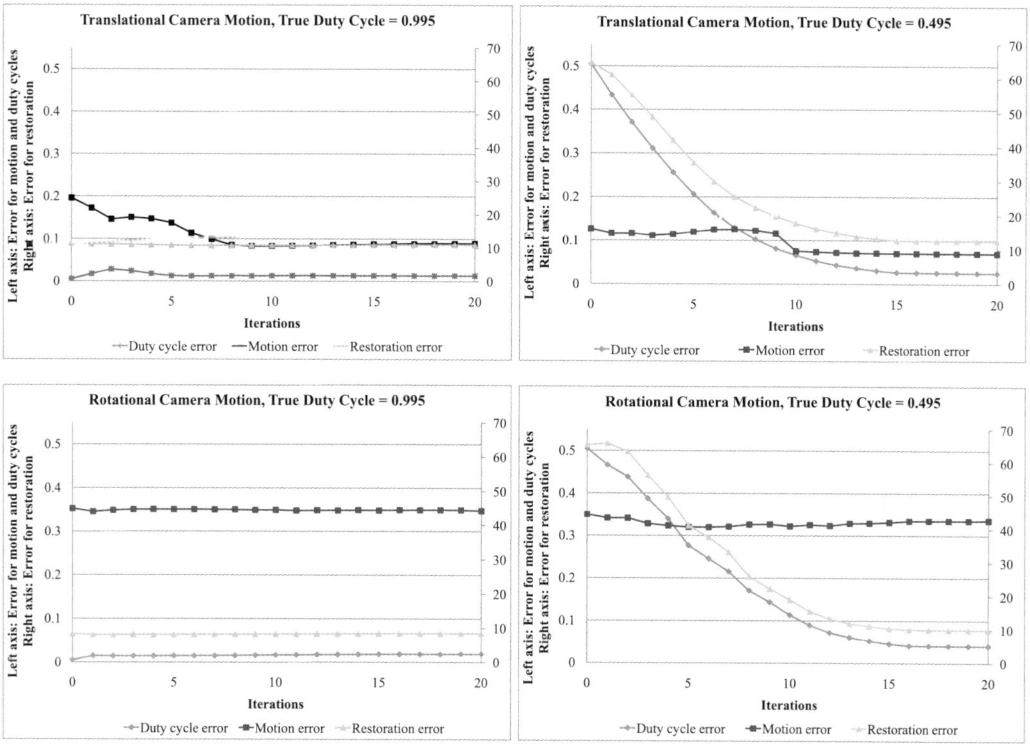

Figure 2.9 Errors of motion, duty cycles, and restoration (i.e. deblur output) on synthetic video sequences, plotted against the number of iterations completed. For the last two, the camera rotation is out-of-plane, hence the perspective change. [Reproduced from Li *et al.* (2010) with permission from IEEE.]

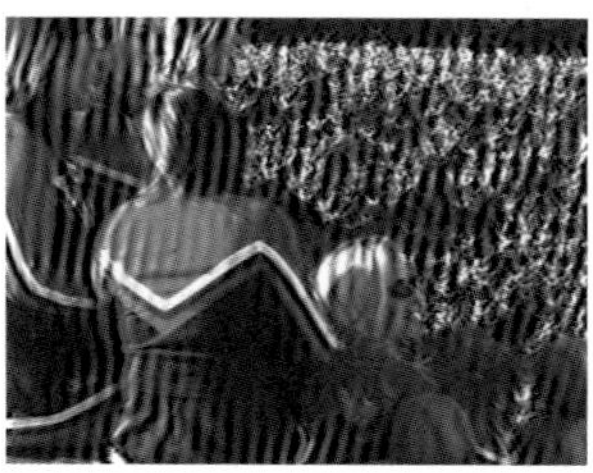
Initial output (iteration 0)

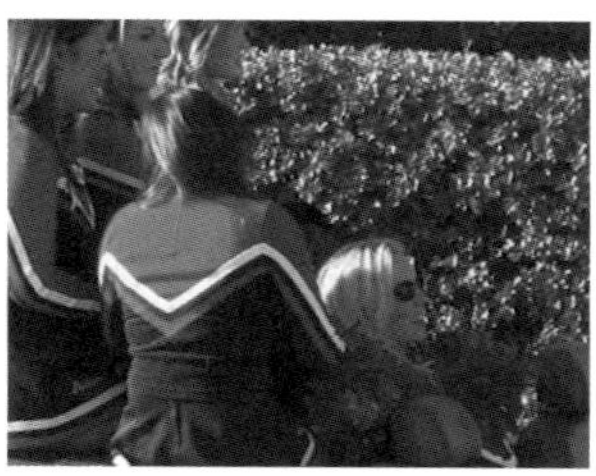
Final output (last iteration)

Figure 2.10 Comparison of sample frames from the initial and the final deblurred output on the synthetic sequence with rotational camera motion and a true duty cycle value of 0.495 (the last plot in Figure 2.9). The initial duty cycle value is 1. [Reproduced from Li *et al.* (2010) with permission from IEEE.]

One may notice that the decrease in motion error is far less than the decrease in duty cycle and restoration errors, and that the sequences with perspective rotation have higher motion error than the translation sequences. This is likely due to the approximations in handling homographies, as described in Section 2.5.1. This, however, does not pose a problem, since the motion estimation is still adequately accurate for reliable restoration.

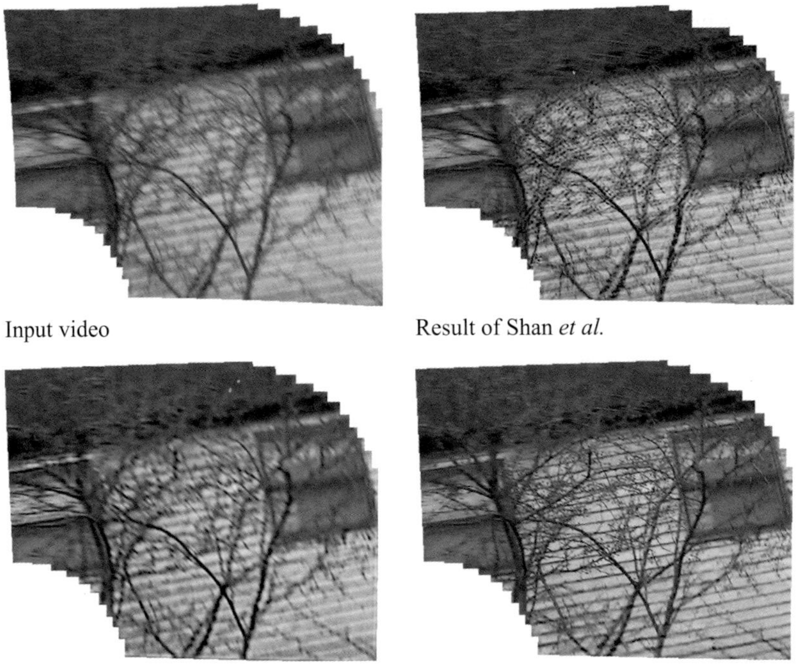

Figure 2.11 Stitched panoramas. The input video was taken by a Canon SD1100. Gamma correction is performed for all three methods. [Reproduced from Li *et al.* (2010) with permission from IEEE.]

2.5.3 Real videos

Real-world video sequences are simply collected using the video mode of a point-and-shoot camera. We used a Canon SD1100, a Canon A540, and a Flip HD for this purpose. For the SD1100 and the A540, we performed gamma correction using published camera response functions (CRF). We were not able to find the CRF for the Flip HD, and thus simply used a gamma value of 1.8. For all the experiments, the initial duty cycle was set to one-quarter, one-half, or one, depending on whether the sequence appears to have a mild, moderate, or high degree of blur, respectively, and a temporal radius of $r = 1$ is used.

Figure 2.11 shows the panorama stitched from a blurry video, and those stitched from the deblurred outputs of Shan *et al.* (2008), Joshi *et al.* (2008), and our method, using Microsoft Image Composite Editor (ICE). Our method produces a far more pleasant result than the competing methods.

2.6 Discussion

In this chapter, we presented a unified model of camera-shake blur that models spatially-varying blur as a function of 6D camera motion. We presented three methods for estimating the spatially-varying blur that are fully automatic and do not require any user input. The methods cover a spectrum of input data from single images, to hybrid image and sensor data, to multiple images.

In all these methods, we have assumed the scene is static and the depth is planar. A logical next step is to allow for some amount of local scene motion and depth variation. In this case, we would down-weight areas that are not consistent with the global motion. This would also allow us to potentially remove moving objects or, alternatively, keep a moving object intact by reinserting it at one selected time.

Acknowledgements

This chapter is based on three recently published papers: Joshi *et al.* (2010), Gupta *et al.* (2010), and Li *et al.* (2010). We would like to thank the other co-authors of these papers, namely, Ankit Gupta, Larry Zitnick, Michael Cohen, Brian Curless, Yunpeng Li, Steven Seitz, and Dan Huttenlocher.

References

Agrawal, A., Xu, Y. & Raskar, R. (2009). Invertible motion blur in video. In *ACM Special Interest Group on Graphics and Interactive Techniques*, **28**(3), 95:1–8.

Baker, S., Scharstein, D., Lewis, J. P., Roth, S., Black, M. J. & Szeliski, R. (2007). A database and evaluation methodology for optical flow. In *IEEE International Conference on Computer Vision*, pp. 1–8.

Bascle, B., Blake, A. & Zisserman, A. (1996). Motion deblurring and super-resolution from an image sequence. In *European Conference on Computer Vision*, pp. 571–82.

Ben-Ezra, M. & Nayar, S. K. (2004). Motion-based motion deblurring. *IEEE Transactions on Pattern Analysis and Machine Intelligence*, **26**(6), 689–98.

Cho, S., Matsushita, Y. & Lee, S. (2007). Removing non-uniform motion blur from images. In *IEEE International Conference on Computer Vision*, pp. 1–8.

Dai, S. & Wu, Y. (2008). Motion from blur. In *IEEE Conference on Computer Vision and Pattern Recognition*, pp. 1–8.

Fergus, R., Singh, B., Hertzmann, A., Roweis, S. T. & Freeman, W. T. (2006). Removing camera shake from a single photograph. *ACM Special Interest Group on Graphics and Interactive Techniques*, **25**(3), 787–94.

Gupta, A., Joshi, N., Zitnick, L., Cohen, M. & Curless, B. (2010). Single image deblurring using motion density functions. In *European Conference on Computer Vision*, pp. 171–84.

Hirsch, M., Sra, S., Schölkopf, B. & Harmeling, S. (2010). Efficient filter flow for space-variant multiframe blind deconvolution. In *IEEE Conference on Computer Vision and Pattern Recognition*, pp. 607–14.

Jia, J. (2007). Single image motion deblurring using transparency. In *IEEE Conference on Computer Vision and Pattern Recognition*, pp. 1–8.

Jin, H., Favaro, P. & Cipolla, R. (2005). Visual tracking in the presence of motion blur. In *IEEE Conference on Computer Vision and Pattern Recognition*, pp. 18–25.

Joshi, N., Kang, S. B., Zitnick, C. L. & Szeliski, R. (2010). Image deblurring using inertial measurement sensors. *ACM Transactions on Graphics*, **29**, 30:1–30:9.

Joshi, N., Szeliski, R. & Kriegman, D. (2008). PSF estimation using sharp edge prediction. In *IEEE Conference on Computer Vision and Pattern Recognition*, pp. 1–8.

Joshi, N., Zitnick, L., Szeliski, R. & Kriegman, D. (2009). Image deblurring and denoising using color priors. In *IEEE Conference on Computer Vision and Pattern Recognition*, pp. 1550–7.

Levin, A. (2007). Blind motion deblurring using image statistics. In *Neural Information Processing Systems Conference*, pp. 841–8.

Levin, A., Fergus, R., Durand, F. & Freeman, W. T. (2007). Image and depth from a conventional camera with a coded aperture. *ACM Special Interest Group on Graphics and Interactive Techniques*, **26**(3), 70:1–10.

Li, Y., Kang, S., Joshi, N., Seitz, S. & Huttenlocher, D. (2010). Generating sharp panoramas from motion-blurred videos. In *IEEE Conference on Computer Vision and Pattern Recognition*, pp. 2424–31.

Likas, A. C. & Galatsanos, N. P. (2004). A variational approach for bayesian blind image deconvolution. *IEEE Transactions on Signal Processing*, **52**(8), 2222–33.

Liu, C., Szeliski, R., Kang, S. B., Zitnick, C. L. & Freeman, W. T. (2008). Automatic estimation and removal of noise from a single image. *IEEE Transactions on Pattern Analysis and Machine Intelligence*, **30**(2), 299–314.

Lucas, B. D. & Kanade, T. (1981). An iterative image registration technique with an application to stereo vision. In *International Joint Conference on Artificial Intelligence*, **81**, 674–9.

Nocedal, J. (1980). Updating quasi-newton matrices with limited storage. *Mathematics of Computation*, **35**, 773–82.

Onogi, M. & Saito, H. (2005). Mosaicing and restoration from blurred image sequence taken with moving camera. In *IEEE International Conference on Advances in Pattern Recognition*, pp. 598–607.

Richardson, W. H. (1972). Bayesian-based iterative method of image restoration. *Journal of the Optical Society of America*, **62**(1), 55–9.

Seitz, S. & Baker, S. (2009). Filter flow. In *IEEE International Conference on Computer Vision*, pp. 143–50.

Shan, Q., Jia, J. & Agarwala, A. (2008). High-quality motion deblurring from a single image. *ACM Special Interest Group on Graphics and Interactive Techniques*, **27**(3), 73:1–10.

Shan, Q., Xiong, W. & Jia, J. (2007). Rotational motion deblurring of a rigid object from a single image. In *IEEE International Conference on Computer Vision*, pp. 1–8.

Stewart, C. V. (1999). Robust parameter estimation in computer vision. *Society for Industrial and Applied Mathematics Review*, **41**(3), 513–37.

Tai, Y., Tan, P., Gao, L. & Brown, M. (2009). *Richardson–Lucy deblurring for scenes under projective motion path*, Technical report, Korea Advanced Institute of Science and Technology.

Tai, Y.-W., Du, H., Brown, M. & Lin, S. (2008). Image/video deblurring using a hybrid camera. In *IEEE Conference on Computer Vision and Pattern Recognition*, pp. 1–8.

Tai, Y.-W., Kong, N., Lin, S. & Shin, S. Y. (2010). Coded exposure imaging for projective motion deblurring. In *IEEE Conference on Computer Vision and Pattern Recognition*, pp. 2408–15.

Weiss, Y. & Freeman, W. T. (2007). What makes a good model of natural images? In *IEEE Conference on Computer Vision and Pattern Recognition*, pp. 1–8.

Whyte, O., Sivic, J., Zisserman, A. & Ponce, J. (2010). Non-uniform deblurring for shaken images. In *IEEE Conference on Computer Vision and Pattern Recognition*, pp. 491–8.

Yuan, L., Sun, J., Quan, L. & Shum, H.-Y. (2007). Image deblurring with blurred/noisy image pairs. *ACM Transactions on Graphics*, **26**, pp. 1–10.

3 Hybrid-imaging for motion deblurring

Moshe Ben-Ezra, Yu-Wing Tai, Michael S. Brown and Shree K. Nayar

3.1 Introduction

This chapter introduces a hybrid-imaging system for motion deblurring, which is an imaging system that couples two or more cameras that function differently to perform a unified task. The cameras are usually selected to have different specialized functions. For example, a hybrid stereo camera presented by Sawhney *et al.* (Sawhney, Guo, Hanna, Kumar, Adkins & Zhou 2001) utilizes two cameras with different spatial resolutions to obtain high resolution stereo output.

In the context of this chapter, a hybrid-imaging system refers to a standard high resolution camera, which we call the *primary detector* with an auxiliary low resolution camera called the *secondary detector*. The secondary detector shares a common optical path with the primary detector, but operates at a significantly higher frame rate.

The primary detector produces a high resolution, high quality colour image but is susceptible to motion blur, whereas the secondary detector output is a *sequence* of low resolution, often monochromatic and noisy images of the same scene taken during the exposure time of the primary detector. An example of the primary and secondary detectors' output is shown in Figure 3.1.

The image sequence produced by the secondary detector is of little visual use. However, it contains information about the *motion* of the camera during the exposure, or more precisely, the motion flow field of the image during integration time. While the camera motion and the observed flow field are not identical (e.g. the observed flow field includes information about moving objects in the scene as well as their depth), the idea of hybrid-imaging motion deblurring is that given the image sequence from the secondary detector, it would be possible to compute the blur function (or the PSF) at every point of the high resolution image taken by the primary detector. Using this PSF it is possible to subsequently deblur the primary detector's image to obtain the sought-after result – a high resolution non-blurred image.

3.2 Fundamental resolution tradeoff

An image is formed when light energy is integrated by an image detector over a time interval. The total light energy received by a pixel during integration must be above a minimum level for the light to be detected. This minimum level is determined by the

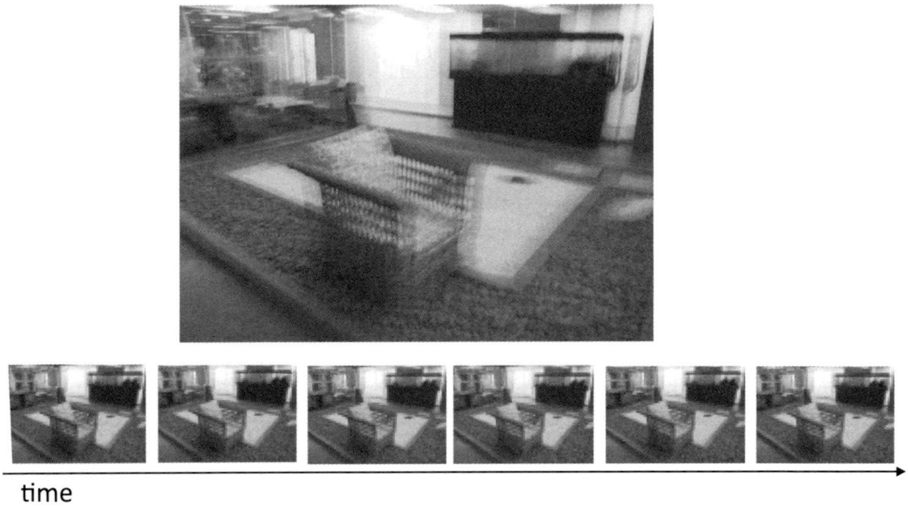

Figure 3.1 Tradeoff between resolution and frame rates. Top: image from a high resolution, low frame-rate camera; bottom: images from a low resolution, high frame-rate camera. [Adopted from Tai *et al.* (2010).]

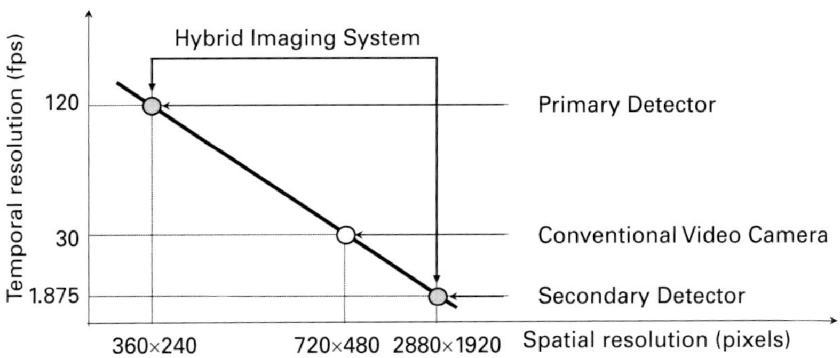

Figure 3.2 The fundamental tradeoff between spatial resolution and temporal resolution of an imaging system. While a conventional video camera (white circle) is a single operating point on the tradeoff line, a hybrid-imaging system uses two different operating points (grey circles) on the line, *simultaneously*. This feature enables a hybrid-imaging system to obtain the additional information needed for non-blind motion deblurring. [Adopted from Ben-Ezra & Nayar (2004).]

signal-to-noise characteristics of the detector. Therefore, given such a minimum level and an incident flux level, the exposure time required to ensure detection of the incident light is inversely proportional to the area of the pixel. In other words, exposure time is proportional to spatial resolution. When the detector is linear in its response, the above relationship between exposure and resolution is also linear. This is the fundamental tradeoff between the spatial resolution (number of pixels) and the temporal resolution (number of images per second).

This tradeoff is illustrated by the solid line in Figure 3.2. The parameters of this line are determined by the characteristics of the materials used by the detector, its efficiency, and the incident flux. Different points on the line represent cameras with different spatio-temporal characteristics. For instance, a conventional video camera (shown as a white circle) has a typical temporal resolution of 30 fps and a spatial resolution of 720×480 pixels.

Instead of relying on a single point on this tradeoff line, we could use two very different operating points on the line to *simultaneously* obtain very high spatial resolution with low temporal resolution, and very high temporal resolution with low spatial resolution. This type of hybrid-imaging system is illustrated by the two grey circles in Figure 3.2. This combined hybrid-imaging system is able to capture the necessary additional information needed for non-blind motion deblurring.

3.3 Hybrid-imaging systems

We now describe three conceptual designs for a hybrid-imaging system. The simplest design, which is illustrated in Figure 3.3(a), uses a rigid rig of two cameras: a high resolution still camera as the primary detector and a low resolution video camera as the secondary detector. Since the secondary detector is used for obtaining motion information it can be monochromatic to collect more light energy. While a simple design, the geometric calibration between the primary and secondary detectors is more complex because the two cameras do not share the same optical axes. Additionally, this design

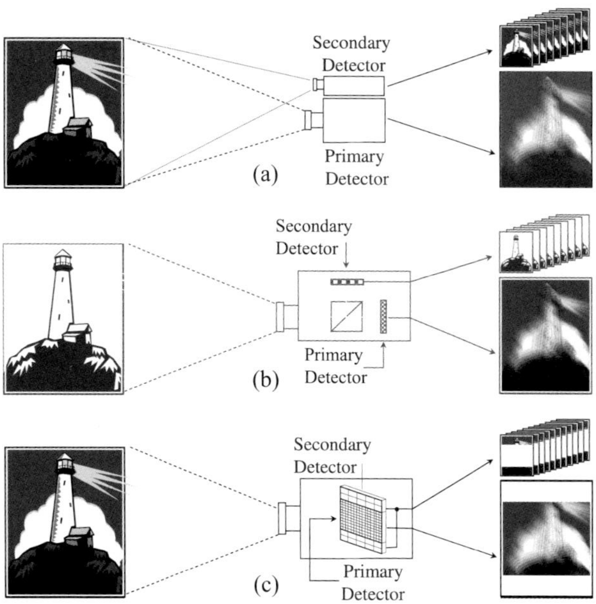

Figure 3.3 Three conceptual designs of a hybrid-imaging system. (a) The primary and secondary detectors are essentially two separate cameras; (b) the primary and secondary detectors share the same lens by using a beamsplitter; (c) the primary and secondary detectors are located on the same chip with different resolutions (pixel sizes). [Adapted from (Ben-Ezra & Nayar 2004).]

requires recalibration when the main lens changes or its zoom setting varies. This problem is addressed by the following two designs.

The second design shares the same lens for both detectors by splitting the incoming light using a beamsplitter. This design, which is shown in Figure 3.3(b), requires simpler calibration than the previous one due to the shared optics. An asymmetric beamsplitter that passes most of the visible light to the primary detector and reflects non-visible wavelengths towards the secondary detector, for example a 'hot mirror', would be preferred.

A third conceptual design, which is illustrated in Figure 3.3(c), uses a special chip layout that includes the primary and the secondary detectors on the same chip. This chip has a high resolution central area (the primary detector) and a low resolution periphery (the secondary detector). Clearly, in this case, the primary and the secondary detectors would not have the same field of view. However they will have sufficient data to compute parametric motion models such as translation, similarity, or homographic models. Such a chip can be implemented using binning technology now commonly found in CMOS (and CCD) sensors. Binning allows the charge of a group of adjacent pixels to be combined before digitization. This enables the chip to switch between a normal full-resolution mode (when binning is off) and a hybrid primary–secondary detector mode (when binning is activated).

Figure 3.4(a) shows a prototype hybrid-imaging system using the camera rig design. The primary detector of the system is a 3M pixel (2048 × 1536) Nikon CoolPix digital camera equipped with a ×6 Kenko zoom lens. The secondary detector is a Sony DV camcorder. The original resolution of the camcorder (720 × 480) was reduced to 360 × 240 to simulate a low resolution detector. The two sensors were calibrated using a calibration target when the camera was static. The sensors are synchronized using a point

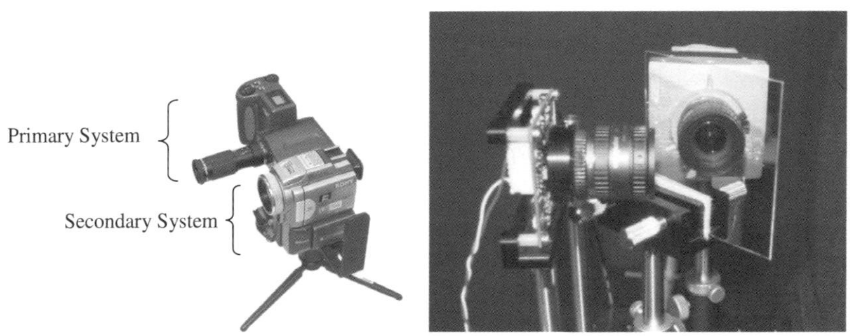

(a) (b)

Figure 3.4 The two implemented hybrid-imaging system prototypes. (a) A rig of two cameras: the primary system consists of a 3M pixel Nikon CoolPix camera equipped with a ×6 Kenko zoom lens; the secondary system is a Sony DV camcorder. The Sony images were reduced in size to simulate a low resolution camera. [Adapted from (Ben-Ezra & Nayar 2004).] (b) A Point Grey Dragonfly 2 camera, which captures images of 1024 × 768 resolution at 25 fps (6.25 fps for image deblurring examples), combined with a Mikrotron MC1311 camera that captures images of 128 × 96 resolution at 100 fps. A beamsplitter is used to align their optical axes and respective images. Video synchronization is achieved using a 8051 microcontroller. [Adapted from (Tai, Du, Brown & Lin 2010).]

of light feedback from the flash unit of the primary detector via a small reflector (not shown).

The second design, shown in Figure 3.4(b), shows another prototype hybrid-imaging system which is a variant of the beamsplitter design. The two cameras share the same optical path and have well-aligned pixel arrays but have different lenses. Camera synchronization is achieved using an external 8051 microcontroller.

3.4 Shift-invariant PSF image deblurring

The secondary detector captures a sequence of images at discrete time intervals that code information regarding the motion flow and blur function of the primary detector. In order to be able to use this information for motion deblurring we must first obtain the discrete motion information from the image sequence and then compute a continuous blur function that will be used at the motion deblurring stage.

3.4.1 Parametric motion computation

The first step is computing a 2D rigid (translation + rotation) parametric motion between successive frames. We compute this motion using a multi-resolution iterative algorithm that minimizes the following optical flow based error function (Lucas & Kanade 1981):

$$\arg\min_{(u,v)} \sum \left(u\frac{\partial I}{\partial x} + v\frac{\partial I}{\partial y} + \frac{\partial I}{\partial t} \right)^2 \tag{3.1}$$

where $\frac{\partial I}{\partial x}, \frac{\partial I}{\partial y}, \frac{\partial I}{\partial t}$ are the spatial and temporal partial derivatives of the image, and (u, v) is the instantaneous motion at time t. This motion between the two frames is defined by the following global rigid motion model:

$$\begin{bmatrix} u \\ v \end{bmatrix} = \begin{bmatrix} \cos\theta & \sin\theta & \Delta x \\ -\sin\theta & \cos\theta & \Delta y \end{bmatrix} \begin{bmatrix} x \\ y \\ 1 \end{bmatrix} \tag{3.2}$$

where $(\Delta x, \Delta y)$ is the translation vector and θ is the rotation angle about the optical axis.

In practice, the motion vectors are computed relative to a reference frame, typically the middle of the sequence and the instantaneous frame to frame motion is derived from that. This is possible because the overlap between frames is significantly larger than the extent of the motion blur. Computing motion with respect to a reference frame is done to prevent drifting by accumulating small errors between successive frames.

Note that the secondary detector, which has a short but nonzero integration time, may also experience some motion blur, which can violate the constant brightness assumption used in the motion computation. We therefore assume that the computed motion displacement between two (slightly) motion-blurred frames is determined by the centroids of the corresponding blur functions. We refer to this as the *motion centroid assumption*.

3.4.2 Shift-invariant PSF estimation

The simplest blur function that we use is the *shift-invariant point spread function* (SI-PSF), which assumes that the blur function is constant across all pixels of the image. In practice, SI-PSF occurs when a camera equipped with a telephoto lens slightly rotates about its x and y axes, but not about the optical axes, resulting in a nearly constant translational motion flow across the image plane. Such motion is likely to happen with scenic and nature photographs taken by trained photographers, and is less likely to happen with less trained photographers that tend to translate the camera or rotate it about the optical axis, usually by pressing the shutter release button.

The discrete motion samples that are obtained by this motion need to be converted into a continuous point spread function. To do that, we define the constraints that this type of motion blur PSF must satisfy, and then use these constraints in the PSF estimation.

A PSF can be treated as an energy distribution function, which can be represented by a convolution kernel $k : (x, y) \mapsto e$, where (x, y) is a location and e is the energy level at that location. The kernel k must satisfy the following energy conservation constraint:

$$\iint k(x,y)\,dx\,dy = 1, \tag{3.3}$$

which states that energy is neither lost nor gained by the blurring operation, i.e. k is a normalized kernel. In order to define additional constraints that apply to a SI-PSF motion blur, we use a time parameterization of the PSF with a path function $f : t \mapsto (x, y)$ and an energy function $h : t \mapsto e$. Note that the functions f and h define a ridge which belongs to a subset of all possible PSFs. Due to physical speed and acceleration constraints, $f(t)$ should be continuous and at least twice differentiable. By assuming that the scene radiance does not change during image integration, we get the additional constraint:

$$\int_t^{t+\delta t} h(t)\,dt = \frac{\delta t}{t_{\text{end}} - t_{\text{start}}}, \quad \delta t > 0, \quad t_{\text{start}} \leq t \leq t_{\text{end}} - \delta t, \tag{3.4}$$

where $[t_{\text{start}}, t_{\text{end}}]$ is the image integration interval. This constraint states that the amount of energy which is integrated at any time interval is proportional to the length of the interval.

Given these constraints and the motion centroid assumption from the previous section, we can estimate a continuous motion blur PSF from the discrete motion samples, as illustrated in Figure 3.5. First, we estimate the path $f(t)$ by spline interpolation as shown in Figure 3.5(a,b); spline curves are used because of their smoothness and twice differentiability properties, which satisfy the speed and acceleration constraints. In order to estimate the energy function $h(t)$ we need to find the extent of each frame along the interpolated path. This is done using the motion centroid assumption by splitting the path $f(t)$ into frames with a 1D Voronoi tessellation, as shown in Figure 3.5(b). Since the constant radiance assumption implies that frames with equal exposure times integrate equal amounts of energy, we can compute $h(t)$ for each frame as shown in

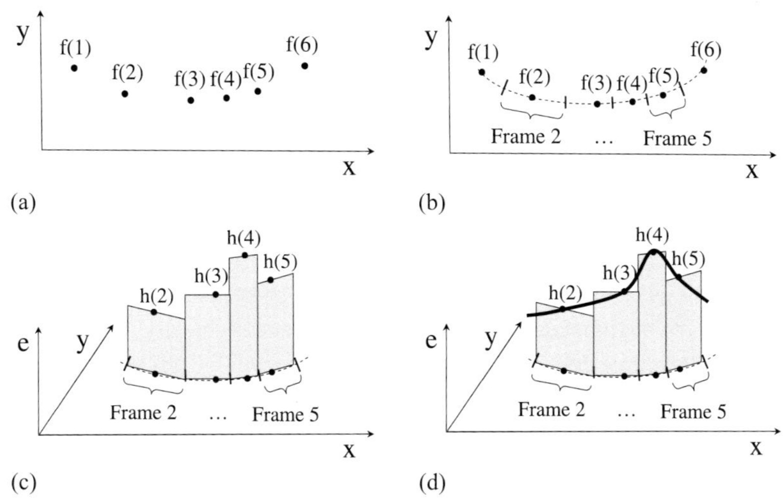

Figure 3.5 The computation of the continuous PSF from the discrete motion vectors. (a) The discrete motion vectors which are samples of the function $f : t \mapsto (x, y)$; (b) interpolated path $f(t)$ and its division into frames by Voronoi tessellation; (c) energy estimation for each frame; (d) the computed PSF. [Adapted from Ben-Ezra & Nayar (2004).]

Figure 3.5(c). Note that all the rectangles in this figure have equal areas. Finally, we smooth $h(t)$ and normalize it to satisfy the energy conservation constraint. The resulting PSF is shown in Figure 3.5(d). The end result of the above procedure is a continuous motion blur PSF that can now be used for motion deblurring.

3.4.3 Image deconvolution

Given the estimated PSF we can deblur the high resolution image that was captured by the primary detector using image deconvolution algorithms. Many such algorithms exist, varying from early algorithms such as the well-known Wiener filter (Wiener 1964), and ratio-based iterative methods (Jansson 1997) to more recent image prior based methods (Krishnan & Fergus 2009, Levin, Fergus, Durand & Freeman 2007, Bioucas-Dias, Figueiredo & Oliveira 2006*a*, Dey, Blanc-Feraud, Zimmer, Roux, Kam, Olivo-Marin & Zerubia 2006, Bioucas-Dias, Figueiredo & Oliveira 2006*b*). The selection of the specific algorithm is application dependent and is orthogonal to the hybrid-imaging framework.

The results reported in this chapter are based on the Richardson–Lucy iterative deconvolution algorithm (Jansson 1997), which is a nonlinear ratio-based method. This method maximizes a Poisson statistics image model likelihood function, yielding the following iteration:

$$\hat{O}^{(k+1)}(x) = \hat{O}^{(k)}(x) \cdot S(-x) \otimes \frac{I(x)}{S \otimes \hat{O}^{(k)}} \tag{3.5}$$

where I is the measured image, $\hat{O}^{(k)}$ is the kth estimation of the result, $\hat{O}^{(0)} = I$, and S is the convolution kernel (the PSG). Given that I and S are positive everywhere, $\hat{O}^{(k)}$ cannot be negative.

3.4.4 Examples – shift-invariant PSF

The basic form of the hybrid-imaging system has been considered so far. Before describing a more advanced form, results are shown for this basic setup. Figures 3.6 and 3.7 show results obtained by the first prototype system shown in Figure 3.4(a). The figures show the input images from the primary and secondary detectors, the recovered PSFs, the resulting deblurred images, and ground truth images taken using a tripod. The ability to handle long exposure times (up to four seconds) and complex PSFs is, as seen in these examples, one of the best advantages of hybrid-imaging framework.

3.4.5 Shift-invariant PSF optimization

While a direct computation of the PSF can already be used for motion deblurring, as seen above, there may still be some inaccuracies in the PSF resulting from noise, insufficient resolution, or insufficient frame rate of the secondary detector.

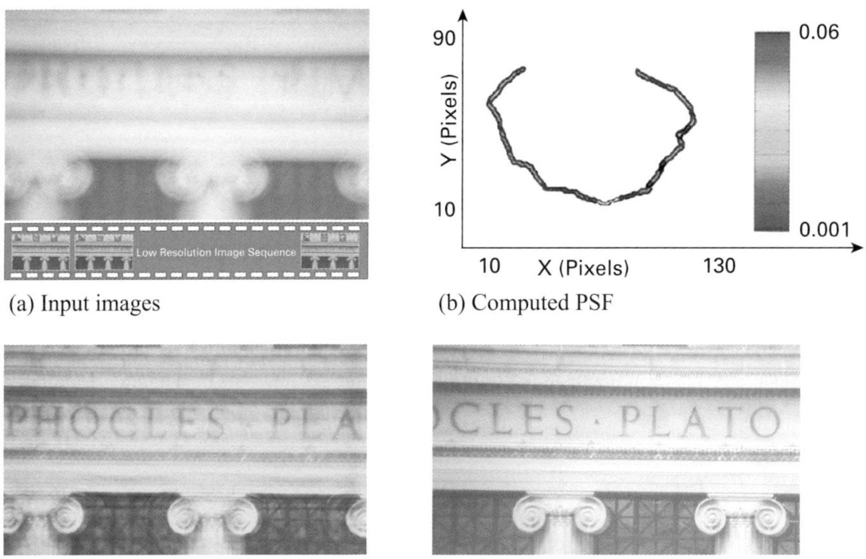

(a) Input images (b) Computed PSF

(c) Computed deblurred image (d) Ground truth image

Figure 3.6 Experimental results for outdoor building scene (focal length = 633 mm, exposure time = 1.0 s). (a) Input images, including the motion-blurred image from the primary detector and a sequence of low resolution frames from the secondary detector; (b) the computed PSF; note the complexity of motion path and the energy distribution; (c) the deblurring result; note the clarity of the text; (d) the ground truth image that was captured without motion blur using a tripod. [Adopted from Ben-Ezra & Nayar (2004).]

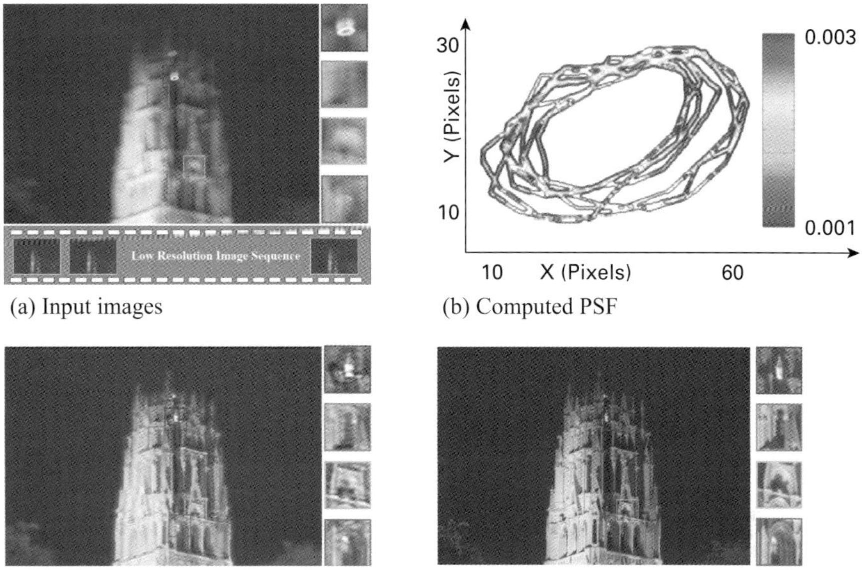

(a) Input images (b) Computed PSF

(c) Computed deblurred image (d) Ground truth image

Figure 3.7 Experimental results for outdoor tower scene (focal length = 884 mm, exposure time = 4.0 s). (a) Input images, including the motion-blurred image from the primary detector and a sequence of low resolution frames from the secondary detector; (b) the computed PSF; note the complexity of the motion path and energy distribution; (c) the deblurring result; (d) the ground truth image that was captured without motion blur using a tripod. [Adopted from Ben-Ezra & Nayar (2004).]

To further refine the PSF, it is possible to jointly use the multiple deconvolution and back-projection constraints available from the hybrid camera input. This approach can be formulated into a maximum a posteriori (MAP) estimation framework as follows:

$$\begin{aligned}
&\arg\max_{I,K} P(I, K|I_b, K_o, I_l) \\
&\quad = \arg\max_{I,K} P(I_b|I, K)P(K_o|I, K)P(I_l|I)P(I)P(K) \\
&\quad = \arg\min_{I,K} L(I_b|I, K) + L(K_o|I, K) + L(I_l|I) + L(I) + L(K)
\end{aligned} \tag{3.6}$$

where I and K denote the sharp images and the PSFs we want to estimate, I_b, K_o, and I_l are the observed blur images, estimated PSFs from optical flows, and the low resolution, high frame-rate images respectively, and $L(\cdot) = -\log(P(\cdot))$. In this formulation, the priors $P(I)$ and $P(K)$ are taken to be uniformly distributed. Assuming that $P(K_o|I, K)$ is conditionally independent of I, that the estimation errors of likelihood probabilities $P(I_b|I, K)$, $P(K_o|I, K)$, and $P(I_l|I)$ follow Gaussian distributions, and that each observation of I_b, K_o, and I_l is independent and identically distributed, we can then rewrite Eq. (3.6) as

$$\arg\min_{I,K} \sum_{i}^{N} \left\| I_{b_i} - I \otimes K_i \right\|^2 + \lambda_B \sum_{j}^{M} \left\| I_{l_j} - d(I \otimes h) \right\|^2 + \lambda_K \sum_{i}^{N} \left\| K_i - K_{o_i} \right\|^2 \tag{3.7}$$

where λ_K and λ_B are the relative weights of the error terms. To optimize the above equation for I and K, an alternating minimization strategy is used, as follows:

1. update $I^{t+1} = I^t + \sum_{i=1}^{N} K_i^t * (I_{b_i} - I^t \otimes K_i^t) + \lambda_B \sum_{j=1}^{M} h \otimes (u(W(I_{l_j}) - d(I^t \otimes h)))$;
2. update $K_i^{t+1} = K_i^t + \widetilde{I}^{t+1} * (I_{b_i} - I^{t+1} \otimes K_i^t) + \lambda_K (K_{o_i} - K_i^t)$

where $\widetilde{I} = I / \sum_{(x,y)} I(x,y)$, $I(x,y) \geqslant 0$, $K_i(u,v) \geqslant 0$, and $\sum_{(u,v)} K_i(u,v) = 1$. The two update steps are processed in alternation until the change in I falls below a specified level or until a maximum number of iterations is reached. The term $W(I_{l_j})$ represents the warped aligned observations. The reference frame to which these are aligned can be any of the M low resolution images. Thus for each deblurred high resolution frame we have up to M possible solutions. In our implementation, we set $N = 3$ in correspondence with the current, previous, and next frames, and M is set according to the relative camera settings (4/16 for video/image deblurring in our implementation). We also initialize I^0 as the currently observed blur image I_b, K_i^0 as the estimated PSF K_{o_i} from optical flows, and set $\lambda_B = \lambda_K = 0.5$.

For more stable and flexible PSF refinement, the PSF is refined in a multi-scale fashion as demonstrated in Fergus, Singh, Hertzmann, Roweis & Freeman (2006), and Yuan, Sun, Quan & Shum (2007). Figure 3.8 illustrates the PSF refinement process. The PSFs are estimated from motion flows of the observed low resolution images and then downsampled to the coarsest level. After refinement at a coarser level, PSFs are then upsampled and refined again. The multi-scale pyramid is constructed using a downsampling factor of $1/\sqrt{2}$ with five levels. The likelihood $P(K_o|K)$ is applied at each level of the pyramid with a decreasing weight, so as to allow more flexibility in refinement at finer levels. Starting at a level coarser than the low resolution images allows the method to recover from some of the errors in the PSF estimation.

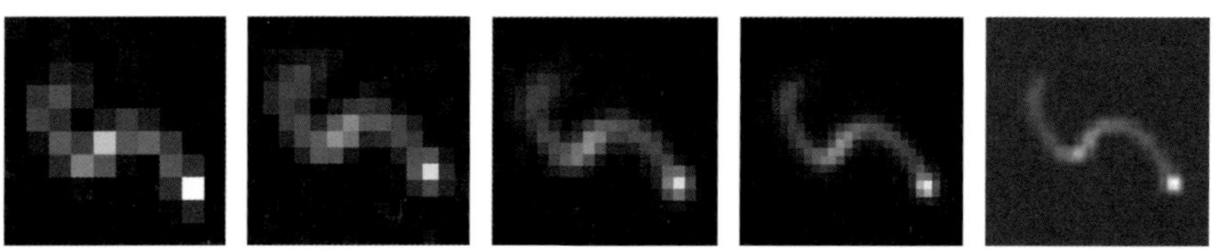

(a) (b) (c) (d) (e)

Figure 3.8 Multi-scale refinement of a motion PSF (a–e) exhibits refined PSFs at progressively finer scales. Our PSF refinement starts from the coarsest level. The result of each coarser level is then upsampled and used as an initial PSF estimate for the next level of refinement. [Adopted from Tai *et al.* (2010).]

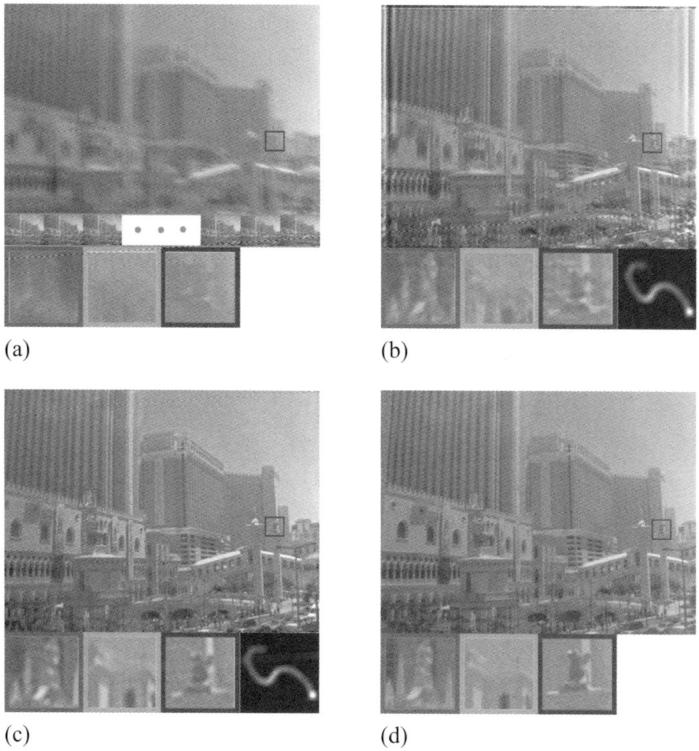

Figure 3.9 Image deblurring using optimized shift-invariant PSF. (a) Input; (b) result generated by a direct (non-optimized) PSF; (c) result generated by an optimized PSF; (d) the ground truth sharp image. PSF and magnified regions are also shown. [Adopted from Tai, Du, Brown & Lin (2010).]

3.4.6 Examples – optimized shift-invariant PSF

Figure 3.9 presents an image deblurring using the hybrid imaging system shown in Figure 3.4(b) and an optimized shift-invariant PSF as described above. In this case the direct estimation of the PSF had a small error, which produces artefacts in the deblurred image. Applying the global optimization strategy, it is possible to refine the PSF to obtain a better result.

3.5 Spatially-varying PSF image deblurring

A sequence of 2D rigid parametric motion described in Section 3.4.1, in the presence of rotations about the optical axis (or possibly homographic or arbitrary optical flow), results in spatially-varying and complex PSFs. These spatially-varying PSFs can be expressed as $K(x, y, u, v)$, where (x, y) is the image coordinate and (u, v) is the PSF coordinate. For large sized PSFs, e.g. 65×65, this representation is impractical due to its enormous storage requirements. One way to reduce the storage size is by constraining

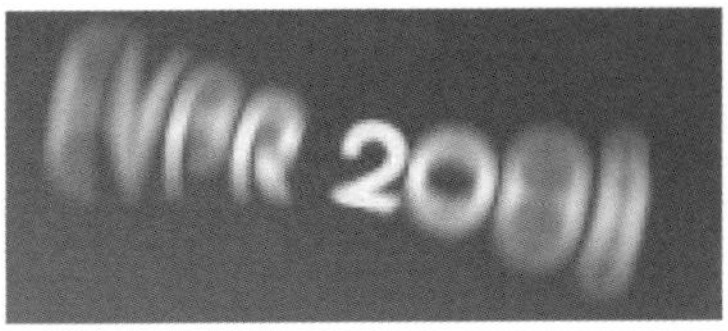

(a)

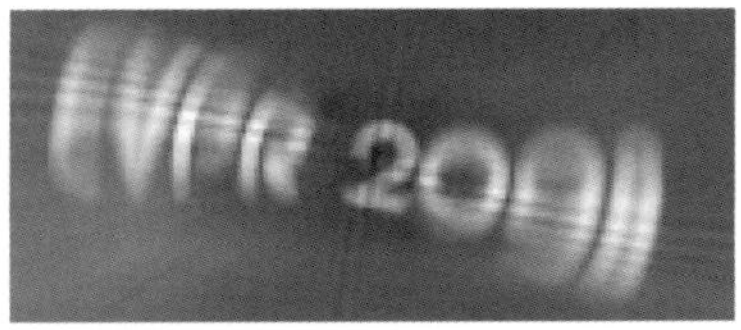

(b)

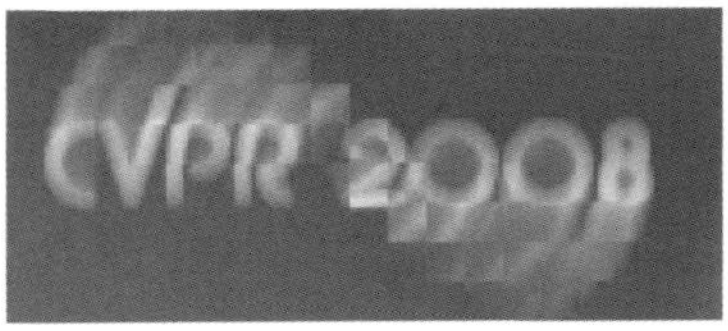

(c)

(d)

Figure 3.10 Convolution with PSF decomposition. (a) Convolution result without PSF decomposition, where full PSFs are generated on-the-fly per pixel using optical flow integration; (b) convolution using 30 PCA-decomposed PSFs; (c) convolution using a patch-based decomposition; (d) convolution using delta function decomposition of PSFs, with at most 30 delta functions per pixel. [Adopted from Tai, Du, Brown & Lin (2010).]

the motion path (Shan, Xiong & Jia 2007); however, in our approach we place no constraints on possible motion. Instead, the spatially-varying PSFs are decomposed into a set of P basis kernels k_l whose mixture weights a_l are a function of image location:

$$K(x, y, u, v) = \sum_{l=1}^{P} a_l(x, y) k_l(u, v). \tag{3.8}$$

The convolution equation then becomes

$$I(x, y) \otimes K(x, y, u, v) = \sum_{l=1}^{P} a_l(x, y)(I(x, y) \otimes k_l(u, v)). \tag{3.9}$$

Alternatively, principal component analysis (PCA) can be used to determine the basis kernels (Lauer 2002). PCA, however, does not guarantee positive PSF values, and may lead to strong ringing artefacts, exemplified in Figure 3.10(b). The ringing artefacts in the convolution result resemble the patterns of basis kernels.

Another method is to use a patch representation which segments images into many small patches such that the local motion PSF is nearly the same within each small patch (Joshi, Szeliski & Kriegman 2008). However, for large and complex motion, the patches need to be very small (generating a large volume of data) to obtain nearly identical motion in each patch. Larger patches will result in inaccurate PSFs leading to discontinuity artefacts as shown in Figure 3.10(c).

Our approach uses a delta function representation, where each delta function represents a position (u, v) within a PSF. From the total 65×65 possible delta-functions in the spatial PSF, we found it possible to select 600 distinct delta functions to provide a sufficient approximation of the spatially-varying PSFs in the convolution process.

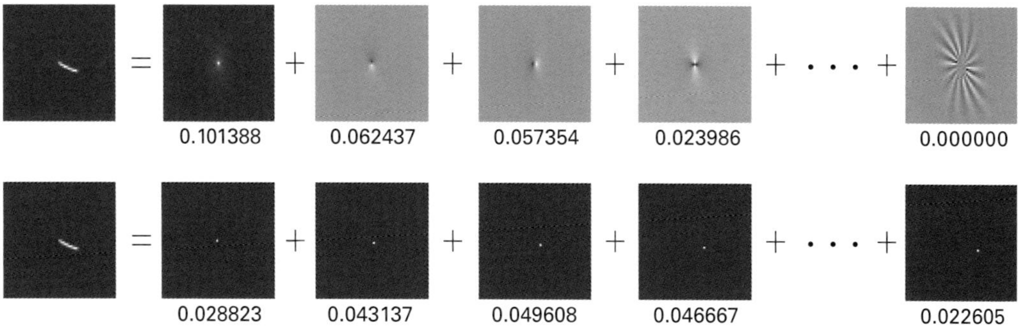

Figure 3.11 PCA versus the delta function representation for PSF decomposition. The top row illustrates the PSF decomposition using PCA, and the bottom row shows the decomposition using the delta function representation. The example PSF is taken from among the spatially-varying PSFs of Figure 3.10, from which the basis kernels are derived. Weights are displayed below each of the basis kernels. The delta function representation not only guarantees positive values of basis kernels, but also provides more flexibility in PSF refinement. [Adopted from Tai, Du, Brown & Lin (2010).]

To further reduce space, to represent the PSF at each image pixel, it was sufficient to use only 40 indexes with associated magnitudes into the selected delta functions.

To show the effectiveness of this approach, examples of basis kernel decomposition using PCA and the delta function representation are shown in Figure 3.11. The delta function representation also offers more flexibility in kernel refinement, while refinements using the PCA representation are limited to the PCA subspace.

Combining Eqs. (3.9) and (3.7), our optimization function becomes

$$\arg\min_{I,K} \sum_i^N \left\| I_{b_i} - \sum_l^P a_{il}(I \otimes k_{il}) \right\|^2 + \lambda_B \sum_j^M \left\| I_{l_j} - d(I \otimes h) \right\|^2 + \lambda_K \sum_i^N \sum_l^P \left\| a_{il}k_{il} - a_{o_{il}}k_{il} \right\|^2 . \tag{3.10}$$

The corresponding iterative update rules are then

1. update $I^{t+1} = I^t + \sum_{i=1}^N \sum_l^P a^t_{il}k_{il} * \left(I_{b_i} - \sum_l^P a^t_{il}(I^t \otimes k_{il})\right) + \lambda_B \sum_{j=1}^M h \otimes \left(u\left(W(I_{l_j})\right.\right.$ $\left.\left.-d(I^t \otimes h)\right)\right)$;
2. update $a^{t+1}_{il} = a^t_{il} + \left(\widetilde{I'}^{t+1} * \left(I'_{b_i} - \sum_l^P a^t_{il}(I'^{t+1} \otimes k_{il})\right)\right) \cdot k_{il} + \lambda_K(a_{o_{il}} - a^t_{il})$

where I' and I'_b are local windows in the estimated result and the blur image. This PSF refinement can be implemented in a multi-scale framework for greater flexibility and stability. The number of delta functions k_{il} stored at each pixel position may be reduced when an updated value of a_{il} becomes insignificant. For greater stability, we process each update rule five times before switching to the other.

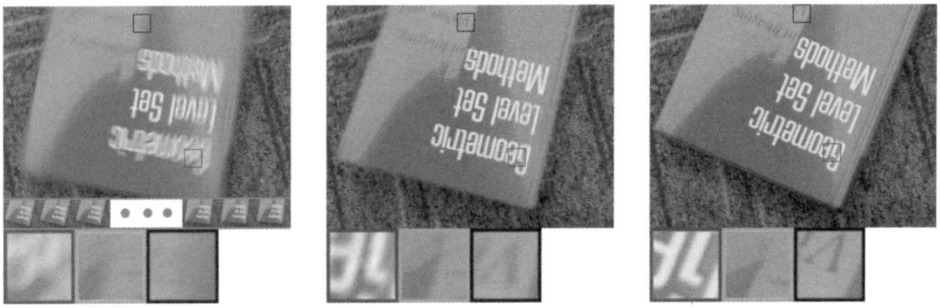

(a) (b) (c)

Figure 3.12 Image deblurring with spatial varying kernels from rotational motion. (a) Input; (b) our spatially-varying PSF results; (c) the ground truth sharp image. Close-ups are also shown. [Adopted from Tai, Du, Brown & Lin (2010).]

3.5.1 Examples – spatially-varying PSF

Figure 3.12 shows an example with in-plane rotational motion of a book. Our approach can successfully deblur the image with spatially-varying PSF. The book title is easily legible after deblurring.

3.6 Moving object deblurring

The hybrid-imaging system can also be used to address the problem of motion blur due to an object moving in front of a stationary (non-blurred) background. This problem is difficult as the moving object 'blends' with the background and therefore it is not enough to know the object's blur function to deblur the image; the blurred object must be separated from the background before it can be deblurred.

Assuming that the blending is linear, the deblurring operation can be expressed as:

$$O = \left(I - (B \cdot \overline{M \otimes S})\right) \otimes^{-1} S + B \cdot \bar{M}, \tag{3.11}$$

where O is the deblurred image, I is the blurred input image, S is the PSF, $\otimes^{-1}$ denotes deconvolution, M is a segmentation mask for the shape of the foreground (non-blurred) object, B is a *clear and non-blurred* background image, $\otimes$ denotes 2D convolution, and $\bar{X}$ is the complement of X. The hybrid-imaging system can provide a good estimation of the PSF for the moving object; this can be done by applying a tracking algorithm to the low resolution (high frame-rate) sequence. Since we assume shift invariance within an object, only a single feature needs to be tracked. It is also possible to extract a low resolution mask (shape) of the foreground object using the secondary detector's image.

Simulation results are shown in Figure 3.13. Figure 3.13(a) shows the original image, which is synthetically blurred using a high resolution ground truth mask (not shown) to form the blurred image shown in Figure 3.13(b). Figure 3.13(c) shows the blending mask computed from the low resolution (simulative) secondary detector image, and Figure 3.13(d) shows the blurred foreground component. The foreground component is

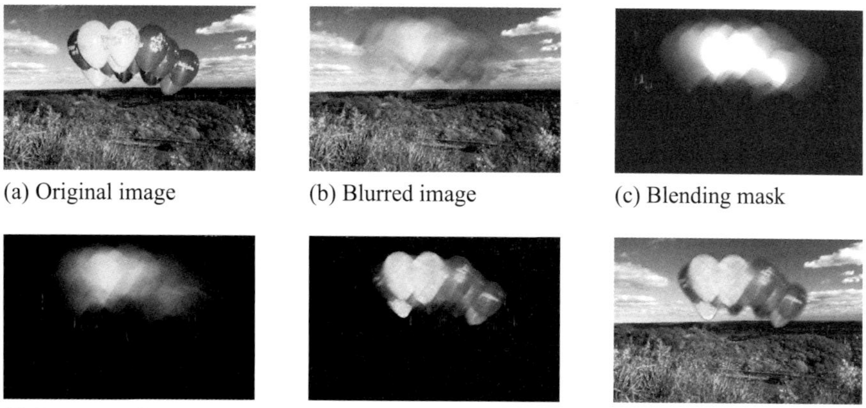

(a) Original image (b) Blurred image (c) Blending mask

(d) Foreground component (e) Deblurred foreground (f) Composite result

Figure 3.13 Object deblurring simulation result. (a) Original non-blurred image; (b) synthetically blurred image; (c) blending (alpha) mask obtained from the secondary detector's image sequence; (d) blurred foreground obtained by subtracting the background component from the primary detector's blurred image; (e) deblurred image of the foreground object; (f) composite of the deblurred foreground and the clear background using a low resolution mask obtained from the secondary detector. [Adopted from Ben-Ezra & Nayar (2004).]

deblurred using PSF computed from the image sequence of the secondary detector, and the deblurred foreground shown in Figure 3.13(e) is composited into a clean background using the same mask to form the final result shown in Figure 3.13(f).

The extension of this method to a blurred background scenario, where it is possible to obtain a clear non-blurred or a clear deblurred image of the background, is straightforward. In this case Eq. (3.11) becomes:

$$O = \big(I - ((B \otimes S_b) \cdot \overline{M \otimes S_f})\big) \otimes^{-1} S_f + B \cdot \bar{M}, \tag{3.12}$$

where S_b and S_f are the PSFs of the background and foreground, respectively.

In many cases it might be difficult to obtain a clear background image of the scene. In such cases, we need to estimate the foreground and background colours, given the mask.

This is a form of the matting problem that can be expressed as:

$$I = \alpha F + (1 - \alpha)B \tag{3.13}$$

where I is the observed image intensity, F, B, and α are the foreground colour, background colour, and alpha value of the fractional occupancy of the foreground, respectively.

The matting problem is an ill-posed problem as the number of unknown variables is greater than the number of observations. Our approach assumes that object motion does not cause motion blur in the high frame-rate camera, so the object appears sharp.

To extract the alpha matte of a moving object, we perform binary segmentation of the moving object in the low resolution images, and then compose the binary segmentation masks with smoothing to approximate the alpha matte in the high resolution image.

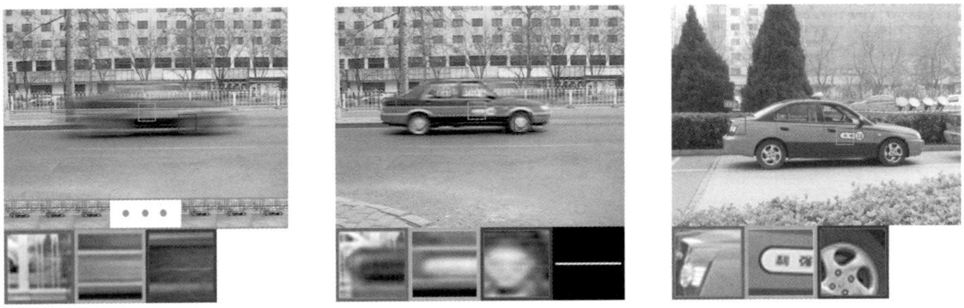

(a) (b) (c)

Figure 3.14 Image deblurring with translational motion. In this example, the moving object is a car moving horizontally. We assume that the motion blur within the car is globally invariant. (a) Input; (b) deblurred image; (c) the ground truth sharp image captured from another car of the same model. Closeup views and the estimated global blur kernels within the motion blur layer are also shown. [Adopted from Tai, Du, Brown & Lin (2010).]

The foreground colour F also must be estimated for deblurring. This can be done by assuming a local colour smoothness prior on F and B and solving for their values with Bayesian matting (Chuang, Curless, Salesin & Szeliski 2001). Given the solution of F and B, the α solution can be refined by solving Eq. (3.13) in closed form. Refinements of F, B, and α can be done in alternation to further improve the result. A recent regularization-based method targeting motion-blurred objects (Lin, Tai & Brown 2011) can also be included to enhance the matting results.

Once moving objects are separated, we deblur each layer separately as mentioned before. An example of this method is shown in Figure 3.14.

3.7 Discussion and summary

This chapter introduced a hybrid-imaging system design targeting image deblurring due to camera shake and, to some extent, moving objects in a scene. In this section, the hybrid-imaging system is discussed in the broader context of motion deblurring.

The idea of the hybrid-imaging system is to allow for non-blind image deblurring, where an additional detector is used to estimate the blur function. As such, it can be considered a hardware assisted deblurring method. Like all hardware assisted methods, the main disadvantage is the need for additional hardware. The cost and complexity of the required hardware can be reduced significantly if the additional hardware is embedded into the camera or the camera sensor.

The main advantage of the hybrid-imaging method over blind and user assisted deblurring methods is robustness. Hybrid-imaging methods do not depend on the content of the image to obtain the blurring function. Hybrid-imaging systems offer the following additional advantages:

1. They can measure motion across inertial systems, for example when taking an image from a moving train.

2. They can detect depth-dependent motion, as well as object motion.
3. By sharing the same optics between the two sensors, a hybrid-imaging system can automatically adjust its accuracy and scale to the focal length of the sensor which can change when the lens is replaced or when a zoom lens is used.
4. As discussed in Section 3.4.5, the additional information from the secondary device can be used in the deblurring processing as additional back-projection constraints to improve PSF estimation and the deblurred result.

Finally, we note that hybrid-imaging systems can be used in conjunction with other deblurring methods by providing an initial estimate used in blind image deblurring. They can also be useful for fusing in information with methods that use other types of sensors, such as inertial sensors, to further improve the deblurring quality and robustness.

Acknowledgements

This chapter is based on the work that appeared in Ben-Ezra & Nayar (2004) and Tai *et al.* (2010) and we gratefully acknowledge IEEE for their permission to reproduce extracts here.

References

Ben-Ezra, M. & Nayar, S. (2004). Motion-based Motion Deblurring. *IEEE Transactions on Pattern Analysis and Machine Intelligence*, **26**(6), 689–98.

Bioucas-Dias, J., Figueiredo, M. & Oliveira, J. (2006*a*). Adaptive total-variation image deconvolution: A majorization–minimization approach. In *Proceedings of the European Signal Processing Conference*, pp. 1–4.

Bioucas-Dias, J., Figueiredo, M. & Oliveira, J. (2006*b*). Total variation-based image deconvolution: a majorization–minimization approach. In *IEEE International Conference on Acoustics, Speech and Signal Processing*, Vol. 2, p. II.

Chuang, Y., Curless, B., Salesin, D. H. & Szeliski, R. (2001). A Bayesian approach to digital matting. In *IEEE Conference on Computer Vision and Pattern Recognition*, pp. 264–71.

Dey, N., Blanc-Feraud, L., Zimmer, C., Roux, P., Kam, Z., Olivo-Marin, J. & Zerubia, J. (2006). Richardson–Lucy algorithm with total variation regularization for 3D confocal microscope deconvolution, *Microscopy Research and Technique*, **69**(4), 260–6.

Fergus, R., Singh, B., Hertzmann, A., Roweis, S. T. & Freeman, W. T. (2006). Removing camera shake from a single photograph. *ACM Transactions on Graphics*, **25**(3), pp. 787–94.

Jansson, P. A. (1997). *Deconvolution of Image and Spectra*, 2nd edn. Academic Press.

Joshi, N., Szeliski, R. & Kriegman, D. (2008). PSF estimation using sharp edge prediction. In *IEEE Conference on Computer Vision and Pattern Recognition*, pp. 1–8.

Krishnan, D. & Fergus, R. (2009). Fast image deconvolution using hyper-Laplacian priors. *Advances in Neural Information Processing Systems*, **22**, 1–9.

Lauer, T. (2002). Deconvolution with a spatially-variant PSF. In *SPIE Proceedings, Astronomical Data Analysis II*, Vol. 4847, pp. 167–73.

Levin, A., Fergus, R., Durand, F. & Freeman, W. (2007). Deconvolution using natural image priors. *Massachusetts Institute of Technology, Computer Science and Artificial Intelligence Laboratory.*

Lin, H., Tai, Y.-W. & Brown, M. S. (2011). Motion regularization for matting motion blurred objects. *IEEE Transactions on Pattern Analysis and Machine Intelligence*, **33**(11), 2329–36.

Lucas, B. & Kanade, T. (1981). An iterative image registration technique with an application to stereo vision. In *Proceedings of the Defense Advanced Research Projects Agency*, pp. 121–30.

Sawhney, H., Guo, Y., Hanna, K., Kumar, R., Adkins, S. & Zhou, S. (2001). Hybrid stereo camera: an IBR approach for synthesis of very high resolution stereoscopic image sequences. In *ACM Proceedings of the 28th Annual Conference on Computer Graphics and Interactive Techniques*, pp. 451–60.

Shan, Q., Xiong, W. & Jia, J. (2007). Rotational motion deblurring of a rigid object from a single image. In *IEEE International Conference on Computer Vision*, pp. 1–8.

Tai, Y., Du, H., Brown, M. & Lin, S. (2010). Correction of spatially varying image and video blur using a hybrid camera. *IEEE Transactions on Pattern Analysis and Machine Intelligence*, **32**(6), 1012–28.

Wiener, N. (1964). *Extrapolation, Interpolation, and Smoothing of Stationary Time Series*. MIT Press.

Yuan, L., Sun, J., Quan, L. & Shum, H. (2007). Image deblurring with blurred/noisy image pairs. *ACM Transactions on Graphics*, **26**(3), 1:1–10.

4 Efficient, blind, spatially-variant deblurring for shaken images

Oliver Whyte, Josef Sivic, Andrew Zisserman and Jean Ponce

In this chapter we discuss modelling and removing spatially-variant blur from photographs. We describe a compact global parameterization of camera-shake blur, based on the 3D rotation of the camera during the exposure. Our model uses three-parameter homographies to connect camera motion to image motion and, by assigning weights to a set of these homographies, can be seen as a generalization of the standard, spatially-invariant convolutional model of image blur. As such we show how existing algorithms, designed for spatially-invariant deblurring, can be 'upgraded' in a straightforward manner to handle spatially-variant blur instead. We demonstrate this with algorithms working on real images, showing results for blind estimation of blur parameters from single images, followed by non-blind image restoration using these parameters. Finally, we introduce an efficient approximation to the global model, which significantly reduces the computational cost of modelling the spatially-variant blur. By approximating the blur as locally-uniform, we can take advantage of fast Fourier-domain convolution and deconvolution, reducing the time required for blind deblurring by an order of magnitude.

4.1 Introduction

Everybody is familiar with camera shake, since the resulting blur spoils many photos taken in low-light conditions. Camera-shake blur is caused by motion of the camera during the exposure; while the shutter is open, the camera passes through a sequence of different poses, each of which gives a different view of the scene. The sensor accumulates all of these views, summing them up to form the recorded image, which is blurred as a result. We would like to be able to deblur such images to recover the underlying sharp image, which we would have captured if the camera had not moved.

In general, the problem of restoring an image after it has suffered some degradation can be broken down into three stages: first, a generative model is needed to relate the undegraded, 'ideal' or latent image (that we would like to recover) to the observed image produced by the camera. Second, the parameters of this model must be estimated, and finally the restored image can be reconstructed, given the model and the estimated parameters.

This chapter is principally concerned with the first of these stages: a geometrically motivated model of *spatially-variant* image blur due to camera shake, which we show can be (mostly) attributed to the rotation of the camera during exposure. We develop a global descriptor for the generative model parameters of this non-uniform blur, analogous to (but different from) a convolution kernel, and show that a more general class of blurs can be modelled other than uniform.

Several authors have proposed models for spatially-variant blur, under different assumptions about the scene and the camera, e.g. simple scene models with unconstrained camera motion (Joshi, Kang, Zitnick & Szeliski 2010, Gupta, Joshi, Zitnick, Cohen & Curless 2010, Tai, Tan & Brown 2011), constrained camera motion (Sawchuk 1974, Klein & Drummond 2005, Shan, Xiong & Jia 2007, Tai, Kong, Lin & Shin 2010), or more complex scene models (Šorel & Flusser 2008, Xu & Jia 2012). On the other hand, much of the work on *algorithms* for deblurring camera shake assumes that the blurred image is simply a 2D convolution of a sharp image with a *spatially-invariant* filter (Fergus, Singh, Hertzmann, Roweis & Freeman 2006, Shan, Jia & Agarwala 2008, Cho & Lee 2009), despite the fact that real camera shake does not, in general, cause spatially-invariant blur (Levin, Weiss, Durand & Freeman 2009), as shown in Figure 4.1. In this chapter we aim to bridge this gap with a practical model for spatially-variant camera-shake blur that can also leverage advances in convolution-based deblurring algorithms.

We begin in Section 4.2 by deriving our geometric model, before discussing how it can be implemented in practice in Section 4.3. In Section 4.4 we show how our model can replace convolution in existing deblurring algorithms, with minimal algorithmic changes. In Section 4.5 we describe an efficient approximation to the model, which significantly reduces its computational cost. In Section 4.6 we present results on real images, using our model to replace the uniform (convolution) blur model in an existing algorithm for camera-shake removal. In Section 4.7 we cover some implementation considerations for our model.

Parts of this chapter are based on previously published work (Whyte, Sivic, Zisserman & Ponce 2010, Whyte, Sivic & Zisserman 2011, Whyte *et al.* 2012).

4.2 Modelling spatially-variant camera-shake blur

A camera may move in several different ways, and it is not necessarily obvious which kinds of motion cause large changes in the view (and hence a large blur), and which cause relatively small changes. Furthermore, even if the camera's motion is fully known for a given photograph, a model is needed to translate this physical 3D motion into image-domain motion before we can begin to deblur the photograph.

We begin by considering the relative blurring effect of different camera motions, and deriving a geometric model for camera shake. In Section 4.3 we develop this into a practical model for deblurring real images. In this work we limit our scope to photographs of static scenes, i.e. the blur is solely due to the motion of the camera.

4.2.1 Components of camera motion

The pose of a camera incorporates two components: *position* and *orientation*. Intuitively, the position tells us where the camera is, while the orientation tells us which

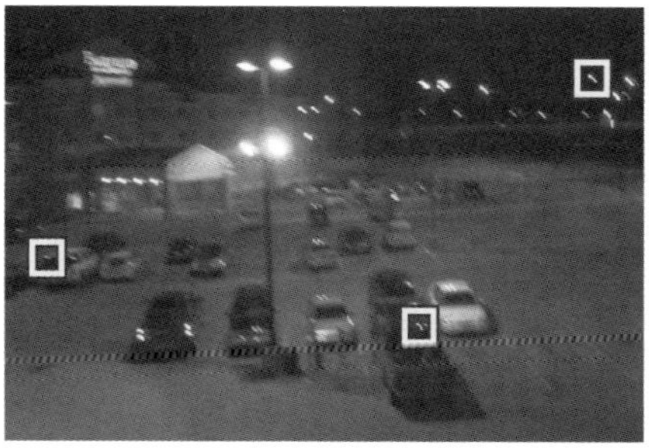

(a) An image blurred by camera shake

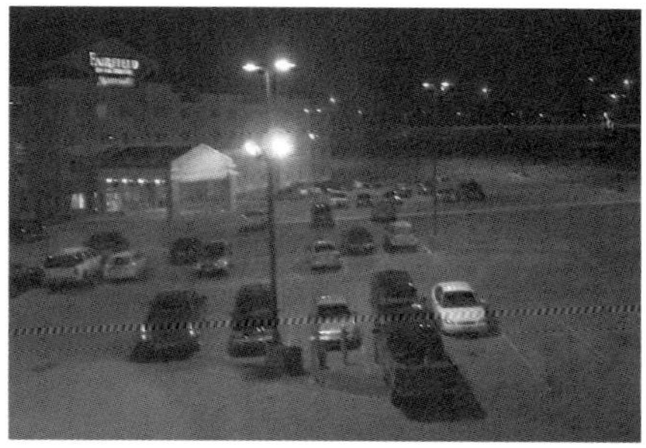

(b) Deblurred using the estimated blur descriptor

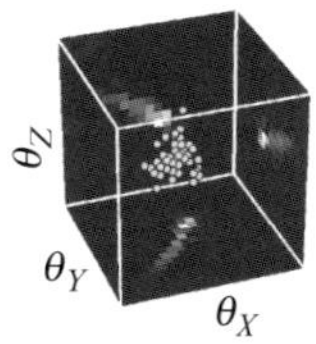

(c) Estimated global blur descriptor

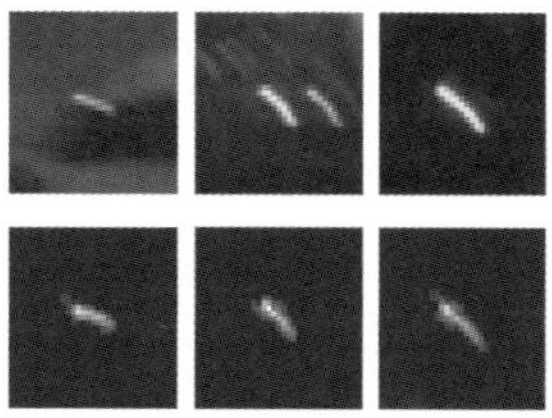

(d) Top: details from blurred image; bottom: corresponding local PSFs generated by (c)

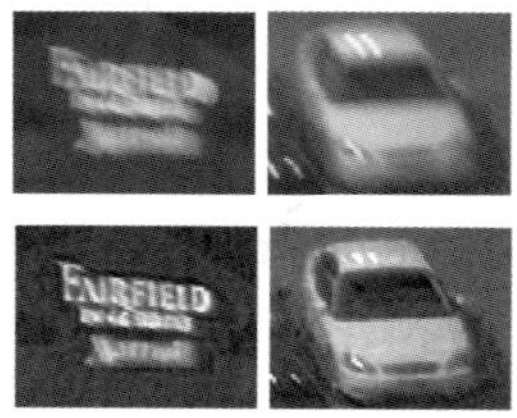

(e) Details from the blurred (top) and deblurred (bottom) images

Figure 4.1 Modelling non-uniform blur in a shaken image. The blurred image (a) clearly exhibits blur which is non-uniform, as highlighted at different locations in the image. Using the model proposed in this work, we can describe this blur using a single global descriptor (c), which in this case has been estimated from the blurred image itself, simply by modifying existing algorithms for blind deblurring (see Section 4.4 for details). Having estimated the blur, the standard Richardson–Lucy algorithm is used to estimate the sharp image. Close-ups of different parts of the image (d) show the variation in the shape of the blur, which can be accurately reproduced using our model, as shown by the local point spread functions generated from it. As can be seen in the deblurred image in (b) and the close-ups in (e), different parts of the image, blurred in different ways, can be deblurred to recover a sharp image. Reproduced from Whyte *et al.* (2012) with permission, © Springer 2012.

way it is pointing, and both may vary while the camera's shutter is open. In this section, we discuss the contribution of each component to the image blur, and conclude that in most cases of camera shake, the changes in orientation (rotation) of the camera during exposure have a significantly larger blurring effect than the changes in position (translation).

Consider the simplified case shown in Figure 4.2 of a scene point P, at a distance D from the camera, being imaged at the centre of the camera's retina/sensor. During the exposure, the camera moves, and the image of the point is blurred through a distance δ pixels. In (a) the camera translates through a distance X parallel to the image plane,

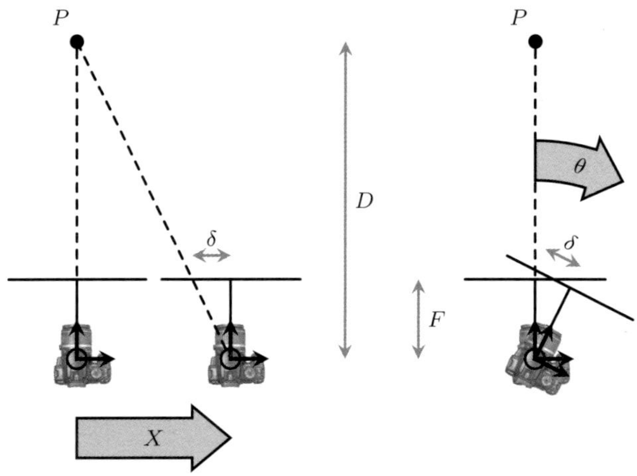

(a) Camera translation (b) Camera rotation

Figure 4.2 Blur due to translation or rotation of the camera. In this simplified example, we consider capturing a blurred image by either (a) translating the camera through a distance X parallel to the image plane, or (b) rotating the camera through an angle θ about its optical centre. We consider the scene point P at a distance D from the camera, whose image is blurred by δ pixels as a result of either of the two motions. In most cases, for a given blur size δ the rotation θ constitutes a significantly smaller motion of the photographer's hands than the translation X (see text for details). Reproduced from Whyte *et al.* (2012) with permission, © Springer 2012.

while in (b) the camera rotates through an angle θ about its optical centre. By simple trigonometry, we can see that in (a) the camera must translate by

$$X = \frac{\delta}{F} D, \tag{4.1}$$

where F is the camera's focal length, while in (b) the camera must rotate through an angle

$$\theta = \tan^{-1}\left(\frac{\delta}{F}\right). \tag{4.2}$$

If we make the common assumption that the camera's focal length F is approximately equal to the width of the sensor, say 1000 pixels, then to cause a blur of $\delta = 10$ pixels by translating the camera, we can see from Eq. (4.1) that $X = \frac{1}{100}D$. Thus the required translation grows with the subject's distance from the camera, so for a subject just 1 m away, we must move the camera by $X = 1$ cm to cause the blur. When photographing a subject 30 m away, such as a large landmark, we would have to move the camera by 30 cm!

To cause the same amount of blur by rotating the camera, on the other hand, we can see from Eq. (4.2) that we would need to rotate the camera by $\theta = \tan^{-1}\left(\frac{1}{100}\right) \simeq 0.6°$, independent of the subject's distance from the camera. To put this in terms of the motion of the photographer's hands, if the camera body is 10 cm wide, such a rotation could be caused by moving one hand just 1 mm forwards or backwards relative to the other. Provided the subject is more than 1 m from the camera, this motion is at least an order of magnitude smaller than for a translation of the camera causing an equivalent amount of blur.

In reality, both the position and orientation of the camera vary simultaneously during the exposure. However, if the camera only undergoes small changes in position (translations), then following the discussion above, we can assert that the variations in the camera's orientation (rotations) are the only significant cause of blur.

From now on, we assume that the translational component of camera motion *does not cause any blur*. Furthermore, we assume that all rotations occur about the camera's optical centre. Note, however, that a camera rotation about a centre that is not the optical centre can be written as a rotation about the optical centre composed with a translation; these translations will generally be small if the centre of rotation is not far from the optical centre. As we shall see in the following section, this model of camera motion leads to spatially-variant blur.

4.2.2 Motion blur and homographies

Under a pinhole camera model, and assuming that the scene being photographed is static, rotations of a camera about its optical centre induce projective transformations of the image being observed. In other words, the image observed at one camera orientation is related to the image at any other by a 2D projective transformation, or homography. For an uncalibrated camera, this is a general 8-parameter homography, but for a camera with known internal parameters, the homography $\mathbf{H}$ is specified by three parameters and is given by

$$\mathbf{H} = \mathbf{K}\mathbf{R}\mathbf{K}^{-1}, \tag{4.3}$$

where the 3×3 matrix $\mathbf{R}$ is a rotation matrix describing the orientation of the camera, and $\mathbf{K}$ is the camera's internal calibration matrix (Hartley & Zisserman 2004). In this work, we assume that the calibration matrix $\mathbf{K}$ is known (see Section 4.2.3).

The rotation matrix $\mathbf{R}$ has only three parameters. We adopt here the 'angle-axis' parameterisation, in which a rotation is described by the angle θ moved about an axis $\mathbf{a}$ (a unit-norm 3-vector). This can be summarised in a single 3-vector $\boldsymbol{\theta} = \theta\mathbf{a} = (\theta_X, \theta_Y, \theta_Z)$. $\mathbf{R}$ is then given by the matrix exponential

$$\mathbf{R}_{\boldsymbol{\theta}} = e^{[\boldsymbol{\theta}]_\times}, \tag{4.4}$$

where

$$[\boldsymbol{\theta}]_\times = \begin{bmatrix} 0 & -\theta_Z & \theta_Y \\ \theta_Z & 0 & -\theta_X \\ -\theta_Y & \theta_X & 0 \end{bmatrix}. \tag{4.5}$$

We fix our 3D coordinate frame to have its origin at the camera's optical centre. The axes are aligned with the camera's initial orientation, such that the *XY*-plane is aligned with the camera sensor's coordinate frame and the *Z*-axis is parallel to the camera's optical axis, as shown in Figure 4.3(a). In this configuration, θ_X describes the 'pitch' of the camera, θ_Y the 'yaw', and θ_Z the 'roll', or in-plane rotation, of the camera.

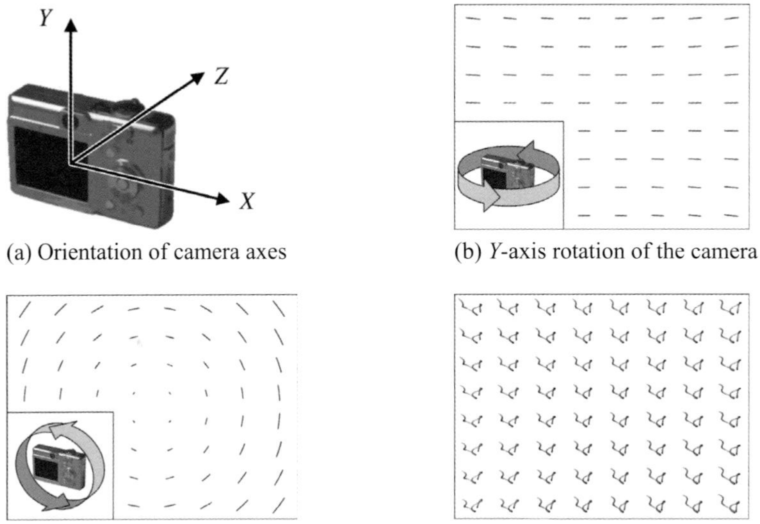

(a) Orientation of camera axes

(b) Y-axis rotation of the camera

(c) Z-axis rotation of the camera

(d) Arbitrary sequence of rotations

Figure 4.3 Our coordinate frame with respect to initial camera orientation, and the paths followed by image points under different camera rotations. We define our coordinate frame (a) to have its origin at the camera's optical centre, with the X and Y axes aligned with those of the camera's sensor, and the Z axis parallel to the camera's optical axis. Under single-axis rotations of the camera, for example about its Y-axis (b), or its Z-axis (c), the paths traced by points in the image are visibly curved and non-uniform across the image. This non-uniformity remains true for general camera shakes (d), which do not follow such simple single-axis rotations, but rather take arbitrary paths through camera pose space. The focal length of the camera in this simulation is equal to the width of the image, the principal point is at the image's centre, and the pixels are assumed to be square. Reproduced from Whyte *et al.* (2012) with permission, © Springer 2012.

Having defined the type of image transformations we expect to occur while the shutter is open, we can write out the image degradation model. Let T denote the exposure time of the photograph. While the shutter is open, the camera passes through a sequence of orientations $\boldsymbol{\theta}_t$, $t \in [0, T]$. As discussed above, at each pose $\boldsymbol{\theta}_t$, the sensor is exposed to a projectively transformed version of the sharp image f, where the projective transformation $\mathbf{H}_t$ is given by Eqs. (4.3)–(4.5). The noiseless blurred image g^* is then modelled as the integral over the exposure time T of all the transformed versions of f:

$$g^*(\mathbf{x}) = \int_0^T f\left(\mathbf{H}_t\mathbf{x}\right) \mathrm{d}t, \tag{4.6}$$

where, with a slight abuse of notation, we use $g^*(\mathbf{x})$ to denote the value of g^* at the 2D image point represented by the homogeneous vector $\mathbf{x}$, and similarly for f.

Under this model, the apparent motion of scene points may vary significantly across the image. Figure 4.3 demonstrates this, showing the paths followed by points in an image under a Y-axis rotation, a Z-axis rotation, and an arbitrary sequence of rotations of the camera. Under the (in-plane) Z-axis rotation, the paths vary significantly across the image. Under the (out-of-plane) rotation about the Y-axis, the paths, while varying considerably less, are still non-uniform. It should be noted that the degree of non-uniformity of this out-of-plane motion is dependent on the focal length of the camera, decreasing as the focal length increases. However, it is typical for consumer cameras to have focal lengths of the same order as their sensor width, as is the case in Figure 4.3. In addition, it is common for camera shake to include an in-plane rotational motion. From this, it is clear that modelling camera shake as a convolution with a spatially-invariant kernel is insufficient to fully describe its effects (see also Figure 4.1).

In general, a blurred image has no temporal information associated with it, so it is convenient to replace the temporal integral in Eq. (4.6) by a weighted integral over a set of camera orientations:

$$g^*(\mathbf{x}) = \int f\left(\mathbf{H}_{\boldsymbol{\theta}}\mathbf{x}\right)w(\boldsymbol{\theta})\,\mathrm{d}\boldsymbol{\theta}, \tag{4.7}$$

where the weight function $w(\boldsymbol{\theta})$ encodes the camera's trajectory in a time-agnostic fashion. The weight will be zero everywhere except along the camera's trajectory, and the value of the function at a point $\boldsymbol{\theta}$ along the trajectory corresponds to the duration the camera spent at the orientation $\boldsymbol{\theta}$.

4.2.3 Camera calibration

In order to compute the homography in Eq. (4.3) that is induced by a particular rotation of the camera, we need to know the camera's calibration matrix **K**. For the results shown in this chapter, we assume that **K** takes the standard form

$$\mathbf{K} = \begin{bmatrix} F & 0 & x_0 \\ 0 & F & y_0 \\ 0 & 0 & 1 \end{bmatrix}. \tag{4.8}$$

This corresponds to a camera whose sensor has square pixels, and whose optical axis intersects the sensor at (x_0, y_0), referred to as the principal point. We recover the focal length of the camera from the image's EXIF tags, and assume that the principal point is at the centre of the image (which is typically sufficient for modelling homographies caused by camera rotations (Szeliski 2004)).

The radial distortion present in many digital cameras can represent a significant deviation from the pinhole camera model. Rather than modelling the distortion explicitly, we pre-process images with the commercially available PTLens software[1] to remove it. A second distortion present in many digital images comes from a nonlinear intensity mapping applied by the camera storing the image, sometimes referred to as 'gamma correction'. In this work, we either use raw camera output images which do not have this nonlinearity, or pre-process the blurred images with the inverse mapping.

4.2.4 Uniform blur as a special case

One consequence of our model for camera shake is that it includes spatially-invariant blur as a special case, and thus gives the conditions under which such a blur model

[1] http://epaperpress.com/ptlens/

is applicable. Given the definitions of **R**, **K** and **H** in Eqs. (4.3), (4.8) and (4.4), and assuming $\theta_Z = 0$, it can be shown (Whyte 2012) that for small θ_X, θ_Y, as $F \to \infty$,

$$\mathbf{H}_{\theta} \to \begin{bmatrix} 1 & 0 & F\theta_Y \\ 0 & 1 & -F\theta_X \\ 0 & 0 & 1 \end{bmatrix}, \tag{4.9}$$

which simply amounts to a translation of the image. Thus we can see that if the focal length of the camera is large (e.g. if the camera is zoomed-in) and there is no in-plane rotation, the blur which results from camera motion will be approximately spatially-invariant.

4.3 The computational model

Real cameras are equipped with a discrete set of pixels, and output a discrete set of samples of the discrete image, denoted by the vector $\mathbf{g} \in \mathbb{R}_+^N$, where $N = H \times W$ pixels for an image with H rows and W columns. We consider the sharp image also to be discrete: $\mathbf{f} \in \mathbb{R}_+^N$. We use i to index into the discrete image $\mathbf{g}$, i.e. $g_i = g(\mathbf{x}_i)$, where $\mathbf{x}_i$ is the coordinate of the ith pixel. Likewise, we use j to index into the sharp image $\mathbf{f}$, such that $f_j = f(\mathbf{x}'_j)$ for a coordinate $\mathbf{x}'_j$. Finally, we note that to evaluate an image at arbitrary (sub-pixel) locations, we interpolate from nearby pixels. In this work, we use bilinear interpolation, whereby sub-pixel values of an image, say $g(\mathbf{x})$, are interpolated as a linear combination of the four nearest pixels.

In this discrete setting, we write the blurred image as a linear function of the sharp image:

$$g_i^* = \sum_j A_{ij} f_j, \tag{4.10}$$

or in matrix-vector notation,

$$\mathbf{g}^* = \mathbf{Af}, \tag{4.11}$$

where the $N \times N$ matrix **A** captures the discrete point spread function (PSF). Each row of the matrix **A** corresponds to a single blurred pixel in **g**, and captures the fact that each blurred pixel is a weighted sum of multiple sharp pixels. In most cases of blur, the light received by each pixel in **g** comes from a relatively small number of nearby pixels in **f**. As a result, the PSF matrix **A** for an image is usually sparse (i.e. contains a relatively small number of nonzero values).

We discretise the camera orientation space into a 3D volumetric grid of size $N_X \times N_Y \times N_Z$, and assign each orientation $\boldsymbol{\theta}^{(k)}$ a weight w_k, for $k \in \{1, \ldots, K\}$, where $K = N_X N_Y N_Z$. The set of weights **w** forms a global descriptor for the camera-shake blur in an image, and by analogy with convolutional blur, we refer to **w** as the *blur kernel*.

Figure 4.1(c) shows a visualisation of **w**, where the cuboidal volume of size $N_X \times N_Y \times N_Z$ is shown, with the points inside representing the nonzero elements of **w** in 3D. The kernel has also been projected onto the three back faces of the cuboid to aid visualisation, with white areas corresponding to a large value, and black corresponding to zero.

Each element w_k corresponds to a camera orientation $\boldsymbol{\theta}^{(k)}$, and consequently to a homography $\mathbf{H}_k$, so that in the discrete setting, the blurred image $\mathbf{g}^*$ is modelled as a weighted sum of a set of projectively-transformed versions of **f**:

$$\mathbf{g}^* = \sum_k w_k \mathbf{T}^{(k)} \mathbf{f}, \tag{4.12}$$

where $\mathbf{T}^{(k)}$ is the $N \times N$ matrix which applies homography $\mathbf{H}_k$ to the sharp image **f**. The matrix $\mathbf{T}^{(k)}$ is very sparse. For example, if bilinear interpolation is used when transforming the image, each row has only four nonzero elements corresponding to the interpolation weights for each pixel. By writing out Eq. (4.12) for a single pixel, we obtain the discrete analogue of Eq. (4.7):

$$g_i^* = \sum_k w_k \left(\sum_j T_{ij}^{(k)} f_j \right), \tag{4.13}$$

where i and j index the pixels of the blurred image and the sharp image, respectively. For a blurred pixel g_i^* with coordinate vector $\mathbf{x}_i$, the sum $\sum_j T_{ij}^{(k)} f_j$ interpolates the value of the sub-pixel location $f(\mathbf{H}_k \mathbf{x}_i)$. We will return to how $\mathbf{H}_k$ and the corresponding matrices $\mathbf{T}^{(k)}$ are sampled in Section 4.7.

Due to the bilinear form of Eq. (4.13), note that when either the blur kernel or the sharp image is known, the blurred image is linear in the remaining unknowns, i.e.

$$\text{given } \mathbf{w}, \quad \mathbf{g}^* = \mathbf{A}\mathbf{f}, \quad \text{where } A_{ij} = \sum_k T_{ij}^{(k)} w_k, \tag{4.14}$$

$$\text{given } \mathbf{f}, \quad \mathbf{g}^* = \mathbf{B}\mathbf{w}, \quad \text{where } B_{ik} = \sum_j T_{ij}^{(k)} f_j. \tag{4.15}$$

In the first form, $\mathbf{A} \in \mathbb{R}^{N \times N}$ is a large sparse matrix, whose rows each contain a local blur filter acting on **f** to generate a blurred pixel. In the second form, when the sharp image is known, each column of $\mathbf{B} \in \mathbb{R}^{N \times K}$ contains a projectively transformed copy of the sharp image. We will use each of these forms in the following sections.

In contrast to this model, when the PSF is spatially invariant, we denote the discrete convolution kernel by **a**, and write

$$\mathbf{g}^* = \mathbf{a} * \mathbf{f}. \tag{4.16}$$

4.4 Blind estimation of blur from a single image

In this section, we demonstrate the effectiveness of the spatially-variant blur model presented in Section 4.3 by using it for blind deblurring of images acquired under camera shake. We consider single-image deblurring, where only a blurred image is available, and use our model to replace the spatially-invariant blur model in an existing algorithm for blind deblurring; specifically, applying our model within the algorithm proposed by Cho & Lee (2009).

'Blind' estimation of PSFs directly from images has a long history (Gull & Skilling 1984, Ayers & Dainty 1988, Fish, Brinicombe, Pike & Walker 1995), however the particular case of camera-shake blur has attracted significant attention recently (Fergus *et al.* 2006, Shan *et al.* 2008, Cai, Ji, Liu & Shen 2009, Cho & Lee 2009, Levin *et al.* 2009, Xu & Jia 2010, Gupta *et al.* 2010, Harmeling, Hirsch & Schölkopf 2010, Krishnan, Tay & Fergus 2011).

The method of Cho & Lee (2009) is similar in spirit to many other MAP-type algorithms recently proposed (Shan *et al.* 2008, Cai *et al.* 2009, Xu & Jia 2010, Krishnan *et al.* 2011), and proceeds by alternately updating the blur kernel and latent image, in a multi-scale framework. The algorithm iterates over three main steps. In the first step, a set of nonlinear filters is applied to the current estimate $\hat{\mathbf{f}}$ of the sharp image, in order to predict the locations and magnitudes of step edges in the sharp image. This is done using a bilateral filter (Tomasi & Manduchi 1998) to denoise the image, followed by a shock filter (Osher & Rudin 1990) to enhance sharp edges, producing an image $\hat{\mathbf{f}}'$. Finally, the derivatives of $\hat{\mathbf{f}}'$ are computed, and thresholded based on their orientation and magnitude, to produce a set of sparse edge maps $\{\mathbf{p}^{(\mathbf{q})}\}$.

In the second step, these predicted edges are used to estimate the blur kernel by solving a linear least-squares problem. Having found the kernel $\hat{\mathbf{a}}$ that minimises this problem, any elements whose values are below a threshold are set to zero. This encourages sparsity in the kernel, and ensures that all the elements are positive. In the third step the sharp image is estimated, using the current estimate of the blur kernel $\hat{\mathbf{a}}$ to deconvolve the blurred image and obtain an improved estimate of the sharp image $\hat{\mathbf{f}}$. This step is also performed by solving a linear least-squares problem.

These three steps are applied iteratively, working from coarse to fine in a multi-scale framework. The iterative process generally converges quickly at each scale, and 5–7 iterations are typically sufficient. The kernel-update and image-update sub-problems involved in the algorithm of Cho & Lee (2009) are linear least-squares by virtue of the fact that convolution is a bilinear operation on the sharp image and the blur kernel. The fact that our blur model (given in Eq. (4.13)) is bilinear in the sharp image and blur kernel is the key feature that allows it to be applied within this and other deblurring algorithms.

Although we have chosen the algorithm of Cho & Lee (2009) to apply our blur model, we note that many MAP-style blind deblurring algorithms consist of the same basic blocks, alternating between updating the kernel and the latent image by solving linear least-squares sub-problems, often in a multi-scale framework (Shan *et al.* 2008, Xu & Jia 2010, Krishnan *et al.* 2011). As such, the following discussion can apply equally well to a number of algorithms.

4.4.1 Updating the blur kernel

The general form of the spatially-invariant kernel-update problem (the second step) in the algorithm of Cho & Lee (2009) is

$$\min_{\mathbf{a}} \; \|\mathbf{a} * \mathbf{f} - \mathbf{g}\|_2^2 + \beta \, \|\mathbf{a}\|_2^2 , \tag{4.17}$$

where the first data-reconstruction term measures how well the blur kernel $\mathbf{a}$ applied to the current estimate of the latent image $\mathbf{f}$ reconstructs the observed blurred image $\mathbf{g}$, and the second term, weighted by a scalar parameter β, is the ℓ_2 regularisation of the blur kernel $\mathbf{a}$. Note that in Cho & Lee's (2009) formulation, the data term replaces $\mathbf{f}$ with the sparse edge maps $\mathbf{p}^{(q)}$ produced by the nonlinear filtering of the current estimate of the latent image $\hat{\mathbf{f}}$, i.e. has the form $\|\mathbf{a} * \mathbf{p}^{(q)} - \mathbf{d}^{(q)} * \mathbf{g}\|_2^2$, where q indexes over the first- and second-order spatial derivatives, and $\mathbf{d}^{(q)}$ are spatial derivative filters.

When applying our spatially-variant blur model, we modify the problem given by Eq. (4.17) to update the spatially variant kernel $\mathbf{w}$ as

$$\min_{\mathbf{w}} \ \|\mathbf{B}\mathbf{w} - \mathbf{g}\|_2^2 + \beta \sum_k w_k, \quad \text{s.t. } \forall k \ w_k \geqslant 0, \tag{4.18}$$

where $\mathbf{B}$ is given by Eq. (4.15). Here we have substituted our blur model into the data-reconstruction term, and replaced the ℓ_2 Tikhonov regularisation with ℓ_1 regularisation and non-negativity constraints on the kernel. Now, instead of simply a linear least-squares problem to update $\mathbf{w}$, we have an instance of the *Lasso* problem (Tibshirani 1996), for which efficient optimisation algorithms exist (Efron, Hastie, Johnstone & Tibshirani 2004, Kim, Koh, Lustig, Boyd & Gorinevsky 2007, Mairal, Bach, Ponce & Sapiro 2010). Similarly to Cho & Lee (2009) we use sparse edge maps $\mathbf{p}^{(q)}$ which result, after expanding $\mathbf{B}$ using Eq. (4.15), in a data term of the form $\|\sum_k w_k \mathbf{T}^{(k)} \mathbf{p}^{(q)} - \mathbf{d}^{(q)} * \mathbf{g}\|_2^2$.

The need for the ℓ_1 (instead of the simpler ℓ_2) regularisation arises from the differences between our kernel and convolution kernels. Fundamentally, our kernels cover a larger space of image transformations than convolution kernels (3-parameter homographies instead of 2D image translations), and we must estimate more parameters from the same amount of data. As a result, the PSF estimation process is liable to become poorly conditioned, due to an increased amount of ambiguity in the data-reconstruction term. We have observed that when simply replacing the convolution in Eq. (4.17) with our model, but without changing the regularisation, the resulting 3D kernels contain many nonzeros spread smoothly throughout, and do not produce good deblurred outputs (see Figure 4.4). On the other hand, with ℓ_1 regularisation, the optimisation is more likely to choose between ambiguous camera orientations than spreading nonzero values across all of them. In the remainder of the chapter, we refer to the original algorithm of Cho & Lee as MAP-ℓ_2, and our ℓ_1 regularised version as MAP-ℓ_1.

In addition to the use of ℓ_1 regularisation, we note that in order to constrain the 3D kernel adequately, we require data from all regions of the image. This can be seen by considering a vertical blur at the left- or right-hand side of the image. Such a blur could be explained by either a rotation of the camera about its X axis, or a rotation about its Z axis. In order to resolve this ambiguity, we would need to look at other regions of the image. To ensure that we use observations from all parts of the image when updating the kernel, we modify the construction of the edge-maps $\mathbf{p}^{(q)}$ from the filtered sharp image $\hat{\mathbf{f}}'$. We simply subdivide the image into 3×3 regions, and apply the gradient thresholding step independently on each.

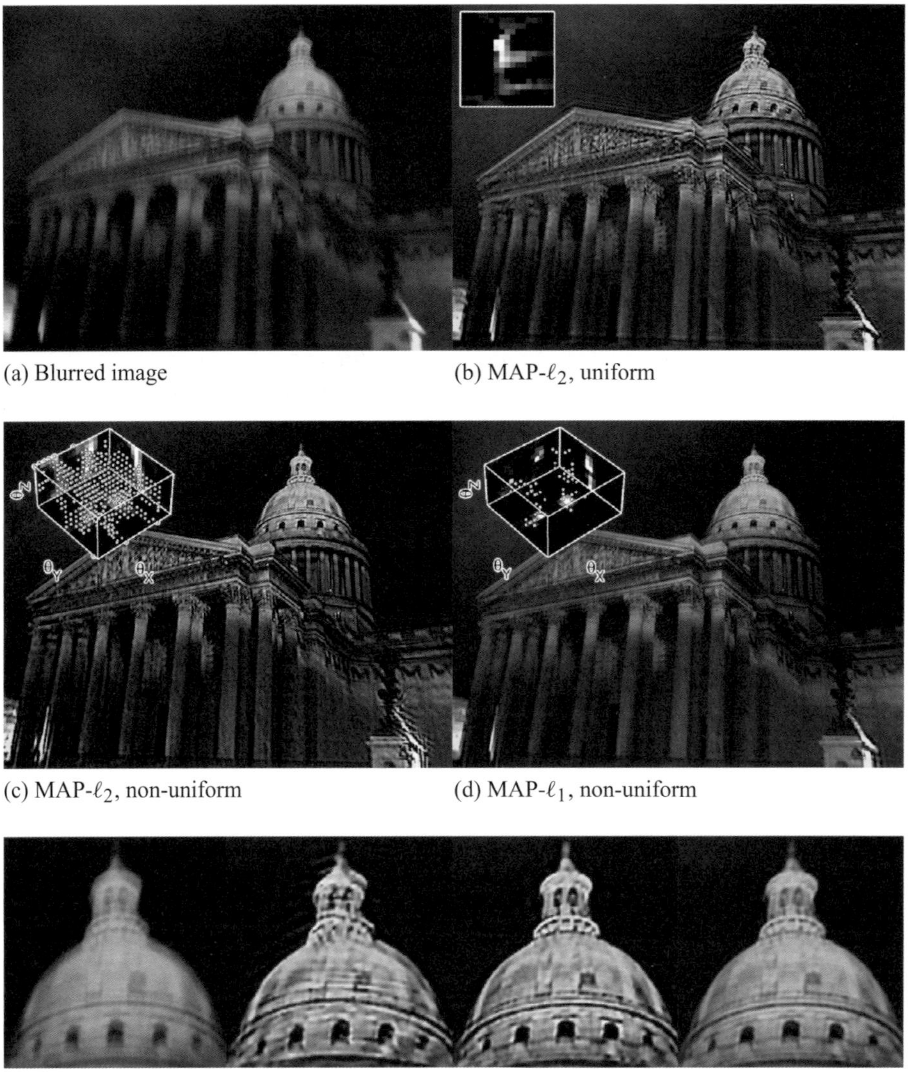

(a) Blurred image (b) MAP-ℓ_2, uniform

(c) MAP-ℓ_2, non-uniform (d) MAP-ℓ_1, non-uniform

(a) (b) (c) (d)

Figure 4.4 Blind deblurring of real camera shake. A hand-held image with camera shake (a), captured with a shutter speed of 1 s, with the results of blind deblurring using the algorithm of Cho & Lee with (b) a uniform blur model and (c–d) our spatially-varying blur model. The estimated kernels are shown inset in the deblurred results. The result using our blur model in the MAP-ℓ_1 algorithm (d) shows more detail and fewer artefacts than those using the uniform blur model, as can be seen in the zoomed-in portions shown in the last row. Also shown is the result when using our blur model with ℓ_2 regularisation on the kernel (c). As can be seen, the ℓ_2 regularisation is not sufficient to produce a good estimate of the kernel, and results in a deblurred output containing many artefacts. The rotational blur kernels in (c–d) cover $\pm 0.7°$ in θ_X and θ_Y and $\pm 1.4°$ in θ_Z.

4.4.2 Updating the latent image

The latent image update (the third step) in the algorithm of Cho & Lee (2009) is performed in a similar way to many non-blind deblurring algorithms (Xu & Jia 2010, Krishnan & Fergus 2009) by solving a linear least-squares problem of the form

$$\min_{\mathbf{f}} \ \|\mathbf{a} * \mathbf{f} - \mathbf{g}\|_2^2 + \alpha \left\|\mathbf{d}^x * \mathbf{f}\right\|_2^2 + \alpha \left\|\mathbf{d}^y * \mathbf{f}\right\|_2^2, \tag{4.19}$$

where $\mathbf{d}^x$ and $\mathbf{d}^y$ represent horizontal and vertical derivative filters, and α is a regularisation weight. Note that in Cho & Lee's (2009) formulation, the data term also takes into account the derivatives of $\mathbf{f}$ and $\mathbf{g}$ as well as the intensities (i.e. $\left\|\mathbf{a} * (\mathbf{d}^{(q)} * \mathbf{f}) - \mathbf{d}^{(q)} * \mathbf{g}\right\|_2^2$ is also penalised, for various derivative filters $\mathbf{d}^{(q)}$).

To apply our blur model, we simply replace the convolution in the data-reconstruction term and update the latent image as

$$\min_{\mathbf{f}} \|\mathbf{A}\mathbf{f} - \mathbf{g}\|_2^2 + \alpha \left\|\mathbf{d}^x * \mathbf{f}\right\|_2^2 + \alpha \left\|\mathbf{d}^y * \mathbf{f}\right\|_2^2, \tag{4.20}$$

where $\mathbf{A}$ is given by Eq. (4.14). We minimise Eq. (4.20) using conjugate-gradient descent (Shewchuk 1994).

Note that at this point, we are not able to take full advantage of the speed optimisations proposed by Cho & Lee (2009), due to their use of Fourier transforms to compute convolutions and to perform direct minimisation of Eq. (4.19) in the frequency domain. However, in the following section we will describe an efficient approximation of the spatially-variant blur model which enables this.

4.5 Efficient computation of the spatially-variant model

Due to the additional computational expense incurred by using a spatially-variant blur model instead of a spatially-invariant one, both blind and non-blind deblurring under this model can be very time consuming. As seen in the previous section, MAP-type deblurring algorithms typically involve solving linear least-squares problems, minimising $\|\mathbf{A}\mathbf{f} - \mathbf{g}\|_2^2$ with respect to $\mathbf{f}$, and $\|\mathbf{B}\mathbf{w} - \mathbf{g}\|_2^2$ with respect to $\mathbf{w}$. As defined in Eqs. (4.14) and (4.15), the matrices $\mathbf{A}$ and $\mathbf{B}$ involve sums over a set of homographies $\{\mathbf{T}^{(\mathbf{k})}\}$, where the size $\mathbf{k}$ of the set can be large, of the order of hundreds or thousands.

In iterative minimisation algorithms for solving such least-squares problems, we must repeatedly compute arbitrary matrix-vector multiplications involving $\mathbf{A}$ and $\mathbf{B}$. For images with millions of pixels these matrices are generally too large to fit into memory, and the matrix-vector products must be computed on the fly, by explicitly warping the image for each homography $\mathbf{T}^{(k)}$. This is by far the biggest bottleneck in the use of our model, and means that both blind PSF estimation and non-blind deblurring are significantly slower than for spatially-invariant blur.

To reduce the running time of the whole deblurring process, in this section we propose an efficient approximation to the blur model from Eq. (4.12), based on the locally-uniform 'efficient filter flow' proposed by Hirsch, Sra, Schölkopf & Harmeling (2010). We begin by describing the approximation in Section 4.5.1, before demonstrating in

Section 4.5.2 how it can be used to compute the matrix-vector products necessary for the kernel update step (Eq. (4.18)) very quickly. In Section 4.5.3 we describe how this approach allows us to update the sharp image (Eq. (4.20)) directly in the frequency domain (instead of using iterative methods). The approximation allows blind deblurring to be performed an order of magnitude faster than when using the exact forward model. Note that concurrently with our work, Hirsch, Schuler, Harmeling & Schölkopf (2011) proposed a similar model for efficiently computing spatially-variant camera-shake blur.

4.5.1 A locally-uniform approximation for camera shake

Hirsch *et al.* (2010) observe that in some cases of spatially-variant image blur, the blur may vary slowly and smoothly across the image. In these cases, it is reasonable to approximate spatially-variant blur as locally-uniform. Following this observation, they propose a model for spatially-variant blur, whereby the sharp image $\mathbf{f}$ is covered with a coarse grid of P overlapping patches, each of which is modelled as having a spatially-invariant blur. The overlapping patches ensure the blur varies smoothly, while allowing the forward model to be computed using P small convolutions. Hirsch *et al.* (2010) assign each patch r a spatially-invariant blur filter $\mathbf{a}^{(r)}$, and their model is given by:

$$\mathbf{g}^* = \sum_{r=1}^{P} \mathbf{C}^{(r)^\top} \left(\mathbf{a}^{(r)} * \left(\mathbf{m} \circ \mathbf{C}^{(r)} \mathbf{f} \right) \right), \tag{4.21}$$

where $\mathbf{C}^{(r)}$ is a matrix that crops the rth patch of the image $\mathbf{f}$ (thus $\mathbf{C}^{(r)^\top}$ reinserts it via zero-padding). The vector $\mathbf{m}$ is the Bartlett–Hann window, and $\cdot \circ \cdot$ represents the Hadamard (element-wise) product. Note that this model can be computed very efficiently by computing the discrete Fourier transforms of each patch and filter using the Barlett–Hann window fast Fourier transform (FFT), multiplying them element-wise in the frequency domain, and then taking the inverse discrete Fourier transform of the result. Under varying assumptions, different authors have also proposed locally-uniform models of spatially-variant blur, which take similar forms to Eq. (4.21) (Nagy & O'Leary 1998, Vio, Nagy, Tenorio & Wamsteker 2005, Tai, Du, Brown & Lin 2010).

In their original work, Hirsch *et al.* (2010) parameterise the blur using a separate filter $\mathbf{a}^{(r)}$ for each patch r. Likewise Harmeling *et al.* (2010), who apply this model to single-image camera-shake removal, also estimate a separate filter per patch using the MAP algorithm of Cho & Lee (2009). One weakness of this approach is that in textureless regions, the algorithm of Cho & Lee may fail, and so heuristics are needed to encourage similarity between neighbouring filters, and to detect and replace failed local kernel estimates.

Given the forward blur model for camera shake in Eq. (4.12), which is parameterised by a single set of weights $\mathbf{w}$, we can in fact write each $\mathbf{a}^{(r)}$ in terms of $\mathbf{w}$. For each patch r, we choose $\mathbf{a}^{(r)}$ to be the point spread function for the central pixel i_r, which is given by the i_rth row of $\mathbf{A}$. Since $\mathbf{A}$ is linear in $\mathbf{w}$, we can construct a matrix $\mathbf{J}^{(r)}$ such that $\mathbf{a}^{(r)} = \mathbf{C}^{(r)} \mathbf{J}^{(r)} \mathbf{w}$. The elements of each $\mathbf{J}^{(r)}$ are simply a rearrangement of the elements of the matrices $\mathbf{T}^{(k)}$; $J_{jk}^{(r)} = T_{i_r j}^{(k)}$.

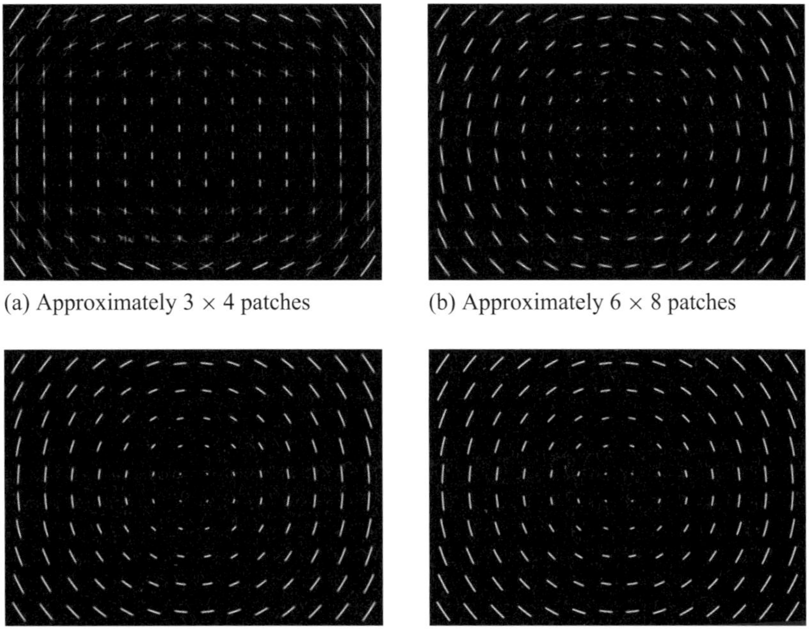

(a) Approximately 3 × 4 patches

(b) Approximately 6 × 8 patches

(c) Approximately 12 × 16 patches

(d) Exact model

Figure 4.5 Approximating spatially-variant blur by combining uniformly-blurred, overlapping patches. Using the model described in Section 4.5.1, we can efficiently compute approximations to the spatially-variant blur model in Eq. (4.12). With a small number of patches (a), the PSF at each pixel is visibly the sum of different blurs from overlapping patches. As more patches are used (b–c), the approximation becomes increasingly close to the exact model (d) – at 12 × 16 patches it is almost indistinguishable.

Having written each filter $\mathbf{a}^{(r)}$ in terms of $\mathbf{w}$, we can then substitute this into Eq. (4.21) to obtain the following approximation of the forward model from Eq. (4.12):

$$\mathbf{g}^* = \mathbf{Af} = \mathbf{Bw} \simeq \sum_{r=1}^{P} \mathbf{C}^{(r)\top} \left(\left(\mathbf{C}^{(r)} \mathbf{J}^{(r)} \mathbf{w} \right) * \left(\mathbf{m} \circ \mathbf{C}^{(r)} \mathbf{f} \right) \right). \tag{4.22}$$

This equation allows the forward model to be computed quickly using only a handful of small convolutions, which can be performed efficiently in the frequency domain. Figure 4.5 shows how the quality of the locally uniform approximation varies with the number of patches being used, compared to the exact model. In all our experiments, we use a grid of 6 × 8 patches.

4.5.2 Updating the blur kernel

In iterative algorithms for estimating the blur kernel $\mathbf{w}$ from Eq. (4.18), we typically need to compute the gradient of $\|\mathbf{Bw} - \mathbf{g}\|_2^2$ with respect to $\mathbf{w}$, which involves computing $\mathbf{B}^\top \mathbf{y}$ for arbitrary $\mathbf{y}$ and $\mathbf{B}^\top \mathbf{B}$. In addition to being able to compute the forward model quickly, Eq. (4.22) provides us with a fast approximate way of computing these products using one convolution per patch:

$$\mathbf{B}^\top \mathbf{y} \simeq \sum_{r=1}^{P} \mathbf{J}^{(r)\top} \mathbf{C}^{(r)\top} \Big(\big(\mathbf{m} \circ \mathbf{C}^{(r)}\mathbf{f}\big) \otimes \big(\mathbf{C}^{(r)}\mathbf{y}\big) \Big) \tag{4.23}$$

$$\mathbf{B}^\top \mathbf{B} \simeq \sum_{r=1}^{P} \mathbf{J}^{(r)\top} \mathbf{C}^{(r)\top} \mathrm{XCorrMatrix}\Big(\big(\mathbf{m} \circ \mathbf{C}^{(r)}\mathbf{f}\big) \otimes \big(\mathbf{C}^{(r)}\mathbf{f}\big) \Big) \mathbf{C}^{(r)} \mathbf{J}^{(r)}, \tag{4.24}$$

where $\cdot \otimes \cdot$ represents 2D correlation, and the function XCorrMatrix constructs the full cross-correlation *matrix* from a cross-correlation *vector* between two signals by replicating and shifting elements. A cross-correlation matrix $\mathbf{M}$ for a pair of signals $\mathbf{u}$ and $\mathbf{v}$ stores their inner-product at all possible translations of both signals. Assuming appropriate boundary conditions, each row of $\mathbf{M}$ is simply a shifted version of the cross-correlation $\mathbf{u} \otimes \mathbf{v}$.

4.5.3 Updating the latent image: fast, non-iterative, non-blind deconvolution

The locally-uniform approximation allows the forward model and its derivatives to be computed much faster using the FFT, such that we can quickly compute the $\mathbf{A}^\top \mathbf{y}$ and $\mathbf{A}^\top \mathbf{A}$ products needed to perform gradient-based optimisation of the sharp image in Eq. (4.20). However, for spatially-invariant blur, the fastest way of updating the sharp image (in Eq. (4.19)) is not using iterative methods such as conjugate-gradient descent, but rather using non-iterative frequency-domain deconvolution, which we outline below. In this section we extend this method to handle spatially-variant blur, via the locally-uniform approximation.

When blur is spatially invariant, using Parseval's theorem, Eq. (4.19) can be transformed into N independent 1D quadratic minimisations in the frequency domain, allowing the solution to be obtained directly by pixel-wise division in the frequency domain (Gamelin 2001):

$$\hat{\mathbf{f}} = \mathcal{F}^{-1}\left(\frac{\mathcal{F}(\mathbf{a})^* \circ \mathcal{F}(\mathbf{g})}{\mathcal{F}(\mathbf{a})^* \circ \mathcal{F}(\mathbf{a}) + \alpha\left(\mathcal{F}(\mathbf{d}^x)^* \circ \mathcal{F}(\mathbf{d}^x) + \mathcal{F}(\mathbf{d}^y)^* \circ \mathcal{F}(\mathbf{d}^y)\right)} \right), \tag{4.25}$$

where $\mathcal{F}(\cdot)$ takes the 2D discrete Fourier transform (computed using the Barlett–Hann window fast Fourier transform), and $\mathcal{F}^{-1}(\cdot)$ the inverse Fourier transform.

However, for spatially-variant blur, the locally-uniform approximation does not immediately permit this. Even though each patch has a spatially-invariant blur, the fact that the patches overlap means that the reconstruction of any blurred pixel will involve several patches. This can be seen by inserting the locally-uniform model into the non-blind deblurring problem in Eq. (4.20), and checking that the sum over patches lies inside the data-reconstruction term:

$$\min_{\mathbf{f}} \left\| \sum_{r=1}^{P} \mathbf{C}^{(r)\top} \left(\mathbf{a}^{(r)} * (\mathbf{m} \circ \mathbf{C}^{(r)}\mathbf{f})\right) - \mathbf{g} \right\|_2^2 + \alpha \left\| \mathbf{d}^x * \mathbf{f} \right\|_2^2 + \alpha \left\| \mathbf{d}^y * \mathbf{f} \right\|_2^2. \tag{4.26}$$

Thus, this equation cannot be minimised independently for each patch.

In order to be able to estimate each patch independently, one simple solution is to use Jensen's inequality to obtain an upper bound on Eq. (4.26), taking the sum over patches outside the data-reconstruction term, and expanding the regularisation terms to also penalise individual patches. Having done this, we minimise the upper bound instead of the original cost. We define the set of blurred patches $\{\mathbf{g}^{(\mathbf{r})}\}$ and sharp patches $\{\mathbf{f}^{(\mathbf{r})}\}$ such that $\mathbf{g}^{(r)} = \mathbf{m} \circ \mathbf{C}^{(r)}\mathbf{g}$ and $\mathbf{f}^{(r)} = \mathbf{m} \circ \mathbf{C}^{(r)}\mathbf{f}$, and estimate each deblurred patch $\hat{\mathbf{f}}^{(r)}$ by solving

$$\min_{\mathbf{f}^{(r)}} \left\| \mathbf{a}^{(r)} * \mathbf{f}^{(r)} - \mathbf{g}^{(r)} \right\|_2^2 + \alpha \left\| \mathbf{d}^x * \mathbf{f}^{(r)} \right\|_2^2 + \alpha \left\| \mathbf{d}^y * \mathbf{f}^{(r)} \right\|_2^2 \tag{4.27}$$

for each patch. This can be done using the direct frequency-domain method of Eq. (4.25). Next, we estimate the full deblurred image $\hat{\mathbf{f}}$ that best matches the deblurred patches $\{\hat{\mathbf{f}}^{(r)}\}$ by minimising $\sum_{r=1}^{P} \left\| \mathbf{m} \circ \mathbf{C}^{(r)}\mathbf{f} - \hat{\mathbf{f}}^{(r)} \right\|_2^2$. This problem can be solved independently for each pixel, yielding the following solution:

$$\hat{\mathbf{f}} = \frac{\sum_r \mathbf{C}^{(r)\top} (\mathbf{m} \circ \hat{\mathbf{f}}^{(r)})}{\sum_r \mathbf{C}^{(r)\top} (\mathbf{m} \circ \mathbf{m})}. \tag{4.28}$$

In Figure 4.6 we compare this method of non-blind deblurring to other possibilities for solving Eq. (4.20) with spatially-variant camera-shake blur. The fast independent method produces results which are visually very similar to those obtained using the exact model, in a significantly shorter amount of time. Using this direct deconvolution method to update the latent image via Eq. (4.20) during blind deblurring provides an additional speed improvement, compared to the use of conjugate-gradient descent with the approximate forward model.

As a demonstration of the speed-up achievable with this approximation, the examples in Figures 4.4, 4.6 and 4.7 all took more than three hours for blind deblurring using the exact model (implemented in Matlab and C), compared to under six minutes using the approximation presented in this section (implemented in Matlab). The results shown in this chapter were all produced using the approximation. The reduction in computational complexity can be quantified by comparing the exact model in Eq. (4.13) with the approximate model in Eq. (4.22). Evaluating Eq. (4.13) requires $\mathcal{O}(NK)$ operations, where N is the number of pixels in the image and K is the number of nonzeros in $\mathbf{w}$, whereas Eq. (4.22) requires $\mathcal{O}(N \log N)$ operations (due to the use of the FFT to compute the convolution). For a camera-shake blur, $\mathcal{O}(K) \geqslant \mathcal{O}(\sqrt{N})$, and thus the approximation provides a significant reduction in computational complexity. For a complete discussion of the computational complexity, as well as a full derivation of the equations presented here, please refer to Whyte (2012).

Application to other non-blind deblurring algorithms

The fast, non-iterative method for non-blind deblurring presented in this section can be used as a building block in more sophisticated non-blind deblurring algorithms. A number of recent algorithms, which place sparse-gradient priors on the deblurred image (Krishnan & Fergus 2009, Wang, Yang, Yin & Zhang 2008, Shan *et al.* 2008), or which use more robust data-reconstruction terms (Yan, Zhang & Yin 2009, Xu & Jia

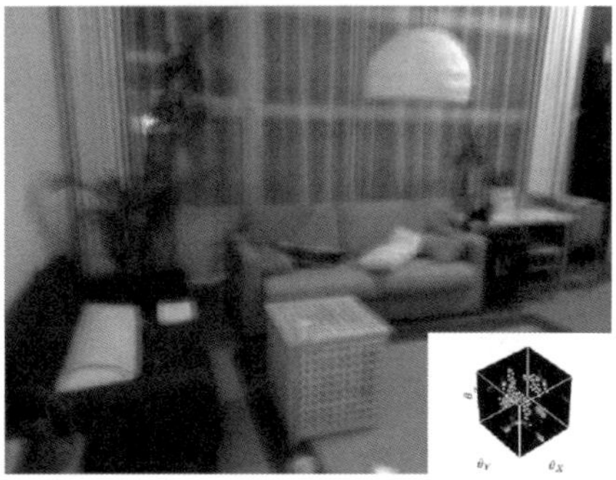

(a) Blurred image and estimated blur kernel

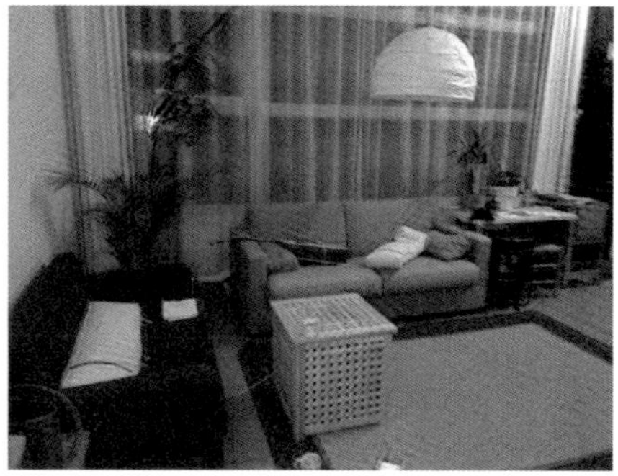

(b) Conjugate gradient descent (20 iterations), exact model, 442 s

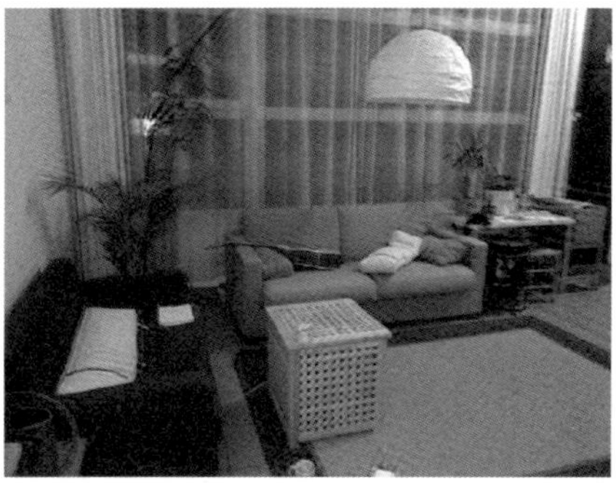

(c) Conjugate gradient descent (20 iterations), approximate model, 137 s

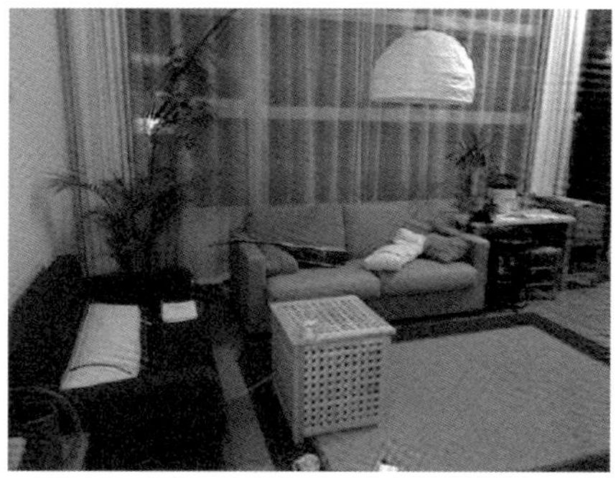

(d) Direct independent deconvolution of patches, 4 s

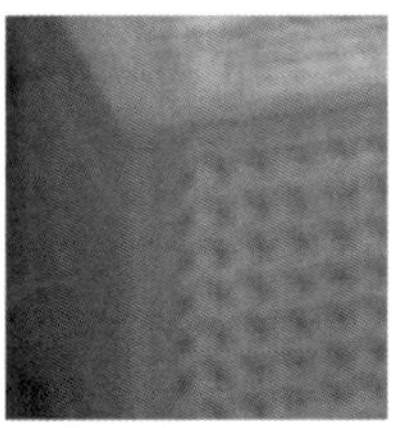

(a)

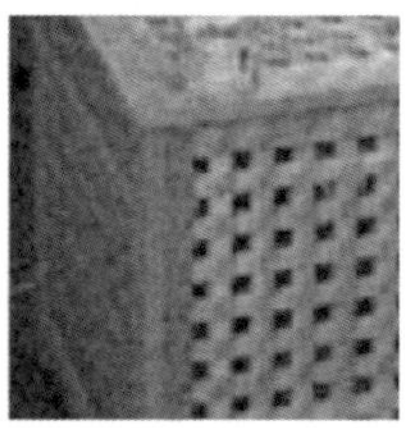

(b)

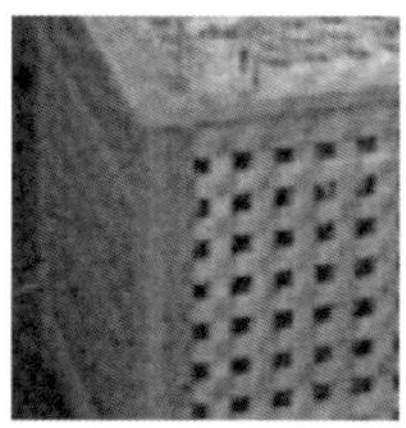

(c)

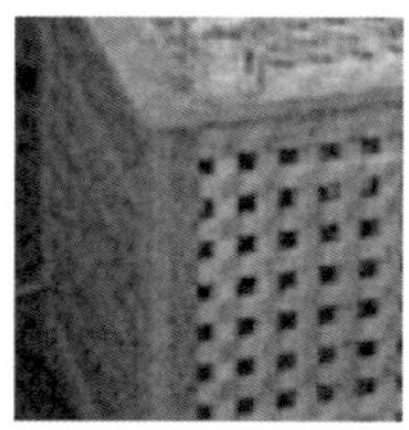

(d)

Figure 4.6 Least-squares non-blind deblurring using the exact and approximate forward models. Given a blurred image of size 1024×768 and blur kernel (a), this figure shows the results and computation times for least-squares deconvolution with ℓ_2 gradient regularisation, using (b) conjugate-gradient descent (CG) with the exact forward model, (c) CG with the approximate forward model, and (d) direct deconvolution using the approach described in Section 4.5.3. The results are visually similar using all three methods. Using CG with the approximate forward model is much faster than with the exact model, however the direct approach takes only a fraction of the time of either of these.

2010), involve solving sub-problems of the form of Eq. (4.19), which has a quadratic data-reconstruction term and quadratic regularisation. As such, the non-iterative method presented in this section can be used to extend these more sophisticated algorithms to handle spatially-variant blur, without a substantial increase in processing time. The

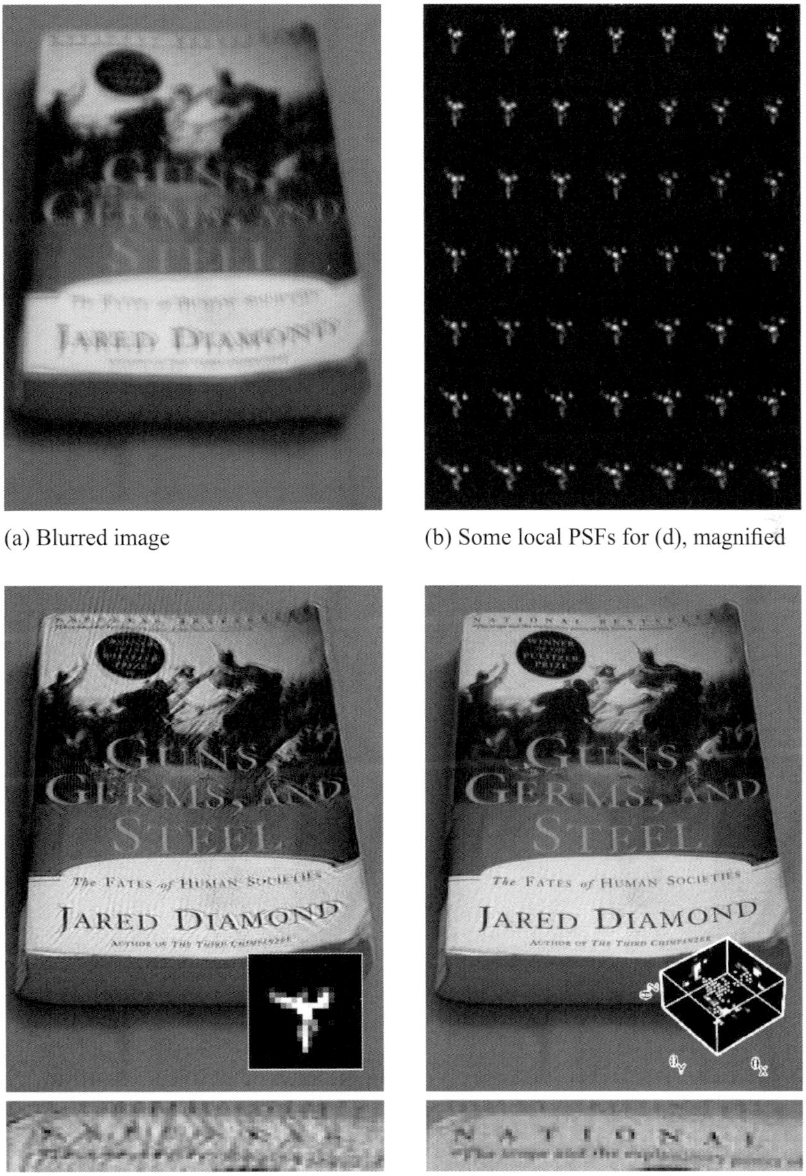

(a) Blurred image (b) Some local PSFs for (d), magnified

(c) MAP-ℓ_2, uniform (d) MAP-ℓ_1, non-uniform

Figure 4.7 Blind deblurring of real camera shake. The result of blind deblurring on a real camera-shake image (a), captured with a shutter speed of 1 s, using the MAP approach of Cho & Lee with both the uniform and non-uniform blur models. Also shown in (b) are some of the local PSFs generated from the blur kernel in (d) at various points in the image. In the blurred image, most of the text on the book cover is too blurred to read. Deblurring the image with the uniform blur model (c) allows some of the text on the cover of the book to be read, however, after deblurring with our non-uniform model (d), all but the smallest text becomes legible.
The estimated kernels for the two models are shown inset in the deblurred results. The blur kernel in (d) covers $\pm 0.4°$ in θ_X and θ_Y, and $\pm 0.9°$ in θ_Z. Reproduced from Whyte *et al.* (2012) with permission, © Springer 2012.

results shown in Section 4.6 use the algorithm of Krishnan & Fergus (2009), modified in this way, to perform the final non-blind deblurring.

4.6 Single-image deblurring results

In this section we present results of single-image deblurring using the MAP-ℓ_1 algorithm to estimate the spatially-variant PSF, with comparisons to results obtained with the original algorithm of Cho & Lee (2009) on real data. Having estimated the PSF, we apply the non-blind deblurring algorithm of Krishnan & Fergus (2009) to estimate the final deblurred image. This algorithm is easily adapted to non-uniform blur since it involves repeated minimisations of quadratic cost functions similar to Eq. (4.20).

Figure 4.4 shows a blind deblurring result on an image blurred by real camera shake. Our algorithm is able to model and remove the blur, while the results with the original algorithm contain visible artefacts. This is explained by both the wide field of view, and the fact that the kernels estimated using our algorithm exhibit significant in-plane rotation. Also shown is the result of using ℓ_2 regularisation on the spatially-variant kernel. As discussed in Section 4.4.1, the kernel produced is not sparse, and as a result the deconvolved output exhibits many artefacts compared to the MAP-ℓ_1 result.

Figure 4.7 shows another example of single-image deblurring, using the MAP algorithm. While the uniform blur kernel provides a reasonable estimate of the true blur, and allows us to resolve some of the text on the book's cover, the use of our non-uniform blur model provides a clear improvement, and renders almost all the text legible.

Additional blind deblurring results of real camera shakes are shown in Figures 4.1 and 4.6.

4.6.1 Limitations and failures

Since the MAP approach to blind deblurring attempts to solve a non-convex minimisation problem, it is not possible to guarantee a globally optimal solution. However, in practice we have found the MAP-ℓ_1 algorithm to be capable of deblurring a wide range of images with large blurs – the blurs removed in this chapter are up to 35 pixels wide (e.g. Figure 4.7) – for both uniform and non-uniform blur. For our model, this corresponds to around 3–5° of rotation around each axis for a photograph whose width and focal length are both 1000 pixels. This is due in large part to the multi-scale approach; by finding a sequence of solutions at increasingly fine resolutions, the large scale structures in the blur kernel and sharp image are resolved before the fine details.

Nevertheless, failures do occur and Figure 4.8 shows such a case. We have observed failures caused by several factors, including a high level of noise in the input images, an excessively large blur, or an unknown nonlinear camera response function. Also, the algorithm may fail when the edge-based heuristics which guide the blind deblurring do not match the particular image being estimated; for example, in an image that contains only fine-scale texture with no large-scale step edges.

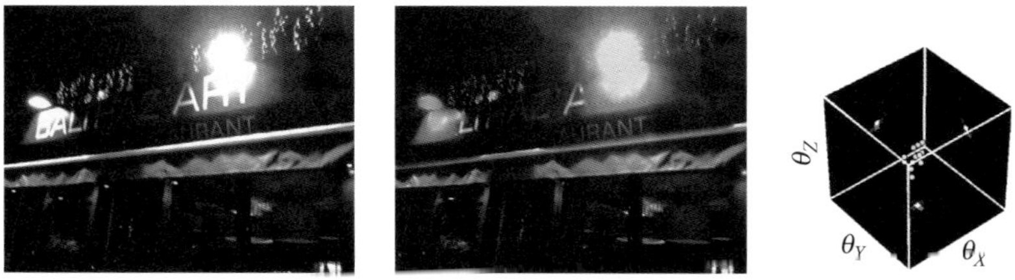

(a) Blurred image

(b) Deblurred result from MAP-ℓ_1 algorithm

(c) Kernel estimated by MAP-ℓ_1 algorithm

Figure 4.8 Blind deblurring failures. A blurred image (a) with the deblurred result (b) and kernel (c) estimated by the MAP-ℓ_1 algorithm. This image contains a large amount of noise, and was captured with a camera-phone whose response function is unknown. As a result, the algorithm fails to estimate a good kernel, and instead returns a kernel which is close to a delta function.

Since we have assumed that camera translation has a negligible blurring effect, our model (and the uniform model too) is unlikely to produce good results on images for which this is not true, due to the depth-dependent blur produced. As discussed in Section 4.2, this is unlikely to be a problem on most shaken images, except for close-up photos where the subject is less than about 1 m from the camera.

4.7 Implementation

The implementation of the algorithm of Cho & Lee (2009) is our own, and we use this implementation for both uniform and non-uniform blur models when comparing results. A binary executable for Cho & Lee's (2009) algorithm is available, however we did not observe an improvement in the results obtained, and thus use our own implementation to permit a fairer comparison between the results from the uniform and non-uniform blur models. The implementation of the non-blind deblurring algorithm of Krishnan & Fergus (2009) is based on Matlab code made available online by the authors[2].

Sampling the set of rotations

One important detail to consider is how finely to discretise the orientation parameter $\boldsymbol{\theta}$. In the discrete case, each grid point $\boldsymbol{\theta}^{(k)}$ corresponds to a transformation matrix $\mathbf{T}^{(k)}$ in the sum in Eq. (4.12). Undersampling the space of orientations will affect our ability to reconstruct the blurred image accurately, but sampling it too finely will lead to unnecessary computation. Since the kernel is defined over the three parameters θ_X, θ_Y and θ_Z, doubling the sampling resolution increases the number of kernel elements by a factor of eight. In practice, we have found that a good choice of grid spacing is that which corresponds to a maximum displacement of 1 pixel in the image. Since we are fundamentally limited by the resolution of our images, reducing the spacing further

[2] http://cs.nyu.edu/~dilip/research/fast-deconvolution/

leads to redundant orientations, which are indistinguishable from their neighbours. We set the size of the 3D kernel in terms of the size of the blur we are attempting to remove, typically a few degrees along each dimension of $\boldsymbol{\theta}$, e.g. $[-5°, 5°]$.

Multi-scale implementation

Most successful blind kernel estimation algorithms here are applied within a multi-scale framework, starting with a coarse representation of image and kernel, and repeatedly refining the estimated kernel at higher resolutions. In the case of single-image deblurring, this is essential to avoid poor local minima; however, in our case, it is also important for computational reasons. At the original image resolution, the kernel may have thousands or tens of thousands of elements, however very few of these should have nonzero values.

Thus, in all of the applications presented in this chapter, which estimate the kernel iteratively, we use our current estimate of the kernel $\hat{\mathbf{w}}_s$ at a scale s to constrain our estimate at the next iteration. To do this, we define an 'active region' where $\hat{\mathbf{w}}_s$ is nonzero, and constrain the nonzeros at the next iteration to lie within this region. By clamping many kernel elements to zero, we eliminate a large amount of computation and memory requirements associated with estimating those elements' values. We first build Gaussian pyramids for the blurred image and, at the coarsest scale $s = 0$, define the active region to cover the full kernel. At each iteration, we find the nonzero elements of our current estimate of the kernel $\hat{\mathbf{w}}_s$, and dilate this region using a $3 \times 3 \times 3$ cube to define the active region for the next iteration. When moving from one scale s to the next scale $s + 1$, we upsample $\hat{\mathbf{w}}_s$ using bilinear interpolation, find the nonzero elements of this upsampled kernel and, as before, dilate this region using a $3 \times 3 \times 3$ cube. This initialises the active region for our next estimate $\hat{\mathbf{w}}_{s+1}$. We repeat this process at each scale until we have found the optimal kernel at the finest scale.

This approach is generally effective at reducing the computational burden of the kernel estimation without reducing accuracy. However, problems may occur if the blur kernel contains long faint structures, as it is possible for these to be clamped to zero at a coarse scale and never be recovered.

Running time

For a 1024×768 image, our C implementation of the exact model in Eq. (4.12) takes approximately 5 s to compute, compared to 2 s for our Matlab implementation of the approximate forward model in Eq. (4.22), on an Intel Xeon 2.93 GHz CPU.

Our implementation, in Matlab and C, of the MAP-ℓ_1 algorithm for spatially-variant blur takes over three hours to deblur a 1 MP (1024×768) image, depending on the size of the blur. This is a significant departure from the spatially-invariant algorithm of Cho & Lee (2009), who report deblurring times of under one minute using their C++ implementation. Using the efficient approximation described in Section 4.5, we are able to perform blind deblurring of the same images, using our spatially-variant blur model in under six minutes.

4.8 Conclusion

In this chapter we have proposed a geometrically-derived model for blur caused by camera shake. For a static scene and a camera with known focal length, we have shown that the blur caused by camera rotation can be modelled using a weighted set of homographies, and have proposed a practical formulation of this model in which the blurred image is bilinear in the sharp image and the weights. We have applied our model for spatially-variant camera-shake blur within an existing camera-shake removal algorithm, and validated the model with experiments demonstrating superior results compared to the spatially-invariant blur model. We have also described how an efficient approximation for spatially-variant blur can be used to reduce the computational cost of computing the spatially-variant forward blur model.

Although we have demonstrated our model only in the algorithm of Cho & Lee (2009) for blind deblurring, it could equally be used with more recent algorithms which have shown superior results, such as those of Xu & Jia (2010) and Krishnan *et al.* (2011). Equally, faster or more sophisticated methods for non-blind deblurring, such as those of Afonso, Bioucas-Dias & Figueiredo (2010) and Zoran & Weiss (2011), could be extended to use our spatially-variant blur model.

Interested readers can access an online demonstration of the blind deblurring algorithm at http://www.di.ens.fr/willow/research/deblurring/.

Acknowledgements

This work was supported in part by the MSR–Inria Laboratory, the EIT ICT Labs, ERC grants VisRec and VideoWorld, and by the Institut Universitaire de France.

References

Afonso, M., Bioucas-Dias, J. & Figueiredo, M. (2010). Fast image recovery using variable splitting and constrained optimization. *IEEE Transactions on Image Processing*, **19**(9), 2345–56.

Ayers, G. R. & Dainty, J. C. (1988). Iterative blind deconvolution method and its applications. *Optics Letters*, **13**(7), 547–9.

Cai, J.-F., Ji, H., Liu, C. & Shen, Z. (2009). Blind motion deblurring from a single image using sparse approximation. In *Proceedings of the 22nd IEEE Conference on Computer Vision and Pattern Recognition*, pp. 104–11.

Cho, S. & Lee, S. (2009). Fast motion deblurring. *ACM Transactions on Graphics (Proceedings of SIGGRAPH Asia 2009)*, **28**(5), 145:1–8.

Efron, B., Hastie, T., Johnstone, L. & Tibshirani, R. (2004). Least angle regression. *Annals of Statistics*, **32**(2), 407–99.

Fergus, R., Singh, B., Hertzmann, A., Roweis, S. T. & Freeman, W. T. (2006). Removing camera shake from a single photograph. *ACM Transactions on Graphics (Proceedings of SIGGRAPH 2006)*, **25**(3), 787–94.

Fish, D. A., Brinicombe, A. M., Pike, E. R. & Walker, J. G. (1995). Blind deconvolution by means of the Richardson–Lucy algorithm. *Journal of the Optical Society of America A*, **12**(1), 58–65.

Gamelin, T. W. (2001). *Complex Analysis*. New York: Springer-Verlag.

Gull, S. & Skilling, J. (1984). Maximum entropy method in image processing. *Communications, Radar and Signal Processing, IEE Proceedings F*, **131**(6), 646–59.

Gupta, A., Joshi, N., Zitnick, C. L., Cohen, M. & Curless, B. (2010). Single image deblurring using motion density functions. In *Proceedings of the 11th European Conference on Computer Vision*, pp. 171–84.

Harmeling, S., Hirsch, M. & Schölkopf, B. (2010). Space-variant single-image blind deconvolution for removing camera shake. In *Advances in Neural Information Processing Systems*, pp. 829–37.

Hartley, R. I. & Zisserman, A. (2004). *Multiple View Geometry in Computer Vision*, 2nd edn. Cambridge University Press.

Hirsch, M., Schuler, C. J., Harmeling, S. & Schölkopf, B. (2011). Fast removal of non-uniform camera shake. In *Proceedings of the 13th International Conference on Computer Vision*, pp. 463–70.

Hirsch, M., Sra, S., Schölkopf, B. & Harmeling, S. (2010). Efficient filter flow for space-variant multiframe blind deconvolution. In *Proceedings of the 23rd IEEE Conference on Computer Vision and Pattern Recognition*, pp. 607–14.

Joshi, N., Kang, S. B., Zitnick, C. L. & Szeliski, R. (2010). Image deblurring using inertial measurement sensors. *ACM Transactions on Graphics (Proceedings of SIGGRAPH 2010)*, **29**(4), 30:1–9.

Kim, S.-J., Koh, K., Lustig, M., Boyd, S. & Gorinevsky, D. (2007). An interior-point method for large-scale ℓ_1-regularized least squares. *IEEE Journal of Selected Topics in Signal Processing*, **1**(4), 606–17.

Klein, G. & Drummond, T. (2005). A single-frame visual gyroscope. In *Proceedings of the 16th British Machine Vision Conference*, pp. 1–10.

Krishnan, D. & Fergus, R. (2009). Fast image deconvolution using hyper-Laplacian priors. In *Advances in Neural Information Processing Systems*, pp. 1033–41.

Krishnan, D., Tay, T. & Fergus, R. (2011). Blind deconvolution using a normalized sparsity measure. In *Proceedings of the 24th IEEE Conference on Computer Vision and Pattern Recognition*, pp. 233–40.

Levin, A., Weiss, Y., Durand, F. & Freeman, W. T. (2009). Understanding and evaluating blind deconvolution algorithms. In *Proceedings of the 22nd IEEE Conference on Computer Vision and Pattern Recognition*, pp. 1964–71.

Mairal, J., Bach, F., Ponce, J. & Sapiro, G. (2010). Online learning for matrix factorization and sparse coding. *Journal of Machine Learning Research*, **11**, 19–60.

Nagy, J. G. & O'Leary, D. P. (1998). Restoring images degraded by spatially variant blur. *SIAM Journal on Scientific Computing*, **19**(4), 1063–82.

Osher, S. & Rudin, L. I. (1990). Feature oriented image enhancement using shock filters. *SIAM Journal on Numerical Analysis*, **27**(4), 919–40.

Sawchuk, A. A. (1974). Space-variant image restoration by coordinate transformations. *Journal of the Optical Society of America*, **64**(2), 138–44.

Shan, Q., Jia, J. & Agarwala, A. (2008). High-quality motion deblurring from a single image. *ACM Transactions on Graphics (Proceedings of SIGGRAPH 2008)*, **27**(3), 73:1–10.

Shan, Q., Xiong, W. & Jia, J. (2007). Rotational motion deblurring of a rigid object from a single image. In *Proceedings of the 11th International Conference on Computer Vision*, pp. 1–8.

Shewchuk, J. R. (1994). *An Introduction to the Conjugate Gradient Method Without the Agonizing Pain*. Technical report, Carnegie Mellon University.

Szeliski, R. (2004). *Image Alignment and Stitching: A Tutorial*, Technical report MSR-TR-2004-92, Microsoft Research.

Tai, Y.-W., Du, H., Brown, M. S. & Lin, S. (2010). Correction of spatially varying image and video motion blur using a hybrid camera. *IEEE Transactions on Pattern Analysis and Machine Intelligence*, **32**(6), 1012–28.

Tai, Y.-W., Kong, N., Lin, S. & Shin, S. Y. (2010). Coded exposure imaging for projective motion deblurring. In *Proceedings of the 23rd IEEE Conference on Computer Vision and Pattern Recognition*, pp. 2408–15.

Tai, Y.-W., Tan, P. & Brown, M. S. (2011). Richardson–Lucy deblurring for scenes under a projective motion path. *IEEE Transactions on Pattern Analysis and Machine Intelligence*, **33**(8), 1603–18.

Tibshirani, R. (1996). Regression shrinkage and selection via the lasso. *Journal of the Royal Statistical Society, Series B (Methodological)*, **58**(1), 267–88.

Tomasi, C. & Manduchi, R. (1998). Bilateral filtering for gray and color images. In *Proceedings of the 6th International Conference on Computer Vision*, pp. 839–46.

Vio, R., Nagy, J., Tenorio, L. & Wamsteker, W. (2005). Multiple image deblurring with spatially variant PSFs. *Astronomy & Astrophysics*, **434**, 795–800.

Šorel, M. & Flusser, J. (2008). Space-variant restoration of images degraded by camera motion blur. *IEEE Transactions on Image Processing*, **17**(2), 105–16.

Wang, Y., Yang, J., Yin, W. & Zhang, Y. (2008). A new alternating minimization algorithm for total variation image reconstruction. *SIAM Journal on Imaging Sciences*, **1**(3), 248–72.

Whyte, O. (2012). Removing camera shake blur and unwanted occluders from photographs. PhD thesis, ENS Cachan.

Whyte, O., Sivic, J. & Zisserman, A. (2011). Deblurring shaken and partially saturated images. In *Proceedings of the IEEE Workshop on Color and Photometry in Computer Vision*, pp. 745–52.

Whyte, O., Sivic, J., Zisserman, A. & Ponce, J. (2010). Non-uniform deblurring for shaken images. In *Proceedings of the 23rd IEEE Conference on Computer Vision and Pattern Recognition*, pp. 491–98.

Whyte, O., Sivic, J., Zisserman, A. & Ponce, J. (2012). Non-uniform deblurring for shaken images. *International Journal of Computer Vision*, **98**(2), 168–86.

Xu, L. & Jia, J. (2010). Two-phase kernel estimation for robust motion deblurring. In *Proceedings of the 11th European Conference on Computer Vision*, pp. 157–70.

Xu, L. & Jia, J. (2012). Depth-aware motion deblurring. In *Proceedings of the IEEE International Conference on Computational Photography*, pp. 1–8.

Yan, J., Zhang, Y. & Yin, W. (2009). An efficient TVL1 algorithm for deblurring multichannel images corrupted by impulsive noise. *SIAM Journal on Scientific Computing*, **31**(4), 2842–65.

Zoran, D. & Weiss, Y. (2011). From learning models of natural image patches to whole image restoration. In *Proceedings of the 13th International Conference on Computer Vision*, pp. 479–86.

5 Removing camera shake in smartphones without hardware stabilization

Filip Šroubek and Jan Flusser

5.1 Introduction

Processing images has become an everyday practice in a wide range of applications in science and technology and we rely on our images with ever growing emphasis. Our understanding of the world is however limited by measuring devices that we use to acquire images. Inadequate measuring conditions together with technological limitations of the measuring devices result in acquired images that represent a degraded version of the "true" image.

Blur induced by camera motion is a frequent problem in photography – mainly when the light conditions are poor. As the exposure time increases, involuntary camera motion has a growing effect on the acquired image. Image stabilization (IS) devices that help to reduce the motion blur by moving the camera sensor in the opposite direction are becoming more common. However, such a hardware remedy has its limitations as it can only compensate for motion of a very small extent and speed. Deblurring the image offline using mathematical algorithms is usually the only choice we have to obtain a sharp image. Motion blur can be modeled by convolution and the deblurring process is called deconvolution, which is a well-known ill-posed problem. In general, the situation is even more complicated, since we usually have no or limited information about the blur shape.

We can divide the deconvolution methods into two categories: methods that estimate the blur and the sharp image directly from the acquired image (blind deconvolution), and methods that use information from other sensors to estimate the blur (semi-blind deconvolution).

The main contribution of this chapter is to illustrate that blur estimation with built-in inertial sensors is possible and to implement image deblurring on a smartphone, which works in practical situations and is relatively fast, making it acceptable for a general user.

5.2 Image acquisition model

In order to perform successful image deblurring, it is important to correctly model the acquisition process. A commonly used model is the following. Let $u(\mathbf{x}) : \Omega \subset \mathbb{R}^2 \to \mathbb{R}$, $\mathbf{x} = [x, y]$, be an original *latent* image describing a real scene, where Ω is

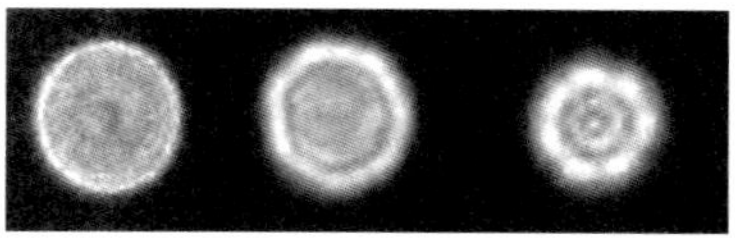

(a) Out-of-focus blurs

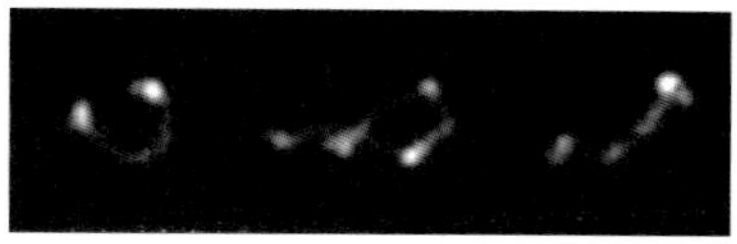

(b) Camera-shake blurs

Figure 5.1 Examples of real camera blurs. (a) Blur caused by an out-of focus lens with different lens parameters (focal length and aperture size); notice the polygonal shape clearly visible in the central image, which corresponds to the aperture opening of a 7-blade diaphragm; (b) blurs caused by camera motion during exposure.

a rectangular image support. A first-order approximation of degradation acting on u is twofold: degradation linear operator H and additive noise n. Then the output of acquisition is a degraded image g given by a relation

$$g = Hu + n\,. \tag{5.1}$$

The difficulty with H is that it is ill-conditioned, which means that during inversion noise n gets amplified and the solution is unstable. We face an *ill-posed inverse* problem that requires special handling.

5.2.1 Space-invariant model

The most common type of degradation, which is considered in this chapter, is *convolution*:

$$[Hu](x,y) = h * u = \int_{\Omega} u(s,t)h(x-s,y-t)\,\mathrm{d}s\,\mathrm{d}t. \tag{5.2}$$

This definition extends to any number of dimensions and not just $\mathbb{R}^2$. For example, in confocal microscopy, convolution is in $\mathbb{R}^3$. Function h is a convolution kernel (or simply blur) and defines the behavior of the convolution operator. It is also called a *point spread function* (PSF), because h is an image the device would acquire after measuring an ideal point source $\delta(\mathbf{x})$ (delta function). Image blur due to camera motion or improper camera focus setting can be in simple cases modeled by convolution. The PSF size influences the degree of blurring and the PSF shape reflects the physical nature of blurring. For example, an out-of-focus camera lens causes PSFs of various shapes depending on the lens aperture, while camera motion causes curvy PSFs, where the curve shape is related to the trajectory of the motion; see Figure 5.1. There is a wide range of imaging devices in which the acquisition process can be modeled by convolution. Apart from devices with classical optical systems, such as digital cameras, optical microscopes or telescopes, convolution degradation occurs also in atomic force microscopy (AFM) or scanning tunneling microscopy (STM), where the PSF shape is related to the measuring tip shape. Media turbulence (e.g. the atmosphere for terrestrial telescopes) can cause blurring that can be modeled by convolution, and there are many more examples.

5.2.2 Space-variant model

To make convolution more general, it is often necessary to allow the PSF to change over the image. This is called *space-variant* convolution, though strictly speaking it is no longer mathematical convolution. Using space-variant convolution we can model more general degradations, such as blur induced by complex camera motion and rotation, out-of-focus blur in a wide-depth scene, or blur due to hot-air turbulence.

The operator H can be written in a form naturally generalizing standard convolution as

$$[Hu](x,y) = \int_{\Omega} u(s,t)h(x-s, y-t, s, t)\, \mathrm{d}s\, \mathrm{d}t \tag{5.3}$$

with h now dependent on the position (x,y) in the image. Convolution in Eq. (5.2) is a special case of Eq. (5.3) with $h(s,t,x,y) = h(s,t)$ for any position (x,y). It is valid to refer to $h(s,t,x,y)$ as a space-variant PSF (or kernel), for the function $h(s,t,x,y)$ taken as a function of two variables with fixed x and y describes how a delta function at (x,y) spreads its energy in space after degradation by H. In many real scenarios h changes slowly with respect to (x,y) as discussed in Section 5.3.3. In this case, the operator H can be locally approximated by convolution with $\tilde{h}(s,t) = h(s,t,x,y)$ in a neighborhood of the point (x,y).

It is worth noting that we can interpret the physical meaning of the PSF in two different ways. The standard view is that the PSF tells us how a point source is spread in space. The blurred image is obtained by spreading every pixel and summing the contributions of all pixels. The other interpretation is that the PSF is a weighted window. The blurred image is calculated by performing weighted averaging around every pixel. In the case of space-invariant convolution both views are equivalent. However in the case of space-variant convolution, this is not true and we assume the first interpretation, which describes the physical nature of the blurring phenomenon.

In practice, we work with a discrete representation, where the same notation can be used with the following differences: PSF h is defined on a discrete set of coordinates, the integral in Eq. (5.3) becomes a sum, operator H corresponds to a matrix and u to a vector obtained by concatenating columns of the image into one long vector. Using the vector–matrix notation, Eq. (5.1) may be written as

$$\mathbf{g} = \mathbf{H}\mathbf{u} + \mathbf{n}\,. \tag{5.4}$$

In the space-invariant case, $\mathbf{H}$ is a convolution matrix (block Toeplitz matrix with Toeplitz blocks) and each column of $\mathbf{H}$ corresponds to the same kernel. In the space-variant case, as each column corresponds to a different position (x,y), it may contain a different kernel $h(s,t,x,y)$.

5.3 Inverse problem

A standard formulation of the image deblurring problem is a stochastic one (Campisi & Egiazarian 2007) which assumes that images and PSFs are random vector fields

(Rosenfeld & Kak 1982) with known prior probability density functions $p(u)$ and $p(h)$, respectively. The Bayesian paradigm dictates that the inference on the latent image and PSF should be based on the posterior probability

$$p(u, h|g) \propto p(g|u, h)p(u)p(h)\,, \tag{5.5}$$

where u and h are assumed to be independent. The conditional density $p(g|u, h)$ is given by our model Eq. (5.1) and is equal to the probability density function of noise n, which is typically normally distributed, $p(n) = N(0, \sigma^2)$. The prior $p(u)$ forces some type of image statistical characteristics – typically we assume sparse distribution (e.g. Laplacian) of image derivatives (Rudin, Osher & Fatemi 1992). The prior $p(h)$ can be similar to $p(u)$, but it is often rectified to force positivity. Estimating the pair $(\hat{U}, \hat{h})$ is equivalent to maximizing the posterior $p(u, h|g)$, which is commonly referred to as the *maximum a posteriori* (MAP) approach. Note that maximization of the posterior is equivalent to minimization of $-\log p(u, h|g)$.

5.3.1 MAP and beyond

The classical solution to the blind deconvolution problem is maximizing the posterior probability $p(u, h|g)$ with respect to both u and h (known as the $\mathrm{MAP}_{u,h}$ approach). However, the posterior has a very uneven shape with many local peaks, and alternating maximization often returns an incorrect solution.

Typically the priors are defined in some other domain than the image domain and are assumed to be independent and sparse, $p(u) = \Pi_i p(\phi_i(u))$, where ϕ_i is a function mapping the image into the other domain. A common choice of ϕ is an image gradient with p favoring sparse distributions, i.e. $\log p(u) = -\sum_i \frac{1}{2\gamma}|\nabla u_i|^p + C$ for $0 \leqslant p \leqslant 1$. In the vector matrix notation this can be expressed as $\log p(u) = -\frac{1}{2\gamma}\mathbf{u}^T\mathbf{L}_u^p\mathbf{u}$, where $\mathbf{L}_u^p$ is a symmetrical sparse matrix with elements related to ∇u. In the case of $p = 2$ (Gaussian prior), $\mathbf{L}_u^2$ is independent of u and is equal to the Laplacian, which we denote simply as $\mathbf{L}$. A detailed derivation of these relations is provided in Šroubek & Flusser (2005).

The most common method of finding $\mathrm{MAP}_{u,h}$ is to alternate between minimization of $-\log p(u, h|g)$ with respect to u and h. It is a well-known fact that this leads to the so-called "no-blur" solution, i.e. the most probable solution is that h is a delta function and $u = g$, which is clearly a solution we want to avoid. It is important to note that the posterior $p(u, h|g)$ has a local extreme for the true u and h, which can be reached if we initialize the alternating minimization close to the true solution. Many authors (Molina, Mateos & Katsaggelos 2006, Fergus, Singh, Hertzmann, Roweis & Freeman 2006, Jia 2007, Shan, Jia & Agarwala 2008, Levin, Weiss, Durand & Freeman 2009) tried to alleviate this problem by searching for more sophisticated priors that would better match the natural image statistics. Unfortunately, mainly due to the fact that blur reduces image variance, most of these priors still favor the blurred image over the sharp one as pointed out by Levin *et al.* (2009). Levin further demonstrated in Levin, Weiss, Durand & Freeman (2011) that a proper estimator matters more than the shape of priors. Marginalizing the posterior with respect to the latent image u and maximizing

$p(h|g) = \int p(u,h|g)du$ leads to the correct solution. This is referred to as the MAP$_h$ approach. Once we estimate $\hat{h}$, we can estimate the sharp image as

$$u^{\mathrm{MAP}} = \underset{u}{\operatorname{argmax}}\, p(u, h = \hat{h}|g). \tag{5.6}$$

The marginalized probability $p(h|g)$ can be expressed in a closed form only for simple priors that are not sparse. Otherwise approximation methods such as variational Bayes (Miskin & MacKay 2000, Fergus *et al.* 2006) or the Laplace approximation (Galatsanos, Mesarovic, Molina & Katsaggelos 2000) must be used. For example, using the Laplace approximation

$$\log p(h|g) \approx \log p(u = u^{\mathrm{MAP}}, h|g) - \frac{1}{2}\log|A| + C, \tag{5.7}$$

where $|A|$ denotes determinant of the variance of $p(u = u^{\mathrm{MAP}}, h|g)$ with respect to u. If we assume a simple Gaussian prior, $\log p(u) = -\sum_i \frac{1}{2\gamma}|\phi_i(u)|^2 + C = -\frac{1}{2\gamma}\mathbf{u}^T\mathbf{L}\mathbf{u} + C$ then the Laplace approximation is exact with $\mathbf{A} = \frac{\mathbf{H}^T\mathbf{H}}{\sigma^2} + \frac{\mathbf{L}}{\gamma}$, u^{MAP} has a closed-form solution and so does $p(h|g)$. In a general case of sparse priors, we can use the above equation as an update equation for $\hat{h} = \operatorname{argmax}_h p(h|g)$ and also iteratively update u^{MAP} in Eq. (5.6). The variational Bayes factorizes the posterior $p(u,h|g)$ and approximates it by $q(u)q(h)$, which can then be solved by alternating maximization. As derived in Molina *et al.* (2006), for example, the variational Bayes leads to a very similar update equation (Eq. (5.7)) for h, where the second term $\frac{1}{2}\log|A|$ is replaced by $\frac{1}{2}\mathbf{h}^T\mathbf{cov}(q(u))\mathbf{h}$. In the case of the simple Gaussian prior, $\mathbf{cov}(q(u)) = \mathbf{A}^{-1}$. In both the Laplace approximation Eq. (5.7) and the variational Bayes, the second term plays a crucial role, since it favors kernels with more blur and penalizes the "no-blur" solution. This brings us to an interesting conclusion that the iterative algorithms of the approximation techniques are very similar to the classical MAP$_{u,h}$ approach with an additional term that penalizes blurs close to delta functions. This conclusion suggests that even MAP$_{u,h}$ should work if a better numerical method is applied.

There are several other modifications (or maybe better to say tweaks) used in blind deconvolution methods, which are often only briefly mentioned in the literature but we believe play a key role and largely improve results in practice. The first one is that in the blur estimation step we use ∇u instead of u. From the perspective of the data model this formulation is equivalent, since $\nabla g = h * \nabla u$. One possible reason for the advantage of derivatives is that it fits better with our model that the image prior is a product of independently and identically distributed variables. An algebraic explanation is that the deconvolution system solved in every iteration is better conditioned. The second modification is that the current blind deconvolution methods, such as Xu & Jia (2010), start the estimation process with a very high noise variance (σ^2) and slowly decrease it to the correct level. A similar idea is used in simulated annealing and has the effect that low frequencies (overall shape) of blurs are estimated first and then the high frequencies (details). The third modification is that the blur and image are estimated in a coarse-to-fine framework, using pyramidal representation of the input blurred image. This has an effect similar to the second modification. However, numerically, the coarser levels have smaller dimensions and so are easier to solve. It is thus possible that a combination of the coarse-to-fine approach and a gradual variance decrease makes the whole system more robust in practice. Last but not least, it is necessary to constrain h to avoid local minima. Recent methods propose to use prior $p(h)$ such that only positive values are allowed and sometimes support constraints are also added. Blur kernels are in general much smaller than the image and it is therefore wise to constrain the blur support and effectively decrease the dimensionality of the problem.

5.3.2 Getting more prior information

There are various ways to get additional information about the degradation process so that the blind deconvolution problem addressed in the previous Section 5.3.1 is better conditioned and thus easier to solve.

One approach is to use multiple degraded images of the same scene. This is called *multichannel blind deconvolution*. The estimation of blurs from multiple observed images is based on a simple but fundamental idea. Let us assume that we have $k > 1$ images g_k that are derived from the original image u according to our space-invariant model

$$g_k = h_k * u + n \,. \tag{5.8}$$

The PSF h_k is different for every k. Let $\hat{h}_k$ denote our estimated PSFs. In the case of no noise, for every pair of images g_i, g_j, $i \neq j$

$$g_i * \hat{h}_j - g_j * \hat{h}_i = 0 \,, \tag{5.9}$$

if $\hat{h}_k = h_k$. This simple relation, which exists only in the multichannel case, gives us means to estimate PSFs directly from blurred images and forms the key idea of all the multichannel blind methods discussed below.

One of the earliest intrinsic multichannel (MC) blind deconvolution methods (Schulz 1993) was designed particularly for images blurred by atmospheric turbulence. Harikumar *et al.* (Harikumar & Bresler 1999) proposed an indirect algorithm, which first estimates the blur functions (published earlier for 1D signals in Gurelli & Nikias (1995)) and then recovers the original image by standard non-blind methods.

Giannakis *et al.* (Giannakis & Heath 2000) developed a similar algorithm based on Bezout's identity of coprime polynomials which finds restoration filters, and by convolving the filters with the observed images recovers the original image. Pai *et al.* (Pai & Bovik 2001) suggested two MC restoration algorithms that estimate the original image from the null space directly, or from the range of a special matrix. Another direct method based on the greatest common divisor was proposed in Pillai & Liang (1999). Interesting approaches based on the ARMA (autoregressive moving average) model are given in Haindl & Šimberová (2002). MC blind deconvolution, based on the Bussgang algorithm, was proposed in Panci, Campisi, Colonnese & Scarano (2003) and performs well on spatially uncorrelated data, such as binary text images and spiky images. Most of the algorithms lack the necessary robustness since they do not include any noise assumptions in their derivation and miss regularization terms. Šroubek *et al.* proposed an iterative MC algorithm (Šroubek & Flusser 2003) that performs well even on noisy images. It is based on least-squares deconvolution by anisotropic regularization of the image and between-channel regularization of the blurs. Another drawback of the MC methods is that the observed images must be spatially aligned, which is seldom true. A first attempt in this direction was done by Šroubek *et al.* in Šroubek & Flusser (2005), where they proposed MC blind deconvolution of images which are mutually shifted by unknown vectors. The same team extended this idea to super-resolution in Šroubek, Cristóbal & Flusser (2007). In super-resolution, the physical resolution of the image is increased, which is equivalent to considering both convolution and downsampling in the degradation operator H.

Another possibility is to acquire a pair of images: one correctly exposed but blurred and one underexposed (noisy) but sharp. Then we can apply the above MC blind deconvolution methods, which are even better posed, as demonstrated in Tico, Trimeche & Vehvilainen (2006), Yuan, Sun, Quan & Shum (2007), and Šorel & Šroubek (2009).

Blind deconvolution in the MC framework is in general a well-posed inverse problem due to the existence of the relations in Eq. (5.9). However, in many practical situations we do not have multiple observations of the same scene, which would differ only by the convolution kernel, and we must revert to the single-channel methods in Section 5.3.1 or try *parametric methods*.

Parametric methods assume a certain set of plausible blurs that are fully described by a few parameters. Inference is done on the parameters and not on the blur function itself, which would be an intractable problem since the dimensionality of the blur functions is too high. A classical example is the blur caused by camera motion, which is limited by six degrees of freedom of a rigid body motion, most commonly decomposed to three rotations and three translations. The main obstacle when dealing with this type of blur is that for translations, the blur depends on scene distance (depth). As a consequence, under general camera motion, we need to estimate the depth map, which makes the algorithm complicated and time consuming. Nevertheless, there are algorithms that work satisfactorily, assuming certain additional constraints on the camera motion. We refer interested readers to Šorel & Flusser (2008), Šorel, Šroubek & Flusser (2010), Favaro, Burger & Soatto (2004), and references therein.

Here we describe a more practical approach that exploits the fact that certain types of camera motion can be neglected in practice. The most common assumption is that all camera translations can be neglected, making the blur independent of scene depth. Validity of this assumption is discussed in Section 5.4. Many devices, such as modern smartphones, are now equipped with inertial sensors (gyroscopes and accelerometers) that can give us very accurate information about camera motion. If we are able to reconstruct the camera path then we can recover blur and perform non-blind image deblurring. This idea was recently described in Joshi, Kang, Zitnick & Szeliski (2010) using an expensive measuring apparatus consisting of a DSLR camera and a set of inertial sensors, where image deblurring is performed offline on a computer. Another possibility is to attach an auxiliary high-speed camera of lower resolution to estimate the

PSF using optical flow techniques as discussed in Ben-Ezra & Nayar (2004) and Tai, Du, Brown & Lin (2010). Later, in Section 5.5, we demonstrate that image deblurring is feasible on modern smartphones without requiring any other devices.

A versatile approach applied recently in Hirsch, Schuler, Harmeling & Scholkopf (2011) is to express the operator of blurring in a specially chosen set of PSF bases B_i

$$Hu = \sum_i d_i B_i u \tag{5.10}$$

which allows us to work with spatially-varying blur in a manner similar to common convolution. In this case, B_i are operators (in the discrete case, matrices) that perform image warping such that images $B_i u$ correspond to all possible transforms (rotations in our case) within a specified range of motions. Note, however, that unlike common convolution such operators do not commute. The set of weights d_i, which are our unknown parameters, are referred to as kernels or motion density functions.

Whyte *et al.* (Whyte, Sivic, Zisserman & Ponce 2010) consider rotations about three axes up to several degrees and describe blurring by the corresponding three-dimensional kernel. For blind deconvolution, it uses a straightforward analogy of the well-known blind deconvolution algorithm (Fergus *et al.* 2006) based on marginalization over the latent sharp image. The only difference is that it uses Eq. (5.10) instead of convolution. For deblurring, following the kernel estimation phase, it uses the corresponding modification of the Richardson–Lucy algorithm.

Gupta *et al.* (Gupta, Joshi, Zitnick, Cohen & Curless 2010) adopted a similar approach but, instead of rotations about x and y axes, consider translations in these directions. Because of the dependence of translation on depth, they require that the scene is approximately planar and perpendicular to the optical axis. Interestingly, in this case it is not necessary to know the real distance because the corresponding kernel works in pixel units. They first estimate locally valid convolution kernels by the original blind deconvolution algorithm (Fergus *et al.* 2006) and estimate the corresponding sharp image patches. In the second step, they estimate the kernels d_i from Eq. (5.10) using the knowledge of both the observed image and an estimate of the sharp image made up of the uniformly distributed patches from the previous step. They do not use all the patches but choose iteratively a subset of patches and check consistency with the rest by a RANSAC-like algorithm. The image is regularized by standard smoothness priors applied separately on derivatives in the x and y directions. The kernel is regularized to be sparse by a $\|.\|_p$ norm applied on kernel values and to be continuous using a quadratic penalty on the kernel gradient.

An obvious advantage of the kernel model Eq. (5.10) is that it is very robust with respect to local non-existence of texture as well as local inaccuracies of the model used, such as sensor saturation for example. On the other hand, it may be considerably more time consuming than algorithms based on local approximation by convolution described in the next section. Another disadvantage is that the actual motion may be more complicated and it is difficult to combine Eq. (5.10) with other models.

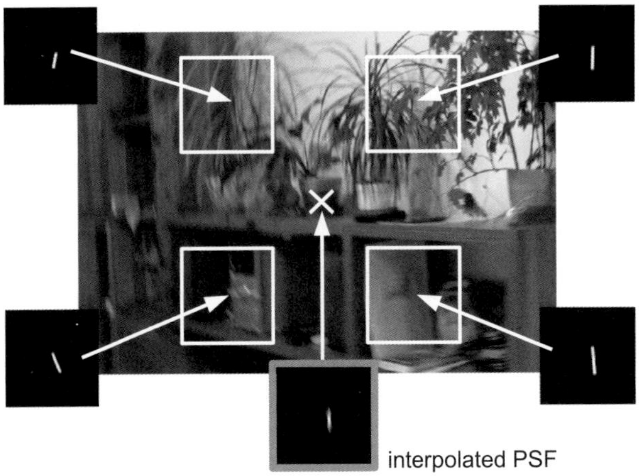

Figure 5.2 Kernel interpolation. If image blur varies slowly, we can estimate convolution kernels on a grid of positions and approximate kernels in the rest of the image by interpolation of four adjacent kernels.

5.3.3 Patch based

There are several types of blur that can be assumed to change slowly with position, which allows the approximation of blurring locally by convolution. Under this assumption we can estimate locally valid convolution kernels that give us a local estimate of the PSF. This usually holds for motion blurs caused by camera shake and optical aberrations. For out-of-focus blur it holds only for approximately planar scenes.

The kernels are usually estimated at a set of regularly spaced positions. For estimation we can use one of many available blind deconvolution methods, working either with a single image (Fergus *et al.* 2006, Xu & Jia 2010) or, using more precise methods, working with multiple images (Šroubek & Flusser 2005). If it is possible to change camera settings, we can also fuse information from one blurred and one noisy/underexposed image (Tico *et al.* 2006, Yuan *et al.* 2007, Šorel & Šroubek 2009).

Once the local convolution kernels are computed, they can be used to estimate the PSF for an arbitrary position (x, y). The simplest possibility is to divide the image into a set of regularly spaced patches, each with an assigned PSF. The patch size is chosen such that the patches overlap and the reconstructed image is generated by blending the reconstructed patches. Blending must be tuned to minimize blocking artifacts that tend to appear around patch boundaries.

A more elaborate but only slightly more computationally demanding method is to assign the estimated kernels just to patch centers (instead of the whole patch) and approximate the PSF h in intermediate positions by bilinear interpolation as indicated in Figure 5.2. An advantage of this solution is that the PSF changes smoothly, thus avoiding the blocking artifacts. Moreover, the corresponding operator H can be computed in time comparable with the time of standard convolution using the fast Fourier transform (FFT) (Stockham 1966, Nagy & O'Leary 1998). The main reason this works is due to simple formulae for blur operators being expressed as a linear combination of a finite set of convolution kernels. Indeed, if the PSF h is defined as a combination of kernels $\sum_i w_i h_i$, then

$$Hu = \sum_i (w_i u) * h_i \,. \tag{5.11}$$

For linear interpolation, the weight functions $w_i(x, y)$ satisfy the constraint $\sum_i w_i(x, y) = 1$ for an arbitrary position (x, y), and $w_i(x, y) = 1$ in the center (x, y) of the window where the kernel h_i was estimated. In Šorel & Šroubek (2009) the authors use this model to deblur an image, provided that another underexposed image taken with sufficiently short shutter time is available. The same model in a more versatile setup working with only one input image and using a recent blind deconvolution method (Cho & Lee 2009) is shown in Harmeling, Michael & Scholkopf (2010).

If the blur changes faster across the image, linear interpolation is not realistic and produces false PSFs, which consequently leads to artifacts in the reconstructed image. More advanced techniques, such as shape-preserving PSF morphing proposed in Šroubek, Šorel, Horackova & Flusser (2013), should then be used instead.

5.4 Pinhole camera model

Let us consider the image a camera captures during its exposure window. Light from a scene point $\mathbf{x}_w = [x_w, y_w, z_w]^T$ projects on the image plane at a location $\mathbf{x} = [x, y]^T$; see Figure 5.3. Using homogeneous coordinates in the image plane $\bar{\mathbf{x}} = [d\mathbf{x}^T, d]^T$, the relation to $\mathbf{x}_w$ is given by

$$\bar{\mathbf{x}} = \mathbf{K}[\mathbf{R}\mathbf{x}_w + \mathbf{t}] \,, \tag{5.12}$$

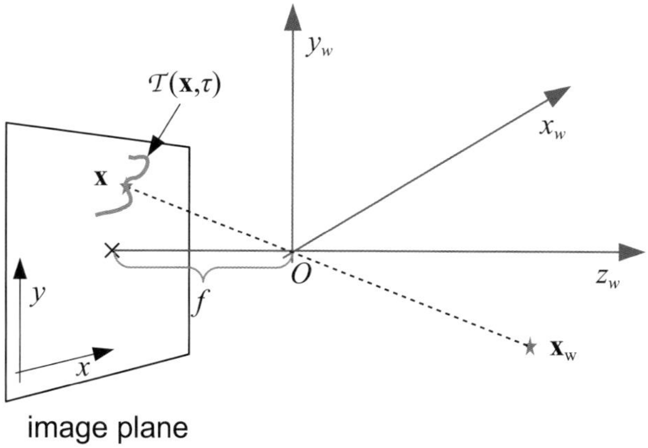

Figure 5.3 Pinhole camera model. A 3D point $\mathbf{x}_w$ in the world coordinate system x_w, y_w, z_w projects into the image plane x, y at the position $\mathbf{x}$. The projection is defined by a camera intrinsic matrix, which is primarily a function of the camera focal length f. Due to camera motion during exposure, the projected point draws a trace (2D curve) $\mathcal{T}(\mathbf{x}, \tau)$, which is a function of time τ.

where $\mathbf{R}$ (3×3) and $\mathbf{t}$ (3×1) are the camera rotation matrix and translation vector, respectively. Upper triangular matrix $\mathbf{K}$ (3×3) is the camera intrinsic matrix. The third element d of the homogeneous coordinates corresponds to distance. The camera intrinsic matrix is typically of the form

$$\mathbf{K} = \begin{bmatrix} fs_x & 0 & c_x \\ 0 & fs_y & c_y \\ 0 & 0 & 1 \end{bmatrix}, \tag{5.13}$$

where f is the camera focal length, $[c_x, c_y]$ is the image center, and $[s_x, s_y]$ is the pixel scale.

During the exposure window the camera position and orientation may change and therefore the extrinsic parameters $\mathbf{R}$ and $\mathbf{t}$ are functions of time τ. The projected point $\mathbf{x}$ moves along a curve parameterized by τ, which we denote as $\mathcal{T}(\mathbf{x}, \tau)$ and call a point trace:

$$\mathcal{T}(\mathbf{x}, \tau): \quad \bar{\mathbf{x}}(\tau) = \mathbf{K}\left[\mathbf{R}(\tau)\mathbf{K}^{-1}\bar{\mathbf{x}}_0 + \frac{1}{d_0}\mathbf{t}(\tau)\right], \tag{5.14}$$

where $\bar{\mathbf{x}}_0 = [\mathbf{x}^T, 1]^T = [x, y, 1]$ is the initial location of the point in the image plane using normalized homogeneous coordinates, and d_0 is the initial distance of the corresponding 3D point $\mathbf{x}_w$ from the camera (see Figure 5.3 for an illustration of the point trace). Assuming a constant illuminance over the exposure period, the light energy emitted from the point is distributed evenly (with respect to time) over the trace $\mathcal{T}$. This effectively gives us a time parameterization of a point spread function for a given point $\mathbf{x}$, which forms the blur operator H. The space-variant PSF $h(s, t, \mathbf{X})$ corresponds precisely to the rendered trace $\mathcal{T}(\mathbf{x}, \tau)$. The point trace as defined in Eq. (5.14) is in the homogeneous coordinates $\mathcal{T}(\mathbf{x}, \tau) = [d(\tau)\mathbf{x}(\tau), d(\tau)]^T$. The space-variant PSF $h(s, t, \mathbf{x})$ can then be expressed as follows:

$$h(s, t, \mathbf{x}) = \frac{1}{T}\int \delta\left(\begin{bmatrix} s \\ t \end{bmatrix} - \mathbf{x}(\tau)\right) \mathrm{d}\tau, \tag{5.15}$$

where T is the exposure time. The point trace in Eq. (5.14) depends on camera intrinsic parameters $\mathbf{K}$ and distance d_0. Unlike the camera parameters (focal length, image center and pixel scale), which are usually known or can easily be estimated for any camera model, the distances of 3D points (the depth map) are not readily available. We cannot assume that the depth map is known in practice. However, only camera translation $\mathbf{t}(\tau)$ is influenced by the distance, and rotation $\mathbf{R}(\tau)$ is independent. The translation term in Eq. (5.14) is $\frac{1}{d_0}\mathbf{K}\mathbf{t}(\tau)$ and after substitution for K we see that the translation vector is multiplied by a factor $\frac{f}{d_0}$ in the x and y directions, and by a factor $\frac{1}{d_0}$ in the z direction. Typical phone cameras have a focal length of several millimeters and objects that we shoot are usually at least a few meters away, thus both factors are small numbers and the influence of camera translation is negligible. Figure 5.4 illustrates influence of a 1 mm camera translation in the x or y direction as a function of the distance of the object from the camera. For typical phone cameras, a scene projected into the camera image plane moves by less than a pixel if the objects are more than 2 m away.

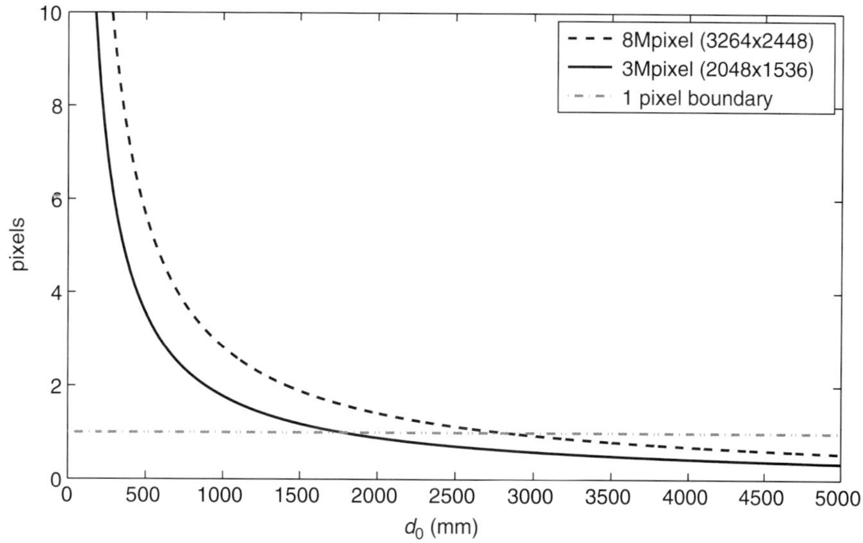

Figure 5.4 Influence of a 1 mm translation in x and y axes as a function of object distance from a camera with a 28 mm focal length (35 mm equivalent). Even for relatively large sensors of 8 MP, if the objects are 3 m away from the camera, a 1 mm translation in the plane perpendicular to the optical axis produces a shift of less than 1 pixel in the image plane.

The above discussion allows us to focus on purely rotational motion of the camera and this provides us with one additional benefit. Smartphones are equipped with both accelerometers and gyroscopes. Accelerometers measure translation acceleration, which is the second derivative of $\mathbf{t}(\tau)$, and gyroscopes measure angular velocity, which is proportional to the first derivative of $\mathbf{R}(\tau)$. To estimate the translation vector $\mathbf{t}$ one has to perform a double integration of accelerometer data along time. On the other hand, estimation of the rotation matrix $\mathbf{R}$ is done by single integration of gyroscope data over time. In double integration, accelerometer noise quickly amplifies and precision deteriorates shortly afterwards. This is not the case for the gyroscope data, which can give a relatively precise estimation of rotational angles throughout the camera exposure time. Since we assume only rotation in our camera motion model, the approximated point trace $\mathcal{T}'$ is given by

$$\mathcal{T}'(\mathbf{x}, \tau): \quad \bar{\mathbf{x}}(\tau) = \mathbf{K}\mathbf{R}(\tau)\mathbf{K}^{-1}\bar{\mathbf{x}}_0 . \tag{5.16}$$

The rotation matrix $\mathbf{R}$ is a function of three angles $\phi_x(\tau)$, $\phi_y(\tau)$, and $\phi_z(\tau)$, which are estimated from gyroscope data – the three angular speeds $\omega_x(\tau)$, $\omega_y(\tau)$, and $\omega_z(\tau)$ – by

$$\phi_x(\tau) = \int_0^\tau \omega_x(t)\,\mathrm{d}t, \quad \phi_y(\tau) = \int_0^\tau \omega_y(t)\,\mathrm{d}t, \quad \phi_z(\tau) = \int_0^\tau \omega_z(t)\,\mathrm{d}t. \tag{5.17}$$

Knowing the camera intrinsic parameters ($\mathbf{K}$) and having the data stream from gyroscopes ($\omega_x(\tau)$, $\omega_y(\tau)$, and $\omega_z(\tau)$), we can generate PSFs $h(s, t, \mathbf{x})$ at any position $\mathbf{x}$ in the image plane using Eqs. (5.15)–(5.17).

5.5 Smartphone application

This section illustrates that blur estimation with built-in inertial sensors is possible and we developed an implementation of image deblurring on a smartphone, which works in practical situations and is relatively fast, so as to be acceptable for a general user.

As a testing platform, we have chosen a *Samsung Galaxy S II* smartphone with an Android OS. It is equipped with all the apparatus needed for our experiments; namely, a relatively high-quality camera, motion sensors, a fast CPU, and enough RAM to perform computations. A block diagram of the deblurring application is shown in Figure 5.5.

We first discuss the space-invariant implementation, which assumes a single PSF in the entire image. Then we show a simple patch-based extension for the space-variant implementation, which applies the space-invariant method in a patch-wise manner, which is still manageable on the tested smartphone platform.

5.5.1 Space-invariant implementation

During the photo acquisition, samples of angular velocity are recorded using the embedded gyroscopes, which are later trimmed to fit the exposure period. An estimation of the PSF is rendered by integrating the curve position from the recorded gyroscope data using Eqs. (5.15) and (5.16) for X in the center of the image.

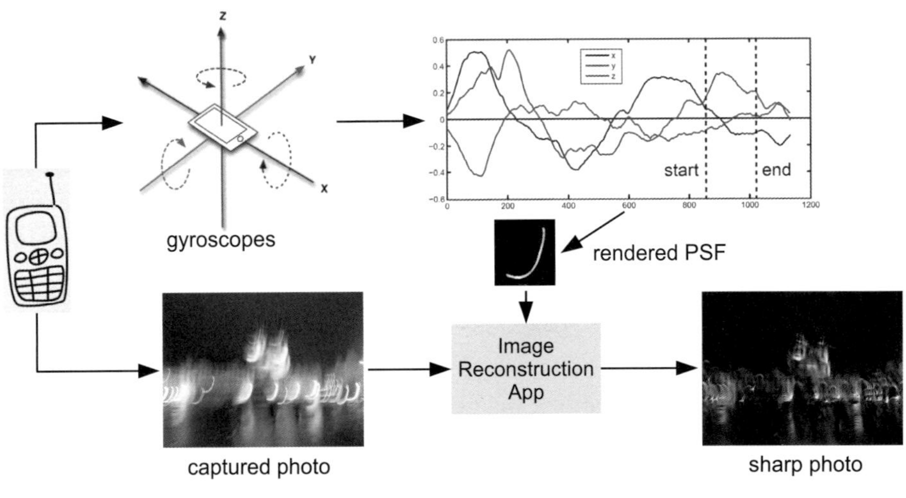

Figure 5.5 The block diagram of the smartphone application. During camera exposure, the application records data from the built-in gyroscopes. The data are processed and PSFs are estimated. The captured photo is divided into overlapping patches; Wiener deconvolution is performed on every patch and the reconstructed patches are blended to generate the sharp photo. The whole process for a 3 MP photo, entirely done on the smartphone, takes around six seconds.

Table 5.1 Computational time (in milliseconds) of the FFT transform of grayscale images with different sizes and different CPU settings.

Resolution	No NEON, no hardware FPU	NEON, 1 core	NEON, 2 cores
1536 × 1152	2900	185	110
2048 × 1536	5300	330	195
2050 × 1538	–	1000	540
3264 × 2448	21200	1450	800

State-of-the-art non-blind deconvolution methods use sparse image priors and the solution is usually found by some iterative minimization algorithms, such as in Shan *et al.* (2008). However the limited computational power of the smartphone prevents us from implementing these sophisticated deconvolution methods. We thus use a simple and fast *Wiener filter* in the form

$$\hat{U} = G\frac{H^*}{|H|^2 + \Phi}, \tag{5.18}$$

where Φ is an estimation of the inverse signal-to-noise ratio (SNR), and G, H, and $\hat{U}$ are discrete Fourier transforms of the observed image g, PSF h and the estimated latent image $\hat{U}$, respectively.

Filtering in the frequency domain treats the image as a periodic function, which causes ringing artifacts around image borders. To overcome this problem, several techniques were proposed in the literature (Liu & Jia 2008, Šorel 2012). We have found it sufficient to preprocess the input image g by blending the opposite image borders at the extremities of the PSF, which creates a smooth transition and eliminates the artifacts.

The intensity values of the output image $\hat{U}$ sometimes lie outside the 8-bit range (0–255), therefore we added optional normalization with clipping of outliers. The normalization is especially useful in the case of large blurs and scenes with high illumination.

To convert images between the frequency and spatial domains, we use the FFT algorithm implemented in the FFTW library. Utilizing a fast ARM Cortex-A9 CPU with two cores and support for a SIMD instruction set (NEON), FFTW proved to be remarkably fast on the tested smartphone; see Table 5.1.

Images are acquired with the native camera with a resolution of 3264 × 2448, which we then rescale to 2048 × 1536 to take advantage of a better performance of the FFTW when the image size is a factor of small primes. Image downsampling has a negligible effect on the image quality, because native camera resolution is unnecessarily high. The optical system of the camera has a very small aperture, which, because of diffraction and optical aberrations, limits the number of pixels that can be effectively captured by the image sensor.

To perform Wiener filtering, the FFT must be applied several times: once for the PSF and twice (forward and backward) for each color channel. That yields a total of 7 FFT operations. With some overhead of bitmap transfers, the deconvolution phase for the image resolution 2048 × 1536 takes about 2.6 s. The whole process, starting from the

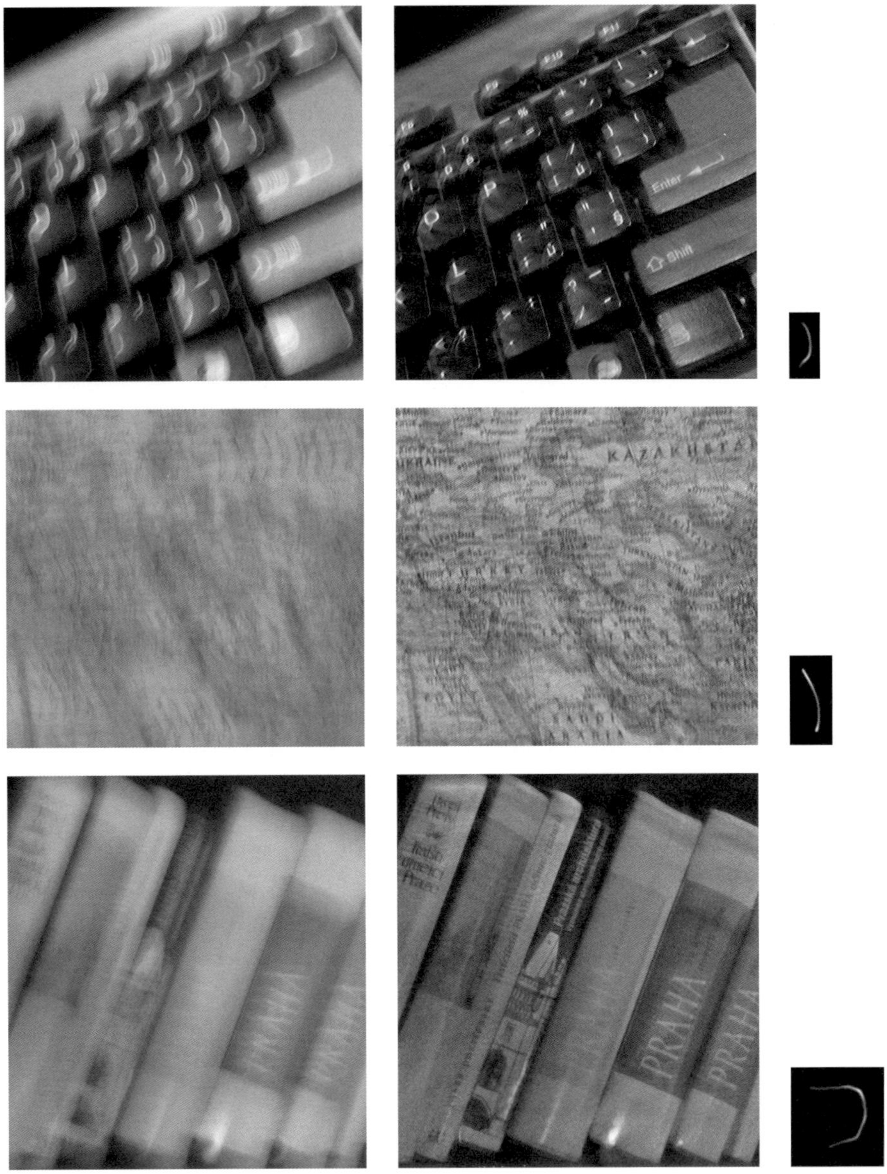

(a) Captured image (b) Reconstructed image (c) PSF

Figure 5.6 Examples of space-invariant deconvolution on the smartphone.

moment the camera shutter closes, takes a little over 6 s. This includes image resizing, PSF estimation, compressing and saving the original and deblurred image files.

In Figure 5.6 we display several of our results together with the PSFs calculated from the gyroscope data. The results were all computed with the signal-to-noise parameter Φ set to 0.01. This value was determined experimentally to provide the best looking

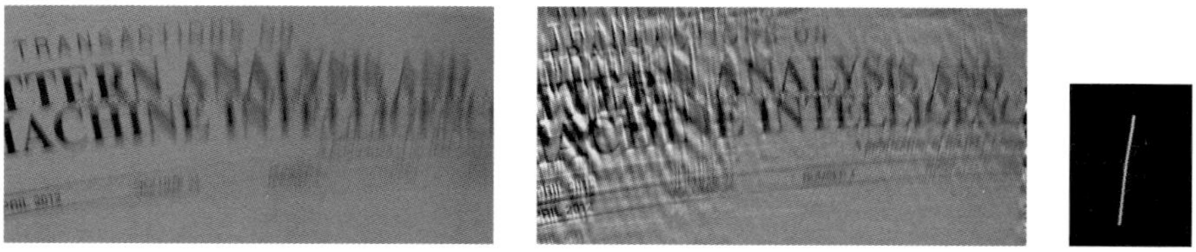

(a) Captured image (b) Reconstructed image (c) PSF

Figure 5.7 An example of reconstruction artifacts in the space-invariant implementation due to the space-variant nature of degradation.

results. The original intention was to set Φ in proportion to the ISO value extracted from EXIF data of the photo, which should determine the amount of noise present in the image. However, we found the dependency of Φ on ISO negligible. This behavior can be explained by the denoising step performed internally by the camera module.

5.5.2 Space-variant implementation

Performance of the space-invariant implementation (as seen in Figure 5.6) is by no means consistent. Deconvolved images are impaired by visual anomalies worsening their appearance. Most often these anomalies manifest as ringing artifacts surrounding sharp edges in the picture, as demonstrated in Figure 5.7.

The main cause is the space-variant nature of the blur that has not been considered so far. The space-variant PSFs are particularly noticeable when rotation around the z axis is significant. Smartphones are equipped with wide-angle lenses with a field of view of approximately 60°. This means that camera projection causes rotation around the x and y axes to produce strongly space-variant blur in areas close to image borders. Using relationships in Eqs. (5.15) and (5.16) allows us to render a correct PSF at every pixel of the image. However to perform image deblurring with space-variant PSFs would be computationally expensive since we cannot use the Wiener filter. Instead we break the image into overlapping patches and generate one PSF in every patch using Eq. (5.16) at the center of the patch. An example of dividing the image into 6×8 patches with 25% overlap and generating the corresponding PSFs is illustrated in Figure 5.8. In every patch we apply the Wiener filter (including edge tapering and normalization) as described in the space-invariant implementation in Section 5.5.1. Due to patch overlaps, we blend the reconstructed patches by weighting them with appropriate Hamming windows, which produces the final image without noticeable seams between patches; see Figure 5.12(d).

Another reason for space-variant blur, which is not connected with camera motion, but inherent to the camera hardware design, is the shutter mechanism. Contrary to systems with a mechanical shutter, CMOS sensors in cheap cameras are continuously illuminated. The readout from the CMOS sensor takes several tens of milliseconds, which results in a picture taken not at a single moment, but with a slight time delay between the first and last pixel row. This process, called *rolling shutter*, is therefore

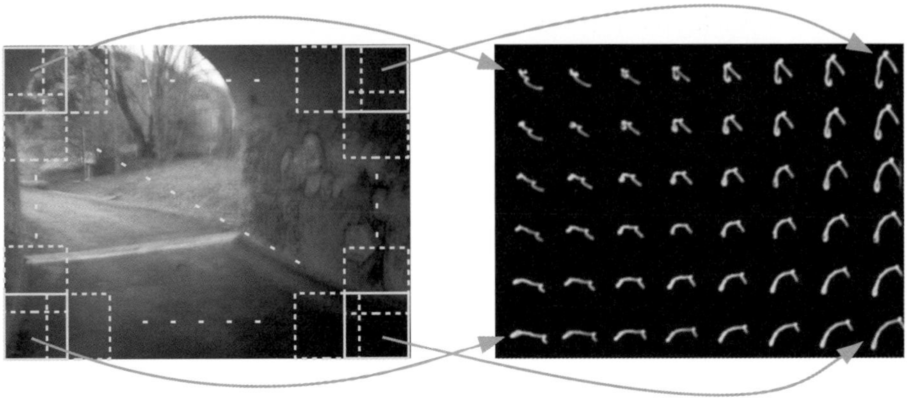

Figure 5.8 Patch-based space-variant reconstruction. The input image is divided into 6×8 patches with 25% overlap. In the center of every patch, we estimate the corresponding PSF using data from smartphone gyroscopes.

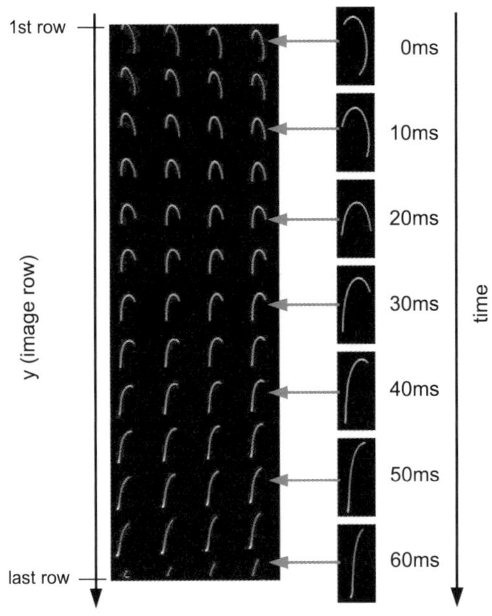

Figure 5.9 A snapshot (exposure time 1/14 s) of a point grid displayed on an LCD screen showing the rolling shutter effect. The middle column shows a series of blur kernels rendered using data from the gyroscope sensor shifted in time. Blurs were created from sensor data starting 0–60 ms after a synchronization timestamp.

another cause of the blur variance as the PSF depends on the vertical position in the image. The correct approach to PSF estimation is thus shifting the inertial sensor data in time according to the vertical position in the image. An example in Figure 5.9 illustrates the rolling shutter effect. We took a snapshot of an LCD screen displaying a grid of white points on a black background. Due to camera motion, the points appear as streaks on the captured image. To accurately model the PSFs at every position, it is necessary to shift the exposure-time window in which the gyroscope data are fetched. Let us assume that we know the precise moment when the camera exposure starts. At the top row of the image there is no delay between the exposure start and the time when we start reading gyroscope data. As we move down towards the bottom rows, the delay increases as these rows are exposed later. As illustrated in Figure 5.9 for the particular testing device, the delay between reading the first and the last row of the image is around 60 ms.

5.6 Evaluation

Removal of camera shake based on gyroscope data, as proposed in the previous section, should give better results than the fully blind deconvolution methods discussed in Section 5.3.1. To illustrate the algorithm's performance, we compare the outcome of the proposed smartphone application in Figure 5.10(b) with the state-of-the-art single-channel blind deconvolution in Figure 5.10(c) proposed by Xu *et al.* in (Xu & Jia 2010). Apart from being far more computationally demanding, Xu's method is unable to estimate the correct blurs in most of the test images we have taken with the smartphone.

The angular speed estimated by gyroscopes is obviously not absolutely precise. To quantify the effect of gyroscope additive noise, we have conducted the following synthetic experiment. A test image was synthetically convolved with a PSF that was rendered using some gyroscope samples recorded previously with our mobile application and an additive noise (SNR = 40 dB) was added to the image. The generated blurred

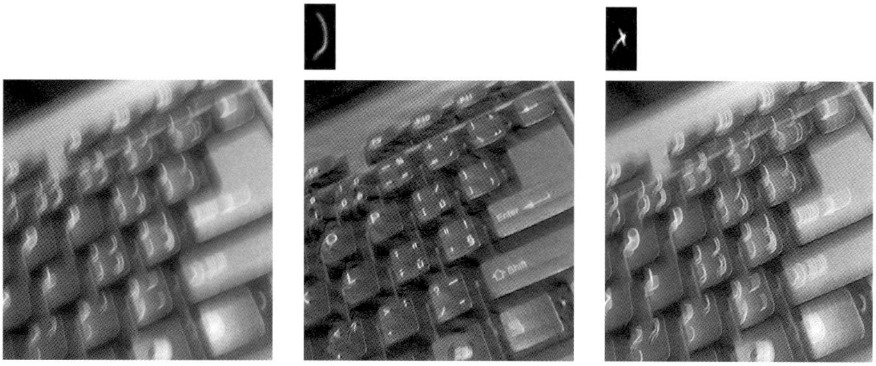

(a) Captured image (b) Reconstructed with gyroscope (c) Reconstructed with blind method

Figure 5.10 Comparison of reconstruction methods. (a) Input blurred image; (b) estimated sharp image and PSF with our smartphone application using gyroscope data; (c) state-of-the-art blind deconvolution method proposed by Xu & Jia (2010).

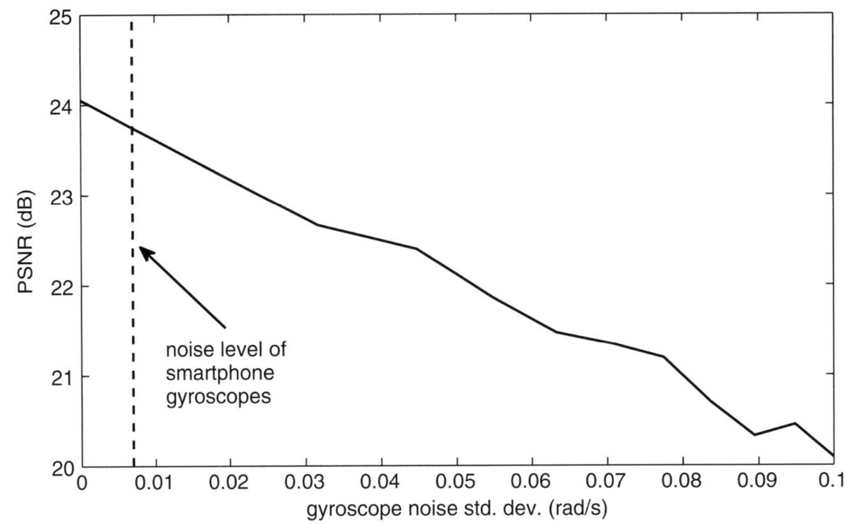

Figure 5.11 PSNR of the original and deblurred image as a function of additive gyroscope noise. Gaussian noise of standard deviation 0–0.1 rad/s was added to gyroscope samples (angular velocity in rad/s). The deconvolution algorithm was then applied using computed blur kernels based on these altered gyroscope measurements. The graph shows mean values of 50 realizations for each of the noise levels.

image imitated a photo that one typically acquires with a smartphone. We then tried to deblur the image using the same gyroscope samples but this time corrupted by additive noise to simulate errors in sensor measurement. The PSNR, where PSNR = $10\log_{10}(|\Omega|255^2/\|u - \hat{U}\|^2)$ (dB), of the results as a function of noise standard deviation (rad/s) in gyroscope samples is summarized in Figure 5.11. The graph clearly shows that the performance of deblurring quickly deteriorates as the noise level increases. Note that a fall of 1 dB in PSNR has a dramatic visual effect on the image quality. The gyroscope noise level typically encountered in motion sensors inside mobile devices (in our case, a Samsung Galaxy S II) is 0.007 rad/s for our sampling rate, which is depicted by the vertical dashed line in the graph. We can see that this noise level is associated with a relatively small drop in PSNR, which is one of the reasons for the satisfactory performance of the proposed smartphone application.

In Figure 5.12 we demonstrate the performance of the space-variant implementation from Section 5.5.2. The smartphone implementation divides the acquired image in Figure 5.12(a) into 6×8 patches of 384×384 pixels with 25% overlap. PSFs generated from gyroscope data in every patch are illustrated in Figure 5.12(c). The reconstruction of the sharp image (Figure 5.12(d)) takes about seven seconds on the testing device. For comparison, we also show the space-invariant implementation in Figure 5.12(b), which assumes only one PSF is generated for the center of the image. In this particular case, the space-variant nature of the PSFs was profound and therefore the image shows noticeable ringing artifacts.

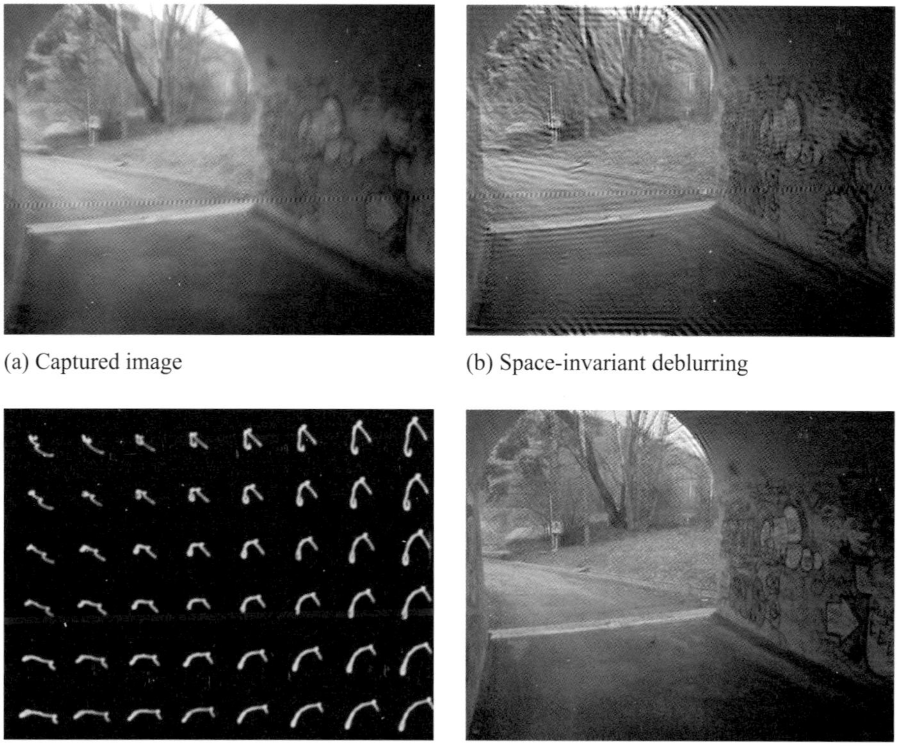

(a) Captured image (b) Space-invariant deblurring

(c) Gyroscope generated blurs (d) Space-variant deblurring

Figure 5.12 An example of space-variant reconstruction. Due to the space-variant nature of the blur, which is clearly visible in (c), the space-invariant deblurring in (b) produces reconstruction artifacts. On the other hand, performing deconvolution patch-wise with correct blurs gives satisfactory results (d).

5.7 Conclusions

The first half of this chapter gives an overview of current trends in blind deconvolution methodology using Bayesian inference. Image deblurring is inherently an ill-posed problem and we need to search for techniques that make it better posed. In general, there are two ways to achieve this. We must either include priors, which penalize inappropriate images and incorrect blur kernels, or use additional information from other sources, such as multiple images of the same scene, parametric models, or information from inertial sensors. Our main consideration is blur caused by camera shake, which is very common in real life and, even though space variant, can be easily parameterized.

The second half of the chapter presents an image deblurring method that can effectively remove blur caused by camera motion using information from inertial sensors – gyroscopes. The proposed method is fully implemented on a smartphone device, which is to our knowledge the first attempt in this direction and renders the method particularly appealing for end users. We have demonstrated that the space-invariant simplification for certain camera motions is plausible, but simultaneously we have uncovered intrinsic sources of space-variant blur. The space-variant implementation of the deblurring algorithm solves the complex camera motion and rolling shutter issues, which can both be modeled by space-variant blur.

There are several topics for future research. Implementing smarter deblurring algorithms that avoid ringing artifacts is viable and clearly a tempting improvement. Gyroscope data are not precise but one can use the calculated PSFs as an initial estimate and apply modified blind deconvolution methods to improve their accuracy.

Acknowledgement

We would like to thank Ondřej Šindelář for implementing the smartphone application and generating the results for this chapter. This work was supported in part by the Grant Agency of the Czech Republic under the project 13-29225S and by the Academy of Sciences of the Czech Republic under the project M100751201.

References

Ben-Ezra, M. & Nayar, S. K. (2004). Motion-based motion deblurring. *IEEE Transactions on Pattern Analysis and Machine Intelligence*, **26**(6), 689–98.

Campisi, P. & Egiazarian, K., eds. (2007). *Blind Image Deconvolution, Theory and Application*. CRC Press.

Cho, S. & Lee, S. (2009). Fast motion deblurring. *ACM Transactions on Graphics (Proceedings of SIGGRAPH Asia 2009)*, **28**(5), 145:1–8.

Favaro, P., Burger, M. & Soatto, S. (2004). Scene and motion reconstruction from defocus and motion-blurred images via anisotropic diffusion. In T. Pajdla & J. Matas, eds., *ECCV 2004, LNCS 3021*. Berlin: Springer-Verlag, pp. 257–69.

Fergus, R., Singh, B., Hertzmann, A., Roweis, S. T. & Freeman, W. T. (2006). Removing camera shake from a single photograph. *ACM Transactions on Graphics (Proceedings of SIGGRAPH 2006)*. New York: ACM, pp. 787–94.

Galatsanos, N. P., Mesarovic, V. Z., Molina, R. & Katsaggelos, A. K. (2000). Hierarchical Bayesian image restoration from partially known blurs. *IEEE Transactions on Image Processing*, **9**(10), 1784–97.

Giannakis, G. & Heath, R. (2000). Blind identification of multichannel FIR blurs and perfect image restoration. *IEEE Transactions on Image Processing*, **9**(11), 1877–96.

Gupta, A., Joshi, N., Zitnick, C. L., Cohen, M. & Curless, B. (2010). Single image deblurring using motion density functions. In *Proceedings of the 11th European Conference on Computer Vision: Part I*. European Conference on Computer Vision. Berlin: Springer-Verlag, pp. 171–84.

Gurelli, M. I. & Nikias, C. L. (1995). EVAM: An eigenvector based algorithm for multichannel blind deconvolution of input colored signals. *IEEE International Conference on Acoustics, Speech and Signal Processing*, **43**, 134–49.

Haindl, M. & Šimberová, S. (2002). Model-based restoration of short-exposure solar images. In L. Jain & R. Howlett, eds., *Frontiers in Artificial Intelligence and Applications*. Vol. 87 of *Knowledge-Based Intelligent Engeneering Systems*. Amsterdam: IOS Press, pp. 697–706.

Harikumar, G. & Bresler, Y. (1999). Perfect blind restoration of images blurred by multiple filters: theory and efficient algorithms. *IEEE Transactions on Image Processing* **8**(2), 202–19.

Harmeling, S., Michael, H. & Scholkopf, B. (2010). Space-variant single-image blind deconvolution for removing camera shake. In J. Lafferty, C. K. I. Williams, J. Shawe-Taylor, R. Zemel & A. Culotta, eds., *Advances in Neural Information Processing Systems 23*, pp. 829–37.

Hirsch, M., Schuler, C. J., Harmeling, S. & Scholkopf, B. (2011). Fast removal of non-uniform camera shake. In *Proceedings of the IEEE International Computer Vision Conference*, pp. 463–70.

Jia, J. (2007). Single image motion deblurring using transparency. In *Proceedings of the IEEE Conference on Computer Vision and Pattern Recognition*, pp. 1–8.

Joshi, N., Kang, S. B., Zitnick, C. L. & Szeliski, R. (2010). Image deblurring using inertial measurement sensors. *ACM Transactions on Graphics*, **29**, 30:1–9.

Levin, A., Weiss, Y., Durand, F. & Freeman, W. (2009). Understanding and evaluating blind deconvolution algorithms. In *Proceedings of the IEEE Conference on Computer Vision and Pattern Recognition*, pp. 1964–71.

Levin, A., Weiss, Y., Durand, F. & Freeman, W. T. (2011). Understanding blind deconvolution algorithms. *IEEE Transactions on Pattern Analysis and Machine Intelligence*, **33**(12), 2354–67.

Liu, R. & Jia, J. (2008). Reducing boundary artifacts in image deconvolution. In *IEEE International Conference on Image Processing*, pp. 505–8.

Miskin, J. & MacKay, D. J. (2000). Ensemble learning for blind image separation and deconvolution. In M. Girolani, ed., *Advances in Independent Component Analysis*, Springer-Verlag, pp. 123–42.

Molina, R., Mateos, J. & Katsaggelos, A. K. (2006). Blind deconvolution using a variational approach to parameter, image, and blur estimation. *IEEE Transactions on Image Processing* **15**(12), 3715–27.

Nagy, J. G. & O'Leary, D. P. (1998). Restoring images degraded by spatially variant blur. *SIAM Journal on Scientific Computing*, **19**(4), 1063–82.

Pai, H.-T. & Bovik, A. (2001). On eigenstructure-based direct multichannel blind image restoration. *IEEE Transactions on Image Processing*, **10**(10), 1434–46.

Panci, G., Campisi, P., Colonnese, S. & Scarano, G. (2003). Multichannel blind image deconvolution using the bussgang algorithm: spatial and multiresolution approaches. *IEEE Transactions on Image Processing*, **12**(11), 1324–37.

Pillai, S. & Liang, B. (1999). Blind image deconvolution using a robust GCD approach. *IEEE Transactions on Image Processing*, **8**(2), 295–301.

Rosenfeld, A. & Kak, A. C. (1982). *Digital Picture Processing*, 2nd edn. Orlando, FL: Academic Press, Inc.

Rudin, L., Osher, S. & Fatemi, E. (1992). Nonlinear total variation based noise removal algorithms. *Physica D*, **60**, 259–68.

Schulz, T. (1993). Multiframe blind deconvolution of astronomical images. *Journal of the Optical Society of America A*, **10**(5), 1064–73.

Shan, Q., Jia, J. & Agarwala, A. (2008). High-quality motion deblurring from a single image. In *ACM Special Interest Group on Graphics and Interactive Techniques*. New York: ACM, pp. 1–10.

Šorel, M. & Šroubek, F. (2009). Space-variant deblurring using one blurred and one underexposed image. In *Proceedings of the IEEE International Conference on Image Processing*, pp. 157–60.

Šorel, M., Šroubek, F. & Flusser, J. (2010). Towards super-resolution in the presence of spatially varying blur. In M. Peyman, ed., *Super-resolution Imaging*. CRC Press, pp. 187–217.

Šroubek, F. & Flusser, J. (2005). Multichannel blind deconvolution of spatially misaligned images. *IEEE Transactions on Image Processing*, **14**(7), 874–83.

Stockham, Jr., T. G. (1966). High-speed convolution and correlation. In *Proceedings of the April 26–28, 1966, Spring Joint Computer Conference*. New York: ACM, pp. 229–33.

Tai, Y.-W., Du, H., Brown, M. S. & Lin, S. (2010). Correction of spatially varying image and video motion blur using a hybrid camera. *IEEE Transactions on Pattern Analysis and Machine Intelligence*, **32**, 1012–28.

Tico, M., Trimeche, M. & Vehvilainen, M. (2006). Motion blur identification based on differently exposed images. In *Proceedings of the IEEE International Conference on Image Processing*, pp. 2021–4.

Šorel, M. (2012). Removing boundary artifacts for real-time iterated shrinkage deconvolution. *IEEE Transactions on Image Processing*, **21**(4), 2329–34.

Šorel, M. & Flusser, J. (2008). Space-variant restoration of images degraded by camera motion blur. *IEEE Transactions on Image Processing*, **17**(2), 105–16.

Šroubek, F., Cristóbal, G. & Flusser, J. (2007). A unified approach to superresolution and multichannel blind deconvolution. *IEEE Transactions on Image Processing*, **16**(9), 2322–32.

Šroubek, F. & Flusser, J. (2003). Multichannel blind iterative image restoration. *IEEE Transactions on Image Processing*, **12**(9), 1094–106.

Šroubek, F., Šorel, M., Horackova, I. & Flusser, J. (2013). Patch-based blind deconvolution with parametric interpolation of convolution kernels. In *Proceedings of the IEEE International Conference on Image Processing*, pp. 1–4.

Whyte, O., Sivic, J., Zisserman, A. & Ponce, J. (2010). Non-uniform deblurring for shaken images. In *IEEE Conference on Computer Vision and Pattern Recognition*, pp. 491–8.

Xu, L. & Jia, J. (2010). Two-phase kernel estimation for robust motion deblurring. In *Proceedings of the 11th European Conference on Computer Vision: Part I*. Berlin: Springer-Verlag, pp. 157–70.

Yuan, L., Sun, J., Quan, L. & Shum, H.-Y. (2007). Image deblurring with blurred/noisy image pairs. In *ACM Special Interest Group on Graphics and Interactive Techniques*, 1:1–10.

6 Multi-sensor fusion for motion deblurring

Jingyi Yu

This chapter presents multi-sensor fusion techniques for motion deblurring. With recent advances in digital imaging, the use of high resolution, high-speed, or high dynamic range cameras has become common practice. However, thus far no single image sensor can satisfy the diverse requirements of all the current industrial camera applications. For example, high-speed (HS) cameras can capture fast motion with little motion blur but require expensive sensors, bandwidth, and storage. The image resolution in HS cameras is often much lower than many commercial still cameras. This is mainly because the image resolution needs to scale linearly with the exposure time (Ben-Ezra & Nayar 2003) to maintain the signal-to-noise ratio (SNR), i.e. a higher speed maps to a lower resolution. In addition, the relatively low bandwidth on the usual interfaces such as USB 2.0 or FireWire IEEE 1394a restricts the image resolution especially when streaming videos at 100–200 fps.

The problem of acquiring high quality imagery with little motion under low light is particularly challenging. To guarantee enough exposures, one can choose to use either a wide aperture or a slow shutter. For example, by coupling a wide aperture with fast shutters, we can capture fast motions of scene objects with low noise. However, wide apertures lead to a shallow depth-of-field (DoF) where only parts of the scene can be clearly focused. In contrast, by coupling a slow shutter with a narrow aperture, one can capture all depth layers in focus. The drawback here is that slow shutters are vulnerable to fast moving objects and will lead to severe motion blurs in the acquired images. Recent advances in computational cameras suggest that it may be possible to fuse heterogeneous types of cameras. In this chapter, we present two multi-sensor fusion techniques for handling fast motions at regular or low light conditions.

6.1 Introduction

Our multi-sensor fusion framework falls in the category of imaging hardware-based solutions. It can be viewed as an extension to the hybrid-imaging system of Ben-Ezra & Nayar (2003). A hybrid-imaging system usually uses a rig of two cameras. The sensor in Ben-Ezra & Nayar (2003) combines a low resolution high-speed video camera and a high resolution low-speed SLR camera. To deblur the SLR image, it tracks the motion of the video camera and then uses it to compute the PSF in the SLR. Their solution can allow parallax between the SLR and the video cameras if the motion is shift invariant, although it does not actively use the parallax for other purposes (e.g. scene geometry recovery). Certain object motions can also be handled by their system (Ben-Ezra & Nayar June 2004). In a similar vein, the design in Tai, Du, Brown & Lin (2010) aligns the video and the SLR cameras to the same viewpoint using a beamsplitter and then applies an iterative optimization algorithm to deblur multiple moving objects.

Our multi-sensor fusion solution (Li, Yu & Chai 2008) differs from these hybrid-imaging systems in several ways. First, we use more than two sensors and assume relatively large baselines between the sensors. As a result, images across the sensors will exhibit strong parallax. Second, we actively use the parallax (e.g. via stereo vision) to simultaneously estimate the moving objects' depths and deblur the images. Finally, we show that it is possible to integrate more than two types of sensors for other computer vision tasks such as enhancing low-light imaging.

6.2 Hybrid-speed sensor

We construct a hybrid-speed sensor that comprises a pair of high-speed color (HS-C) cameras and a single high resolution color (HR-C) camera (Li *et al.* 2008). In our real implementation, we use a pair of Point Grey Dragonfly Express cameras as the HS-C cameras. The Dragonfly Express captures images of resolution 320×240 at 120 fps. We also position a Point Grey Flea2 camera on top of the two Dragonfly cameras. The Flea2 serves as the HR-C camera and captures images of resolution 1024×768 at 7.5 fps. We use the software solution provided by Point Grey to synchronize the HS-C cameras and the HR-C camera so that every HR-C frame synchronizes with 16 HS-C frames. These three cameras are connected to two SIIG FireWire 800 3-Port PCI-E cards. We attach all three camera modules to a wood plate and mount it on two tripods, as shown in Figure 6.1. Each camera uses a micro-lens with a 4.8 mm focal length. The two HS-Cs are positioned about 7.5 cm away from each other and the HR-C camera is placed above the two HS-C cameras.

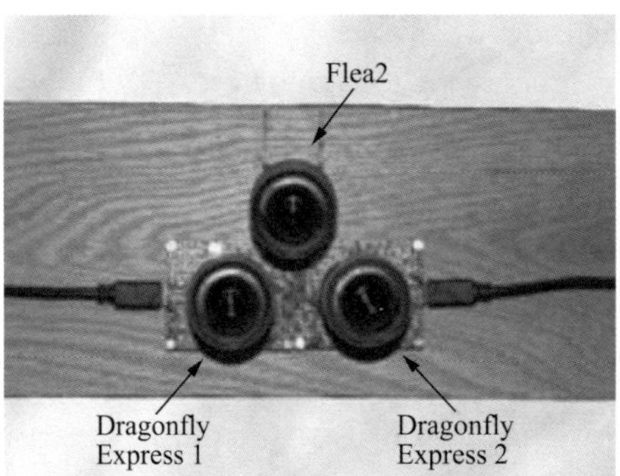

Figure 6.1 Our hybrid-speed sensor for depth estimation and motion deblurring. This sensor combines a Point Grey Flea2 camera that serves as the HR-C camera and two Point Grey Dragonfly Express cameras that serve as the HS-C cameras. [Adopted from Li *et al.* (2008).]

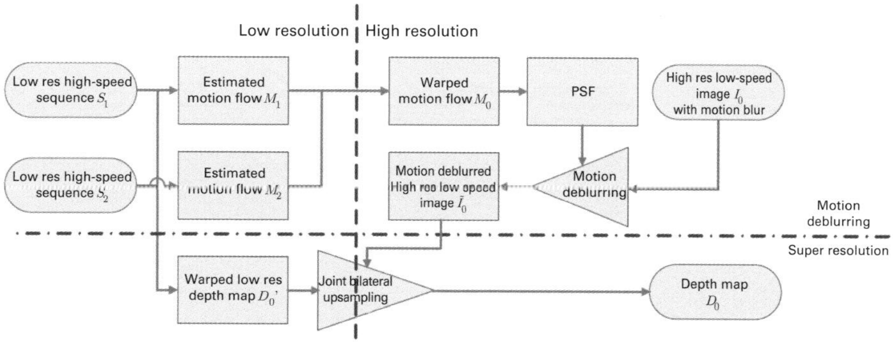

Figure 6.2 The processing pipeline using our hybrid-speed sensor for motion deblurring and depth map super-resolution. We first estimate the motion flows M_1 and M_2 in two HS-C sequences. We then warp the flow to the HR-C camera as M_0. To motion deblur the HR-C image, we estimate the PSF from M_0 and use the Richardson–Lucy algorithm to deblur the image. To construct a high resolution depth map D_0, we warp the low resolution depth map estimated from the HS-C cameras to the downsampled HR-C, and use joint bilateral filters to upsample the D_0'. [Adopted from Li *et al.* (2008).]

An overview of our HR motion deblurring and depth map super-resolution framework is shown in Figure 6.2. We assume each frame I_0 in the HR-C camera maps to two HS-C sequences $S_1(t)$ and $S_2(t)$ ($t = 1 \ldots K$, and in our case, $K = 16$). We first estimate the motion flow fields M_1 and M_2 in S_1 and S_2 and warp them onto I_0 as M_0. Next, we estimate the PSF in I_0 from M_0 and apply the Richardson–Lucy (R–L) algorithm to deblur I_0. Recall that I_0 corresponds to K consecutive HS-C frames, therefore I_0 can be deblurred using K different PSFs, each derived from the relative motions with respect to a specific HS-C frame. We use $\tilde{I}_0(t)$ to represent the deblurred result of I_0 for frame t. To generate the super-resolution depth map, we first compute a low resolution depth map $D_L(t)$ between $S_1(t)$ and $S_2(t)$. We then warp $D_L(t)$ to the downsampled HR-C camera as $D_0'(t)$ and then upsample $D_0'(t)$ using a joint bilateral filter, whose spatial kernel is defined on $D_0'(t)$ and range kernel defined on $\tilde{I}_0(t)$.

6.3 Motion deblurring

6.3.1 Motion flow estimation

We first partition each frame in the HS-C camera into multiple foreground regions Ω_i^f and a background region Ω^b. To do so, we simply take a background image without any foreground objects and then use foreground/background subtraction followed by a graph-cut (Boykov & Funka-Lea 2006) to extract the foreground regions. We then group all foreground pixels into separate connected regions. We assume that each region has homogeneous motion and that the regions do not overlap. We also use the estimated disparity map to establish correspondences between the foreground regions in the two HS-C cameras. This allows us to individually motion-deblur each foreground region using the method described in Section 6.3.3. Notice that since our system captures a sequence of images, it is also possible to directly composite the background image by applying a median filter across the frames.

To estimate the motion flow in each region Ω_i^f, we assume an affine motion model between two consecutive frames although we can concatenate several successive frames to form more complex motions. We apply a multi-resolution iterative algorithm (Bergen, Anandan, Hanna & Hingorani 1992) to minimize the energy function:

$$\arg\min_p \sum_x \left[I(w(x;p)) - I'(x)\right]^2$$

$$w(x;p) = \begin{pmatrix} p_1 x + p_2 y + p_3 \\ p_4 x + p_5 y + p_6 \end{pmatrix} \tag{6.1}$$

where p corresponds to the motion flow parameter, and w is the warping function. In our case, we estimate the motion flow in each HS-C camera, i.e. $I = S_j(t)$ and $I' = S_j(t+1)$, $j = 1, 2$. At each frame, we use the Gauss–Newton method to find the optimal motion (Baker & Matthews 2004).

6.3.2 Motion warping

Once we have estimated the motion flow for every foreground region Ω_i^f in each HS-C camera, we warp the motion flow to the HR-C camera. Denote $M_1(t)$ and $M_2(t)$ as the estimated motion flow samples in S_1 and S_2 at frame t. To compute the motion flow sample $M_0(t)$ in the HR-C image I_0, we first establish the correspondences between $S_1(t)$ and $S_2(t)$. We use scale-invariant feature transform (SIFT) detection (Lowe 2004) to process $S_1(t)$ and $S_2(t)$ and then perform a global matching. To remove outliers, our algorithm uses random sample consensus (RANSAC) with projective transformations as its precondition.

Given a pair of corresponding feature points p_1 and p_2 in $S_1(t)$ and $S_2(t)$, we connect p_1 with S_1 camera's center-of-project (COP) to form a ray r_1. We then project r_1 onto the HR-C image I_0 as a line l_1. We apply a similar process to p_2 to obtain line l_2. Finally we intersect l_1 and l_2 to obtain p_0, as shown in Figure 6.3(a).

To estimate the motion flow of p_0 in the HR-C camera I_0, we assume p_0's corresponding point p_1 in $S_1(t)$ moves to q_1 by motion flow $M_1(t)$, and p_2 in $S_2(t)$ moves to q_2 by $M_2(t)$. Similar to the way in which we find p_0, we then combine q_1 and q_2 to find their corresponding point q_0 in I_0. We use the displacement between p_0 and q_0 as a motion flow sample. To improve robustness, we average multiple motion flow samples to compute $M_0(t)$.

Since our method combines the correspondences and the HS-C's motion flow, our camera is able to model more complex motions than the system proposed by Ben-Ezra and Nayar. In Ben-Ezra & Nayar (2003), the motion flow in the HR-C camera is assumed to be identical to the one estimated by the HS-C camera. This is only valid for planar motions. Our method estimates the motion flow for each foreground region Ω_i^f separately, and warps it using the correspondences. Therefore, we can compute the motion flow for each foreground object in I_0, even if they lie at different depths or move along different trajectories. Notice that the accuracy of our motion warping relies

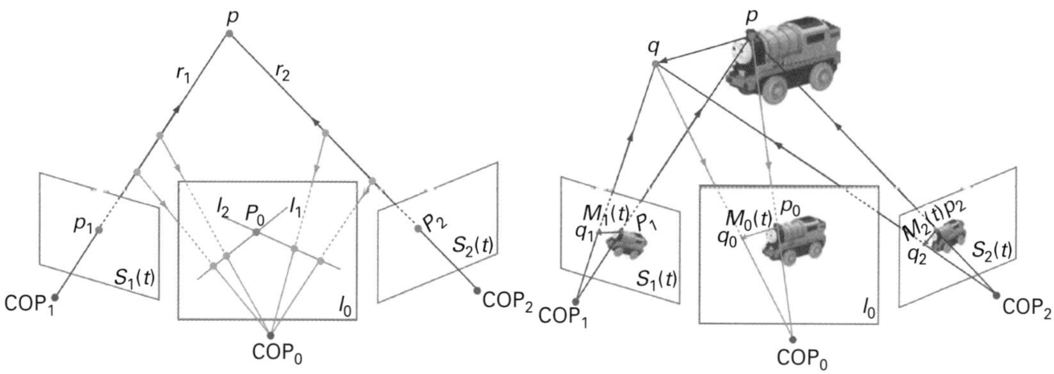

(a) (b)

Figure 6.3 Motion estimation. (a) For each correspondence pair p_1 and p_2 in the HS-C pair, we project the ray that passes through p_1 in $S_1(t)$ onto I_0 as l_1. We similarly project p_2 as l_2 in I_0. Finally, we compute the intersection point of l_1 and l_2 to find the corresponding point p_0; (b) to estimate the motion flow of p_0 in I_0, we use the motion flow in the HS-C cameras to find q_0 and treat p_0q_0 as its motion flow. [Adopted from Li *et al.* (2008).]

heavily on camera calibrations. In our implementations, we use the method discussed in Zhang (2000) to calibrate all three cameras.

6.3.3 PSF estimation and motion deblurring

Recall that every HR-C frame maps to K ($K = 16$) HS-C frames, therefore, we can compute the PSF with respect to each frame t. To do so, we concatenate all K discrete motion samples $M_0(t)$, $t = 1 \ldots K$ to form a single motion path. Ben-Ezra and Nayar proposed to use the motion centroid to construct a Voronoi tessellation for interpolating the motion path smoothly. Notice that our HS-C camera captures at a very high frame rate, therefore each motion sample usually only covers three to five pixels in the HR-C image. To estimate the kernel accurately, first we align the motion path with the center of the kernel appropriately, then resample each $M_0(t)$ into N subpixels ($N = 50$ in our experiment). Next, we count the number of subpixels N_p falling into p for each pixel p covered by $M_0(t)$ in the kernel. Finally, we use this count N_p to estimate the weight of entry p in the kernel and normalize the kernel to arrive at the PSF.

Once we estimate the PSF, we can deblur the HR-C image using existing image deconvolution algorithms (Fergus, Singh, Hertzmann, Roweis & Freeman 2006, Yitzhaky, Mor, Lantzman & Kopeika 1998). In our implementation, we choose to use the R–L iterative deconvolution algorithm. The R–L deconvolution always produces non-negative gray level values and works better than linear methods if the blur kernel is known and the noise level in the image is low. Before applying the R–L algorithm, we first estimate the masks for each of the foreground regions in the HR-C image, obtained by using the technique presented in Section 6.3.1. These masks are used to composite deblurred results into a common background image. In Figures 6.4 and 6.5, we show the motion results on captured dynamic scenes.

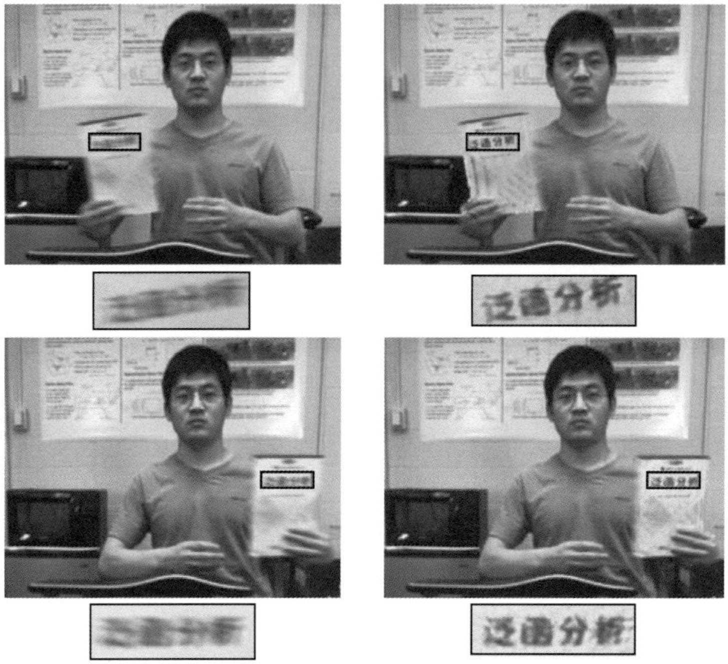

(a) (b)

Figure 6.4 Motion deblurring results in a dynamic scene. The book was quickly passed over from one hand to the other; (a) shows two original HR-C frames, and (b) shows the deblurred results using our approach. [Adopted from Li *et al.* (2008).]

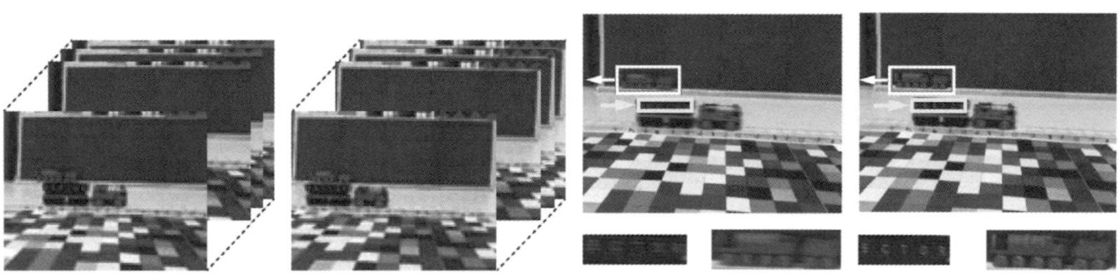

(a) (b) (c) (d)

Figure 6.5 Motion deblurring results with spatially-varying kernels. (a) and (b) show the two HS-C sequences; (c) shows the HR-C image with strong motion blur; (d) shows the motion deblurred results using our method. Notice the front and back train are moving in opposite directions at different speeds. Our method is able to motion deblur both objects. [Adopted from Li *et al.* (2008).]

6.4 Depth map super-resolution

Next, we show how to use the deblurred HR-C image for generating super-resolution depth maps.

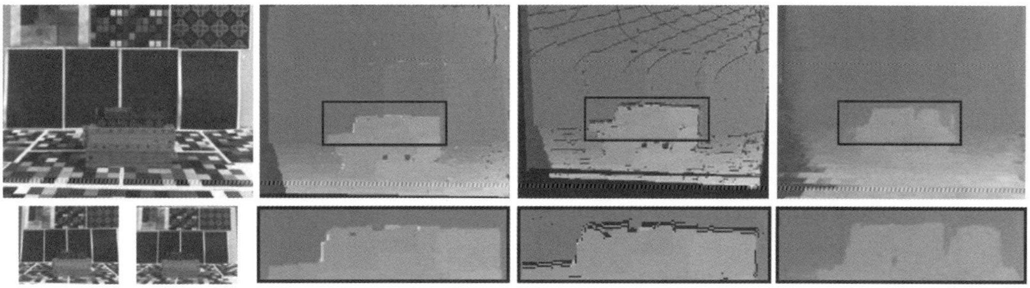

(a) (b) (c) (d)

Figure 6.6 Depth map super-resolution results for a static scene. (a) shows the input images from the HS-C and HR-C cameras; (b) shows the depth map from the HS-C camera pair. The result is upsampled using bilinear interpolation for comparison; (c) shows the reprojected low resolution depth map onto the HR-C camera. The warped map is very noisy and contains holes; (d) shows the upsampled results using the joint bilateral filters. [Adopted from Li *et al.* (2008).]

6.4.1 Initial depth estimation

First we use the two HS-C cameras and apply the graph-cut algorithm (Kolmogorov & Zabih 2002) to estimate a low resolution disparity map with respect to the left HS-C camera. We then convert it to the depth map D_L. Next, we warp D_L to the downsampled HR-C camera as D'_0. We adopt a similar notation as in Zhang (2000): every pixel p_1 in D_L is mapped to its corresponding pixel p_0 in D'_0 as:

$$p_0 = (\mathbf{A}_0\, \mathbf{R}_0)\, (\mathbf{A}_1\, \mathbf{R}_1)^{-1}\, (p_1 - \mathbf{A}_1\, \mathbf{T}_1) + \mathbf{A}_0\, \mathbf{T}_0 \tag{6.2}$$

where p is the homogeneous coordinate of the pixel as $(sx, sy, s)^T$, s represents the depth of the point, $\mathbf{A}_1$ and $\mathbf{A}_0$ are the camera intrinsic matrices of the HS-C camera S_1 and the downsampled HR-C camera. $\mathbf{R}$ and $\mathbf{T}$ are extrinsic to the cameras. Here we downsample the HR-C to have the same spatial resolution as the HS-C camera to reduce holes caused by missing data in the reprojected depth map. A sample depth map D_L is shown in Figure 6.6(b). Notice the depth map is incomplete because of the large baseline between the two HS-C cameras. This is less problematic since we focus on capturing the depth map of the moving foreground objects. However, the foreground can exhibit some serious artifacts. For example, in Figure 6.7, the left boundary of the foreground object in the depth map (c) partially merges with the background and the silhouette of the toy train lacks fine details. We use the joint bilateral upsampling to reduce these artifacts.

We choose to warp the depth map to the HR-C camera instead of warping the HR-C image to the HS-C camera, mainly because the depth-based warping inevitably corrupts the quality of the image by introducing holes near occlusion boundaries. Since our goal is to construct a high resolution depth map, it is more preferable to maintain the quality of the high resolution image than the low resolution depth map.

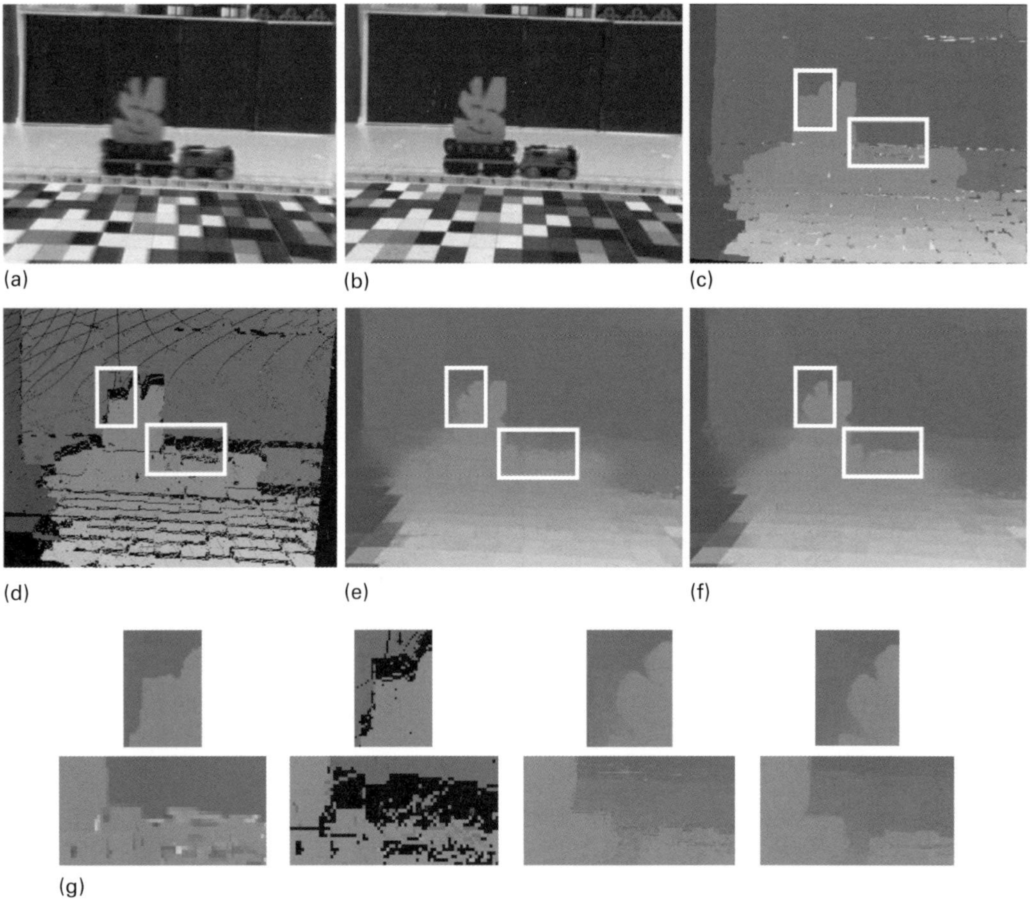

Figure 6.7 Results on a toy train scene using our hybrid-speed sensor. (a) shows the motion-blurred image from the HR-C camera; (b) shows the motion deblurred result using the method described in Section 6.3; (c) shows the low resolution depth map. We upsample the map using bilinear interpolation to compare it with the bilateral filtered results; (d) is the reprojected depth map; (e) shows the upsampling result using joint bilateral filters; (f) shows further improved results using the joint median filter; (g) shows closeup images of the depth maps. [Adopted from Li *et al.* (2008).]

6.4.2 Joint bilateral upsampling

To enhance the warped depth map, we use the joint bilateral filters. The basic bilateral filter is an edge-preserving filter (Tomasi & Manduchi 1998) that uses both a spatial filter kernel and a range filter kernel evaluated on the data values themselves. A joint bilateral filter chooses the range filter from a second guidance image. It can effectively combine flash and no-flash pairs (Petschnigg, Szeliski, Agrawala, Cohen, Hoppe & Toyama 2004, Eisemann & Durand 2004), upsample the low resolution exposure maps, and enhance the low resolution range maps (Kopf, Cohen, Lischinski & Uyttendaele 2007, Yang, Yang, Davis & Nister 2007). In our method, we also use the joint bilateral filter for depth map super-resolution.

Our joint bilateral filter combines the low resolution depth map D'_0 and the motion deblurred high resolution image $\tilde{I}_0$. It computes the value at each pixel p in the high resolution depth map D_0 as:

$$D_0(p) = \frac{1}{W_p} \sum_{q \in \Theta \backslash \Gamma} G_s(\| p - q \|) G_r \left(\left| \tilde{I}_0(p) - \tilde{I}_0(q) \right| \right) D'_0(q)$$

$$W_p = \sum_{q \in \Theta \backslash \Gamma} G_s(\| p - q \|) G_r \left(\left| \tilde{I}_0(p) - \tilde{I}_0(q) \right| \right) \tag{6.3}$$

where G_s is the spatial kernel centered over p and G_r is the range kernel centered at the image value at p in $\tilde{I}_0$; Θ is the spatial support of the kernel G_s. W_p is the normalization factor. Since the warped low resolution depth map D'_0 contains holes, we exclude the points Γ that correspond to the holes.

To emphasize the color difference in the range kernel, we choose

$$\left| \tilde{I}_0(p) - \tilde{I}_0(q) \right| = \max \left(\left| r_p - r_q \right|, \left| g_p - g_q \right|, \left| b_p - b_q \right| \right) \tag{6.4}$$

where r, g, b are the color channels of the pixel in $\tilde{I}_0$. In Figure 6.6(c), we show the reprojected low resolution depth map of resolution 320×240 and the enhanced high resolution depth map in Figure 6.6(d). The joint bilateral filter significantly improves the quality of the depth map by filling in the missing holes and partially correcting the inaccurate boundaries (e.g. the background between the engine and carriage of the toy train in Figure 6.6). However, the resulting boundaries still appear blurry due to the spatial Gaussian. Although we can re-apply the joint bilateral filters on the upsampled depth image to further improve the boundaries, we implement a simple scheme based on the observation that with a large enough spatial support Θ, there should be enough pixels with a similar color and depth to pixel p. To do so, we apply a joint median filter: we first pick N pixels from Θ that have the closest color to p and then apply the median filter on the depth values of these N pixels. In our experiment, we choose N to be 20% of all pixels in Θ.

6.4.3 Results and discussion

In Figures 6.4 and 6.5, we show the motion deblurring results using our hybrid camera. In Figure 6.4(a), we apply a motion deblurring algorithm to a dynamic scene: the book was quickly passed over from the right hand to the left hand. Notice that the texts on the book are unrecognizable in the original HR-C frames, but are clearly readable in the deblurred results shown in Figure 6.4(b). The hand holding the textbook is also effectively deblurred. In Figure 6.5, we use the hybrid camera to deblur spatially-varying kernels. The foreground and the background toy trains are moving in opposite directions at different speeds. Since our method estimates the PSF for each individual train separately, using the method discussed in Section 6.3 we are able to effectively deblur both objects.

In Figure 6.6, we validate our joint bilateral filter method for depth map super-resolution in a static scene. We position a toy train on top of the rails. Notice the

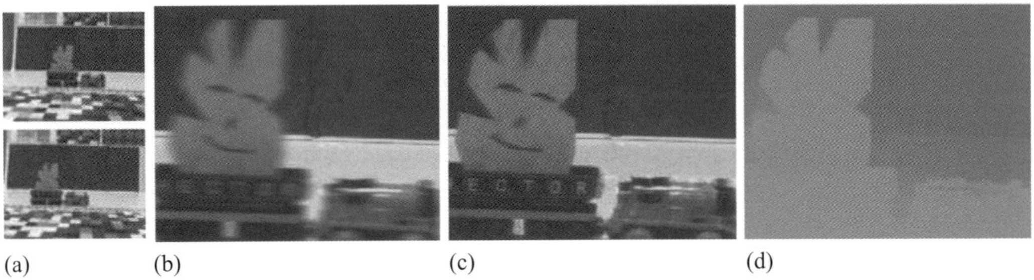

Figure 6.8 Closeup views of a toy train scene. (a) shows two frames from the HS-C cameras; (b) shows a cropped region of the HR-C image; (c) shows the motion deblurred result; (d) shows the reconstructed high resolution depth map. [Adopted from Li *et al.* (2008).]

toy train has complex contours. If we simply upsample the low resolution depth map obtained from the HS-C cameras, the details of the contours disappear. Using joint bilateral upsampling, we are able to recover these fine details and partially correct the erroneous boundaries in the depth map (e.g. the V-shaped contour between the engine and the carriage).

Finally, we show how to use the hybrid camera to motion deblur the HR-C image and reconstruct the high resolution depth map simultaneously. Figure 6.8 shows the results using our system in a toy train scene. We mount a star-shaped object on top of the toy train to emphasize the accuracy of the recovered depth maps. The HS-C and HR-C inputs are shown in Figures 6.8(a) and 6.7(a), and the deblurred results are shown in Figures 6.8(c) and 6.7(b). Our method effectively reduces the motion blur by recovering the text on the toy train and the contour of the star-shaped object, although we also observe some ringing artifacts caused by the R–L deconvolution.

In Figure 6.7, we show the depth super-resolution results using joint bilateral filters. We choose a large enough kernel size σ_S for the spatial Gaussian to fill in the missing data. In this example, we set σ_S to be 35. We normalize the depth map to have intensity between 0 and 1, and we set σ_r of the range Gaussian to be 0.02. Figure 6.7(c) shows the depth map from the HS-C pair and Figure 6.7(d) shows the warped result. The reprojected depth map is very noisy and contains many holes. Using the joint bilateral filter, the depth map is significantly enhanced (Figure 6.7(e)). The seams at the occlusion boundaries are further enhanced (Figure 6.7(f)) using the joint median filter. Closeups are shown at the bottom of Figure 6.7. The typical upsampling rate in our experiments is 9. Due to the camera calibration requirement, we did not further downsample the HS-C camera to demonstrate a higher upsampling rate, although it could easily be achieved if we used a different HR-C camera with a higher resolution.

6.5 Extensions to low-light imaging

Our hybrid-speed sensor assumes sufficient light. However, under low light, the high-speed cameras can incur severe noise due to insufficient exposure and produce incorrect blur kernel estimations. In this section, we briefly discuss our recent work on extending

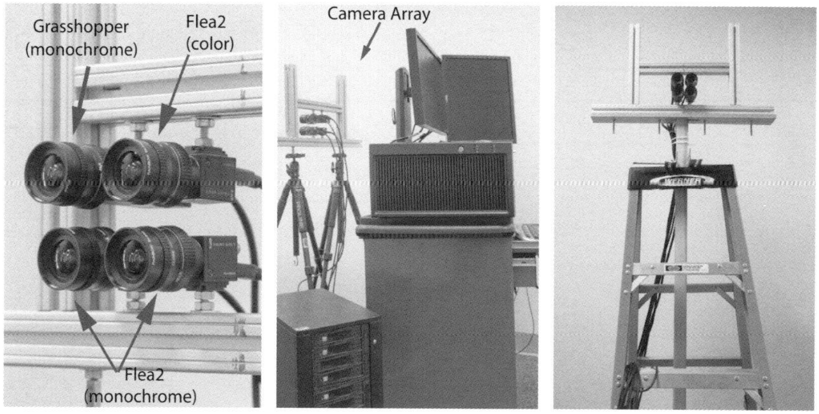

(a) (b) (c)

Figure 6.9 Our heterogeneous sensor for low-light imaging. (a) Our heterogeneous sensor consists of four cameras; (b) and (c) show the mounting of the sensor for indoor and outdoor applications. [Adopted from Li *et al.* (2013).]

the hybrid-speed sensor to low-light imaging. We modify the sensor by combining a single high-speed monochrome (HS-M) camera, a pair of high resolution monochrome (HR-M) cameras, and a single high resolution RGB color (HR-C) camera. The HR-M cameras use a wide aperture and a relatively slow shutter for acquiring images with a low level of noise whereas the HS-M camera uses fast shutters for capturing motion blur-free images on fast moving objects. The HR-C camera is used to capture color information about the scene and it uses slow shutters to reduce the color noise. We call this sensor the heterogeneous sensor.

6.5.1 Sensor construction

Figure 6.9 shows the construction of the heterogeneous sensor: it consists of one Point Grey Grasshopper high-speed monochrome (HS-M) camera (top-left), one Flea2 high resolution color (HR-C) camera (top-right), and two Point Grey Flea2 high resolution monochrome (HR-M) cameras (bottom). All the cameras are equipped with the same Rainbow 16mm C-mount F1.4 lens. We mount the four cameras on a T-slotted aluminum grid, which is then mounted on two conventional tripods for indoor applications. To deal with the long working range of outdoor applications, we also build a giant "tripod" from a six-foot step ladder to hold the camera array grid, as shown in Figure 6.9(c). Since the observable regions of the cameras overlap substantially, we use Zhang's algorithm (Zhang 2000) directly for camera calibration.

We built a single workstation with an Intel SC5400BASE-NA chassis to control and synchronize these cameras, and stream the image data to the storage device. All the cameras are connected to this workstation via PCI-E FireWire adaptors. The use of a FireWire bus allows us to synchronize the cameras using Point Grey software. In our system, the HS-M camera captures images at $640 \times 480 \times 8$ bit at 120 fps, the two HR-M

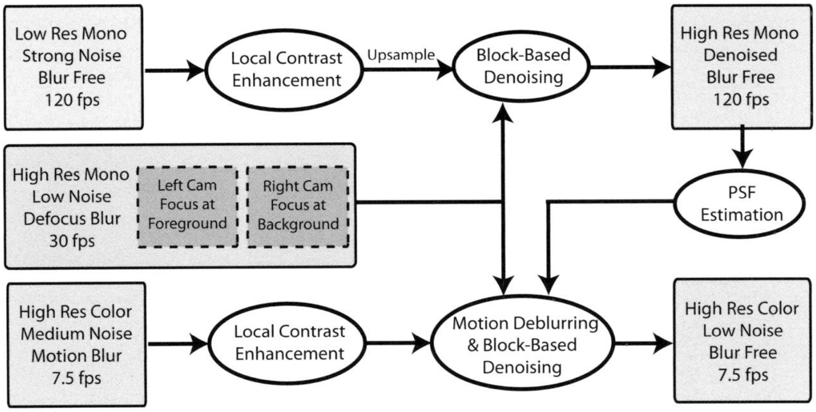

Figure 6.10 Simultaneous deblurring and denoising using our heterogeneous sensor. [Adopted from Li *et al.* (2013).]

cameras work at $1024 \times 768 \times 8$ bit at 30 fps, and the Flea2 color camera captures color images of size $1024 \times 768 \times 24$ bit at 7.5 fps. To stream these $\sim$100 MB/s image data, we connect the workstation to an external SATA disk array as the data storage device. It is capable of writing data at 500 MB/s when configured to RAID 0, as shown in Figure 6.9(b).

To improve the SNR in low-light imaging, we choose to use two HR-M cameras since they gather significantly more light than the HR-C for the same sensor. This is because monochrome sensors do not use the "Bayer array" for color filtering. Bayer arrays can commonly filter out approximately two-thirds of the incoming light at every pixel. Also, monochrome sensors are responsive to a much wider range of spectra and hence have higher light efficiency. In addition, we use wide apertures on the monochrome cameras. However, due to large apertures the resulting images have a shallow DoF, i.e. they exhibit strong blurs for out-of-focus regions. Our solution is to use two HR-M cameras focusing at different parts of the scene to extend the DoF.

6.5.2 Processing pipeline

In our heterogeneous sensor, every color frame region I_i^{hrc} captured by the HR-C camera maps to four pairs of synchronized high resolution grayscale frames $I_p^{hrm_j}$ ($p = 4i, \ldots, 4i+3$; $j = 0, 1$) by the two HR-Ms, and to 16 synchronized low resolution grayscale images I_q^{hsm} ($q = 16i, \ldots, 16i+15$) by the HS-M. Figure 6.10 shows our system pipeline for fusing the imagery data. Specifically, we use the HR-M images as the spatial prior to denoise HS-M and HR-C images. We then estimate motion flow using denoised HS-M sequences to deblur the HR-C images.

The most challenging task in the pipeline is to denoise the HS-M camera using the HR-M camera pair. Recall that the HR-M and HS-M sensors use different exposure settings, therefore we first conduct feature-preserving tone-mapping (Mertens, Kautz & Van Reeth 2007) on the low dynamic range HS-M image to match the intensity level of the HR-M images. The tone-mapped HS-M image has enhanced contrast but still exhibits strong noise. Hence we use the HR-M images to denoise the HS-M ones. However, the three sensors are separated by relatively large baselines and their images exhibit strong parallax. We therefore conduct a patch-based denoising scheme: for each noisy patch in the HS-M image, we first locate its corresponding patches in the HR-M images via multi-view block matching, and then impose them as spatial priors in total variation (TV) based denoising (Li *et al.* 2013).

The analysis of motion flows in the HS-M sequence also enables the separation of static versus moving regions in the HR-C image. For the static region, we only need conduct denoising and we apply the same TV-based denoising scheme by using patches from the HR-M cameras. To deblur the moving regions, we adopt a similar approach to Section 6.3 to estimate the PSF in the HR-C image. The main difference is that the HR-C image is corrupted by both blur and noise under low light. As a result, direct deconvolution can lead to large errors. Our approach is to apply the simultaneous deblur/denoise scheme (Wang, Yang, Yin & Zhang 2008, Krishnan & Fergus 2009, Li *et al.* 2013).

6.5.3 Preliminary results

We have conducted preliminary experiments on our heterogeneous sensor.

Synthetic scenes

To compare our hybrid denoising algorithm with BM3D quantitatively, we generate a pseudo multi-camera array and render the scene using Autodesk® 3ds Max®. Specifically, the HR-M cameras are rendered at a resolution of 1024×768 with aperture $f/8$, the HS-M camera at 640×480 with aperture $f/22$, and the HR-C at 1024×768 also with aperture $f/22$. All cameras have a focal length of 135 mm. We construct a scene with a moving paper cup with constant velocity 66.18 cm/s parallel to the image plane of the sensor. We focus one HR-M camera on the front of the table and the second one on the background orchid. To simulate motion blurs, we render 16 frames from the HR-C camera and average them as the motion-blurred HR-C image. To simulate noisy HS-M and HR-C images, we add zero-mean Gaussian noise of variance 0.09 and 0.03, respectively.

Figure 6.11 shows our denoised HS-M images. We compare our approach with BM3D (Dabov, Foi, Katkovnik & Egiazarian 2007) using the peak signal-to-noise ratio (PSNR) in decibels (dB). Our algorithm not only outperforms BM3D by 0.24 dB but also preserves fine details better, such as the texture of the table, the text on the paper cup, and the contour of the background orchid. In Figure 6.12, we show our deblurring/denoising result of the synthetic HR-C image. We first estimate the motion information of the paper cup from the denoised HS-M image sequence and then compute the PSF to conduct non-blind deconvolution of the HR-C image. Finally, we apply our hybrid denoising technique. The deblurring/denoising process is applied on individual color channels and the results are then combined. Our approach is able to partially recover the text on the moving cup, even though it is corrupted by noise. It also increases the PSNR from 17.69 dB to 22.95 dB on the HR-C image.

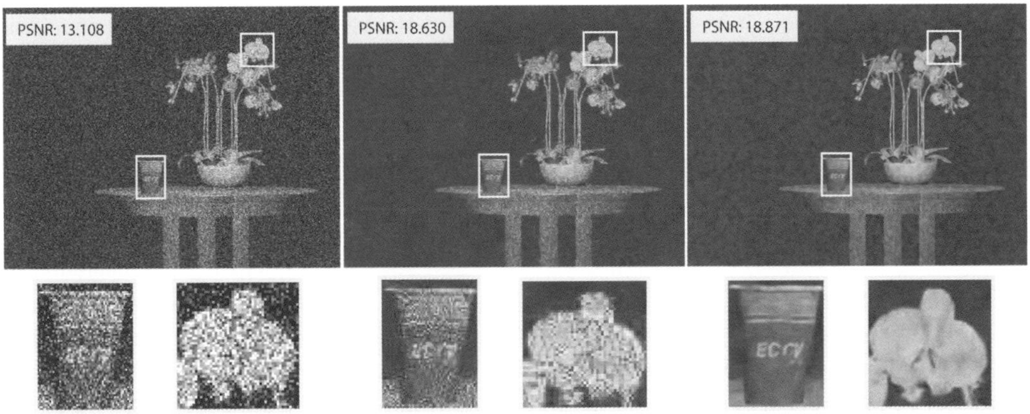

Figure 6.11 Path-based denoising on a sample image in a synthetic scene. From left to right: the synthesized low-light image captured by the HS-M camera, the BM3D denoising result, and our result. [Adopted from Li *et al.* (2013).]

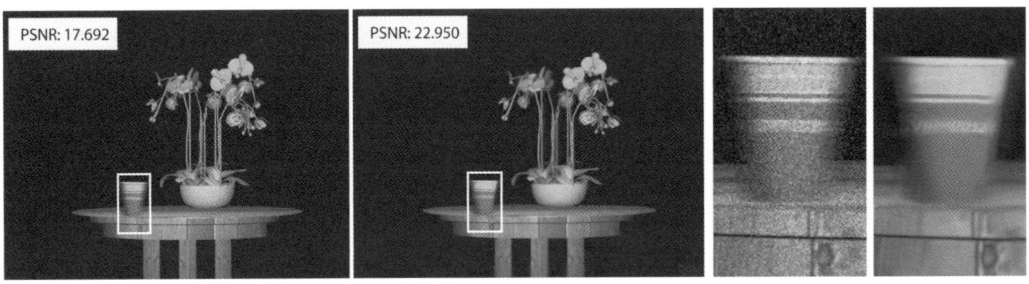

Figure 6.12 Image deblurring/denoising on a sample HR-C image. From left to right: the motion-blurred HR-C image, our deblurring/denoising result, and closeup views of the moving object. [Adopted from Li *et al.* (2013).]

Real scenes

Figure 6.13 shows an indoor scene of a toy train moving in front of a cluttered background. We focus one HR-M camera on the toy train and chocolate box at the front, and the second on the book in the background. Defocus variations due to large apertures can be clearly observed in (b) and (c). The use of large apertures, however, significantly reduces the noise. In fact, the HR-M images exhibit very low noise and hence are suitable for denoising the HS-M and HR-C images using our patch-based denoising. Specifically, we first boost the contrast of the HS-M image (a) via feature-preserving tone mapping and the result is shown in (d). However the closeup views show that the contrast-enhanced result is still severely degraded. By applying our denoising algorithm, we can significantly reduce the noise while preserving local details such as the text on the background book and the contour of the foreground train. In contrast, the results using BM3D appear overly smooth and exhibit artifacts near edge boundaries. For example, the closeup view (e) shows that the BM3D result exhibits a special diagonal scanline noise pattern (possibly caused by the rolling shutter).

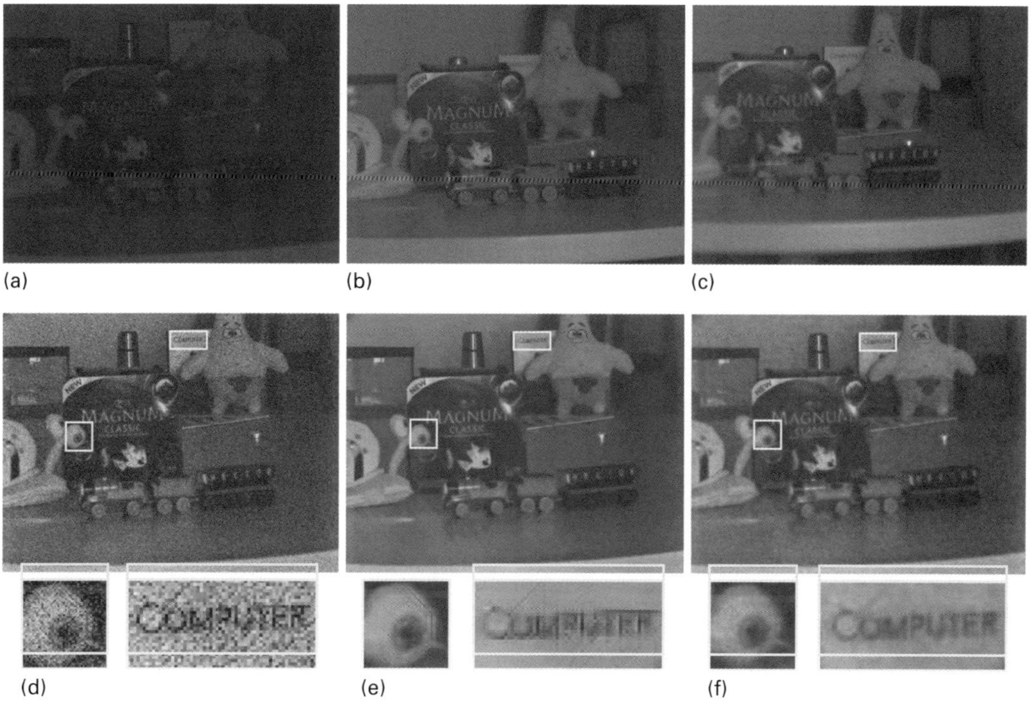

Figure 6.13 Patch-based denoising on a sample image of an indoor scene. (a) Shows the low-light image captured by the HS-M camera; (b) and (c) show the large aperture stereo image pair from HR-M cameras; (d) shows the contrast enhanced (a); (e) and (f) show the denoising results of (d) by BM3D and our method. [Adopted from Li *et al.* (2013).]

In Figure 6.14, we show the deblurred result on the HR-C image. We first use the denoised HS-M sequence to estimate the PSF of the motion-blurred region in HR-C and then simultaneously denoise/deblur the region. In this example, we only apply the multi-view block matching algorithm on the green channel. This is because a typical Bayer array contains twice as many green as red or blue sensors and the green channel is less noisy. Our approach is able to reconstruct the contours of the moving train, partially recover the text on it, and significantly reduce the noise.

6.6 Discussion and summary

We have presented two multi-sensor fusion techniques for reducing motion blurs under sufficient and low-light conditions. Our solution leverages the latest advances on high-speed, high resolution and multi-spectral sensors by integrating multiple types of sensors into a unified imaging system. The multi-sensor fusion solution eliminates the need for co-axial setups or the requirement of shift-invariant kernels in hybrid-imaging systems. It actively exploits the parallax across the sensors for simultaneous depth estimation and image deblurring/denoising. A downside of our solution, however, is that the system tends to be bulky and less portable since three or more sensors are generally needed.

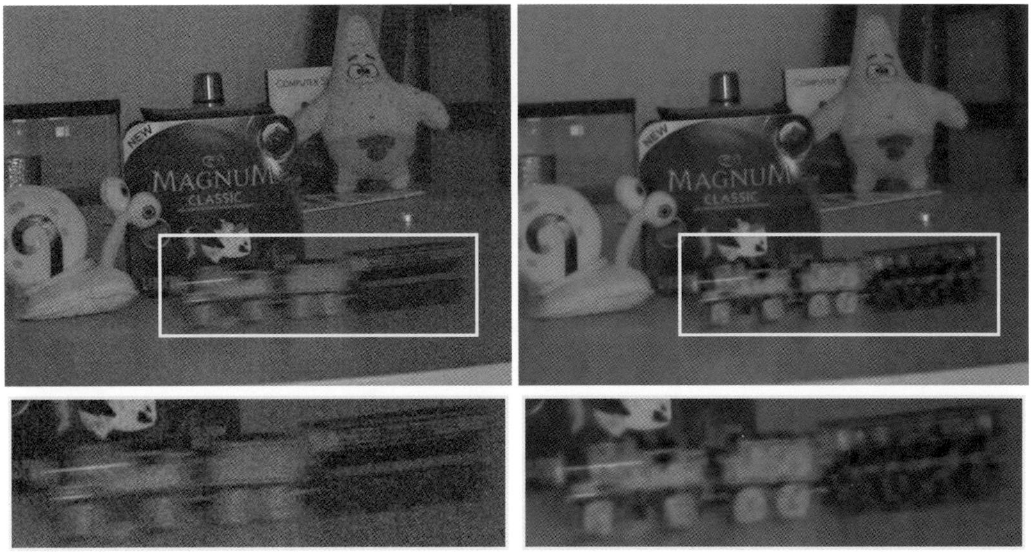

Figure 6.14 Deblurring/denoising result on a sample HR-C frame. Left: the HR-C frame (contrast-enhanced) corresponding to the HS-M/HR-M sequences in Figure 6.13; right: the deblurred/denoised result. [Adopted from Li *et al.* (2013).]

Our hybrid-speed sensor aims to handle fast motions under sufficient lighting conditions robustly. The sensor consists of a pair of high-speed color (HS-C) cameras and a single high resolution color (HR-C) camera. The HS-C cameras are able to capture fast motion with little motion blur. They also form a stereo pair and can estimate the low resolution depth map. We estimate the motion flows in the HS-C cameras and then warp them using the depth map onto the HR-C camera to get the PSFs for motion deblurring. The HR-C image, once deblurred, is then used to super-resolve the depth map.

We have also demonstrated preliminary work on extending our hybrid-speed sensor to low-light imaging. We configured a heterogeneous sensor by combining a high-speed monochrome (HS-M) camera, two high resolution (HR-M) cameras, and a single high resolution RGB color (HR-C) camera. The HR-M cameras use large apertures to gather more light. The HS-M camera captures fast motions without motion blurs but produces noisy images. The HR-C camera provides color information of the scene using a slow shutter but incurs strong motion blurs. We strategically fuse the heterogeneous imagery data from the sensors to conduct simultaneous denoising and deblurring. We are currently conducting more thorough experiments on both synthetic and real scenes.

Our multi-sensor fusion solutions can potentially be applied to a range of imaging tasks beyond deblurring. For example, if we ignore motion blurs, we can simplify the hybrid-speed sensor to a hybrid-resolution stereo camera by coupling one high resolution color camera with one low resolution monochrome camera. Our recent work (Yu, Thorpe, Yu, Grauer-Gray, Li & Yu 2011) has shown that such sensors can be used to recover high quality depth maps at interactive speeds and in addition synthesize dynamic refocusing effects. Our hybrid-speed sensor can also be fine-tuned to acquire high resolution, high-speed 3D videos. In particular, we can adjust the frame rate and the resolution of the sensors to enable more accurate PSF estimation and reliable deblurring. More sophisticated super-resolution techniques such as dictionary learning based methods can be used to generate high resolution depth maps for creating 3D video content.

Acknowledgements

Part of this chapter is based on the work that appeared in Li *et al.* (2008) and we gratefully acknowledge IEEE for their permission to reproduce large portions here.

References

Baker, S. & Matthews, I. (2004). Lucas–Kanade 20 years on: A unifying framework. *International Journal of Computer Vision*, **56**(3), 221–55.

Ben-Ezra, M. & Nayar, S. (June 2004). Motion-based motion deblurring. *IEEE Transactions on Pattern Analysis and Machine Intelligence*, **26**(6), 689–98.

Ben-Ezra, M. & Nayar, S. K. (2003). Motion deblurring using hybrid imaging. In *Proceedings of the 2003 IEEE Computer Society Conference on Computer Vision and Pattern Recognition*, pp. 657–64.

Bergen, J. R., Anandan, P., Hanna, K. J. & Hingorani, R. (1992). Hierarchical model-based motion estimation. In *Proceedings of the Second European Conference on Computer Vision*, pp. 237–52.

Boykov, Y. & Funka-Lea, G. (2006). Graph cuts and efficient N-D image segmentation. *International Journal of Computer Vision*, **70**(2), 109–31.

Dabov, K., Foi, A., Katkovnik, V. & Egiazarian, K. (2007). Image denoising by sparse 3-D transform-domain collaborative filtering. *IEEE Transactions on Image Processing*, **16**(8), 2080–95.

Eisemann, E. & Durand, F. (2004). Photography enhancement via intrinsic relighting. *ACM Transactions on Graphics*, **23**(3), 673–8.

Fergus, R., Singh, B., Hertzmann, A., Roweis, S. T. & Freeman, W. T. (2006). Removing camera shake from a single photograph. *ACM Transactions on Graphics*, **25**(3), 787–94.

Kolmogorov, V. & Zabih, R. (2002). Multi-camera scene reconstruction via graph cuts. In *Proceedings of the 7th European Conference on Computer Vision, Part III*, pp. 82–96.

Kopf, J., Cohen, M. F., Lischinski, D. & Uyttendaele, M. (2007). Joint bilateral upsampling. *ACM Transactions on Graphics*, **26**(3), 96:1–10.

Krishnan, D. & Fergus, R. (2009). Fast image deconvolution using hyper-Laplacian priors. In *Neural Information Processing Systems Conference*, **22**, 1–9.

Li, F., Ji, Y. & Yu, J. (2013). *A Hybrid Camera Array for Low Light Imaging*. University of Delaware Technical Report UD-CIS-2013-01.

Li, F., Yu, J. & Chai, J. (2008). A hybrid camera for motion deblurring and depth map super-resolution. In *Computer Vision and Pattern Recognition*, pp. 1–8.

Lowe, D. G. (2004). Distinctive image features from scale-invariant keypoints. *IEEE International Conference on Computer Vision*, **60**(2), 91–110.

Mertens, T., Kautz, J. & Van Reeth, F. (2007). Exposure fusion. In *IEEE 15th Pacific Conference on Computer Graphics and Applications*, pp. 382–90.

Petschnigg, G., Szeliski, R., Agrawala, M., Cohen, M., Hoppe, H. & Toyama, K. (2004). Digital photography with flash and no-flash image pairs. *ACM Transactions on Graphics*, **23**(3), 664–72.

Tai, Y.-W., Du, H., Brown, M. S. & Lin, S. (2010). Correction of spatially varying image and video motion blur using a hybrid camera. *IEEE Transactions on Pattern Analysis and Machine Intelligence*, **32**(6), 1012–28.

Tomasi, C. & Manduchi, R. (1998). Bilateral filtering for gray and color images. *Proceedings of the 6th International Conference on Computer Vision*, pp. 839–46.

Wang, Y., Yang, J., Yin, W. & Zhang, Y. (2008). A new alternating minimization algorithm for total variation image reconstruction. *SIAM Journal on Imaging Sciences*, **1**(3), 248–72.

Yang, Q., Yang, R., Davis, J. & Nister, D. (June 2007). Spatial-depth super resolution for range images. *Computer Vision and Pattern Recognition*, pp. 1–8.

Yitzhaky, Y., Mor, I., Lantzman, A. & Kopeika, N. S. (1998). Direct method for restoration of motion-blurred images. *Journal of the Optical Society of America A: Optics, Image Science & Vision*, **15**(6), 1512–19.

Yu, Z., Thorpe, C., Yu, X., Grauer-Gray, S., Li, F. & Yu, J. (2011). Dynamic depth of field on live video streams: a stereo solution. *Computer Graphics International*, pp. 1–9.

Zhang, Z. (2000). A flexible new technique for camera calibration. *IEEE Transactions on Pattern Analysis and Machine Intelligence*, **22**(11), 1330–4.

7 Motion deblurring using fluttered shutter

Amit Agrawal

Motion blur is the consequence of relative motion between the camera and the scene within the camera's exposure (shutter) time. While a sharp photo might be restored computationally using deblurring algorithms, can we assist motion deblurring by modifying the imaging process itself? Motion deblurring systems modify the traditional image capture to simplify or help the subsequent deblurring process. Such systems include but are not limited to (a) coded exposure cameras which modulate the light integration pattern so as to make the resulting point spread function (PSF) invertible; (b) auxiliary low spatial resolution high frame-rate cameras to help in PSF estimation; and (c) auxiliary sensors such as gyroscopes/inertial measurement units (IMUs) to assist in PSF estimation. Current digital single lens reflex (SLR) cameras and lenses also incorporate image stabilization for handling limited motion blur, which will detect user hand-shake and shift the image parallel to the image plane appropriately. In this chapter, we describe coded exposure photography for motion deblurring.

7.1 Related work

Capture time solutions

Using a fast shutter speed (short exposure time) can reduce motion blur, but increases noise and penalizes static parts of the scene. High-speed cameras capture fast motion but require expensive sensing, bandwidth and storage, along with brilliant scene lighting. A high-speed camera also fails to exploit the inter-frame coherence. Edgerton (1951–63) has shown visually stunning results for high-speed objects using a modest exposure time but an extremely narrow-duration flash. Flash, however, is impractical in outdoor or distant scenes. In addition, it captures an instant of the action and fails to indicate the general movement in the scene.

Smarter cameras

To overcome camera shake, newer cameras optically stabilize their lenses. Adaptive optical elements controlled by inertial sensors compensate for camera motion to reduce blur (Canon 2006, Nikon 2005). Alternatively, some CMOS cameras perform multiple high-speed frame captures within normal exposure time, enabling multiple image-based motion blur removal (Liu & Gamal 2001). These methods are able to produce clear and crisp images, given a reasonable exposure time. Ben-Ezra and Nayar (Ben-Ezra &

Nayar 2004) proposed a hybrid imaging system that accurately estimates the PSF using an auxiliary low resolution high frame-rate camera, enabling deblurring even after long exposure times. These methods compensate for camera motion but do not respond to object motion within the scene.

Coded sampling

Binary and continuous codes are commonly used in signal processing to modulate signals with a broadband response. These include chirps that sweep the carrier over a wide frequency band during the pulse interval. Maximum length sequences (m-sequences) and modified uniformly redundant arrays (MURAs) are popular choices for coding and decoding by circular convolution. Coded-aperture astronomical imaging uses MURA codes (Gottesman & Fenimore 1989) to improve the signal-to-noise ratio while capturing X-ray and gamma-ray wavelengths unsuitable for conventional lenses. However, motion blur corresponds to linear convolution, which is equivalent to circular convolution with a zero-padded sequence, and MURA codes are thus not optimal. Broadband codes also find wide application in spread spectrum coding for noise-robust communication and in code division multiplexing (CDMA) that minimizes interference with adjacent broadcast channels.

7.2 Coded exposure photography

In conventional imaging, the shutter is opened at the start of the exposure time and remains fully open for the entire exposure duration. The exposure time defines a temporal box filter resulting in a low-pass non-invertible PSF, leading to high spatial-frequency content being lost in the blurred image. Deblurring motion blur due to fast moving objects thus becomes an ill-posed problem.

Coded exposure photography (Raskar, Agrawal & Tumblin 2006) is an image capture method that reduces the loss of high spatial frequencies in blurred images. The coded exposure camera flutters the camera's shutter open and closed during the exposure time with a pseudo-random binary sequence. This results in modulation of the light falling on the sensor, and modifies the PSF to become broadband or invertible. Unlike the temporal box filter, the flutter pattern preserves all spatial frequencies. The deblurring process thus becomes well posed and the sharp photo can be easily recovered.

7.3 Image deconvolution

Let us assume 1D object motion parallel to the sensor plane with constant velocity. For a conventional camera, the PSF $h(x)$ corresponds to a box filter

$$h(x) = 1/k, \qquad 0 < x < k, \tag{7.1}$$

where k is the blur size in pixels. We call such blur *flat blur*. The captured blurred image $i(x)$ can be modeled as a convolution of the sharp image of the object $s(x)$ with the motion PSF $h(x)$:

$$i(x) = s(x) * h(x) + n(x), \tag{7.2}$$

where $*$ denotes the convolution operator and $n(x)$ denotes image noise. Since convolution in the spatial domain corresponds to multiplication in the frequency domain,

$$I(\omega) = S(\omega) \times H(\omega) + N(\omega), \tag{7.3}$$

where ω denotes the frequency and capitalized symbols denote the respective Fourier transforms. Inverse filtering (Jain 1988) can recover an estimate of the sharp image as

$$\hat{S}(\omega) = \frac{I(\omega)H^*(\omega)}{|H(\omega)|^2}. \tag{7.4}$$

However, if $H(\omega)$ contains zeros, inverse filtering becomes ill posed and noise is severely amplified (Tikhonov & Arsenin 1977). The Fourier transform of the box filter is a sinc function, which is zero at certain frequencies, thereby making deblurring ill posed. To overcome this problem, several previous algorithms have chosen their reconstruction from the range of possible solutions using maximum-likelihood estimation approaches such as Wiener filtering (Jain 1988) and the Richardson–Lucy algorithm (Richardson 1972, Lucy 1974). The latter is a nonlinear ratio-based method that produces non-negative gray level values.

Rather than leaving the shutter open for the duration of the exposure, the coded exposure camera "flutters" it open and closed in a rapid irregular binary sequence. The resulting blur pattern is called *coded blur*. The fluttering modulates the integration of motion on and off in such a way that the resultant point spread function has maximum coverage in the Fourier domain. Although the object motion is unknown a priori, the temporal pattern can be chosen so that the convolved (blurred) image preserves the higher spatial frequencies of moving objects and allows us to recover them by a relatively simple decoding process.

7.3.1 Motion model

For 1D motion, the discrete equation for each motion line can be written as

$$\mathbf{i} = A\mathbf{s} + \mathbf{n}, \tag{7.5}$$

where $A_{(n+k-1)\times n}$ denotes the 1D circulant *motion smear matrix* for object size n, and $\mathbf{s}$, $\mathbf{i}$ and $\mathbf{n}$ denote the vectors of the sharp object, the blurred object, and noise intensities along each motion line, respectively. The estimated deblurred image is then given by

$$\hat{\mathbf{s}} = (A^T A)^{-1} A^T \mathbf{i} = \mathbf{s} + (A^T A)^{-1} A^T \mathbf{n}. \tag{7.6}$$

Assuming noise to be zero mean, independently and identically distributed (IID) Gaussian noise with variance σ^2, the covariance matrix of the estimation error, $\hat{\mathbf{s}} - \mathbf{s}$, is equal to

$$\Sigma = (A^T A)^{-1} A^T \sigma^2 A (A^T A)^{-T} = \sigma^2 (A^T A)^{-1}. \tag{7.7}$$

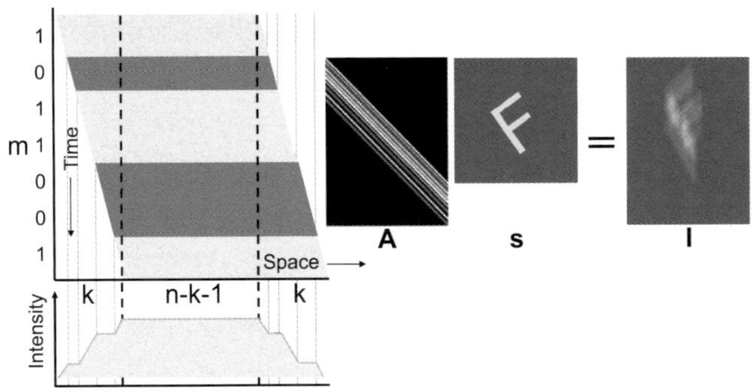

Figure 7.1 The 1D motion blur process. Left: a time–space projection for a moving object chopped with a code. Right: a linear system to transform the unknown image into a blurred photo. Figure courtesy of Raskar *et al.* (2006). Reproduced with permission from ACM.

Given a finite exposure time of T seconds, we subdivide the integration time into m time slices, called *chops*, so that each chop is T/m seconds long. The on/off chop pattern then is a binary sequence of length m. The motion blur process is a time–space projection (Figure 7.1), where motion in T seconds causes a linear blur of k pixels. Hence, within one single chop's duration, the smear covers k/m pixels. Consider the simple case of an object moving vertically in front of a black background and evaluated along a vertical scan line as shown in Figure 7.1. If the object length is n pixels and blur size is k pixels, the total blurred size is $(n + k - 1)$. Our goal is to find the best estimate of the n pixels from the given blurred image.

The smear matrix A can be obtained as follows. Each pixel in the unknown image $\mathbf{s}$ contributes to a total of k pixels after smearing. The first column of the circulant matrix A is the PSF vector of length k followed by $n - 1$ zeros. Each subsequent column is obtained from the previous one by cyclically permuting the entries one step forward. In the case of flat blur, the time–space projection of an input signal of length n creates a response with a trapezoidal intensity profile. The ramps have a span of k pixels each and the plateau is $n - k - 1$ pixels long. For coded blur, the overall intensity profile shape is still trapezoidal, but the shutter's rapid flutter changes the ramps to a more jagged shape as shown in Figure 7.1. Figure 7.2 shows a comparison of traditional imaging with coded exposure imaging for a moving test pattern.

7.4 Code selection

Our goal is to select a temporal code that improves the invertibility of the deconvolution process. We analyze the invertibility by studying the condition number of the coding matrix and the variance of the frequency spectrum of the code. The invertibility of the smearing matrix A, in the presence of uncertainty and noise, can be judged by standard matrix conditioning analysis. The condition number is the ratio of the largest to the smallest singular value of A and indicates the sensitivity of the solution to the noise in the input image. We note that the eigenvalues of a circulant matrix correspond to the

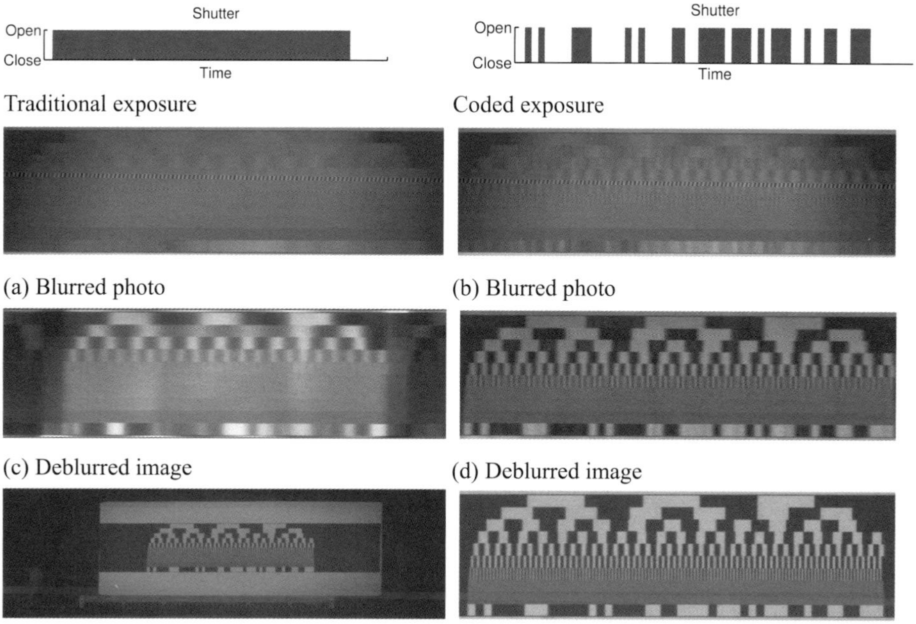

Figure 7.2 The key idea behind coded exposure is to temporally sample the motion with minimum loss of spatial frequencies. (a,b) Photos of a moving target pattern with varying spatial frequencies show that motion blur in conventional exposure loses high spatial frequencies, but coded exposure preserves those frequencies with attenuated magnitudes; (c,d) the deblurred results show recovered spatial frequencies; (e,f) images of the static scene. Figure courtesy of Raskar *et al.* (2006). Reproduced with permission from ACM.

magnitude of the discrete Fourier transform (DFT) of the first column of the circulant matrix and that each column in A is the PSF vector padded with zeros. Based on this observation, we choose a coded sequence with a broadband frequency response so that the corresponding condition number for the smearing matrix is as small as possible.

In theory, we could modulate the opacity of the filter continuously over time to achieve a broadband frequency response, e.g. using a chirp-like function. However, in practice, binary (on/off) opacity switching with fixed chop duration is much easier to implement. Choices for broadband binary codes include Walsh–Hadamard codes, maximum length sequences, and MURA codes. The MURA sequence may seem the obvious choice as its discrete Fourier transform is flat. However, for motion blurring, circular convolution occurs with the PSF vector of length k padded with $n - 1$ zeros, where n is the size of the object in pixels. The DFT of a MURA pattern without zero padding is flat. However, the DFT can resolve only the discrete frequencies with exactness. There is spectral leakage for components falling between the DFT bins. Zero padding results in greater resolution in the frequency components and shows the weakness of MURA patterns. Thus, a MURA is not optimal for zero-padded patterns, and prompted our search for the best possible code.

Because the decoding involves inversion of the frequency spectrum, we also add a smoothness constraint to our search for the best binary chop pattern. The frequency

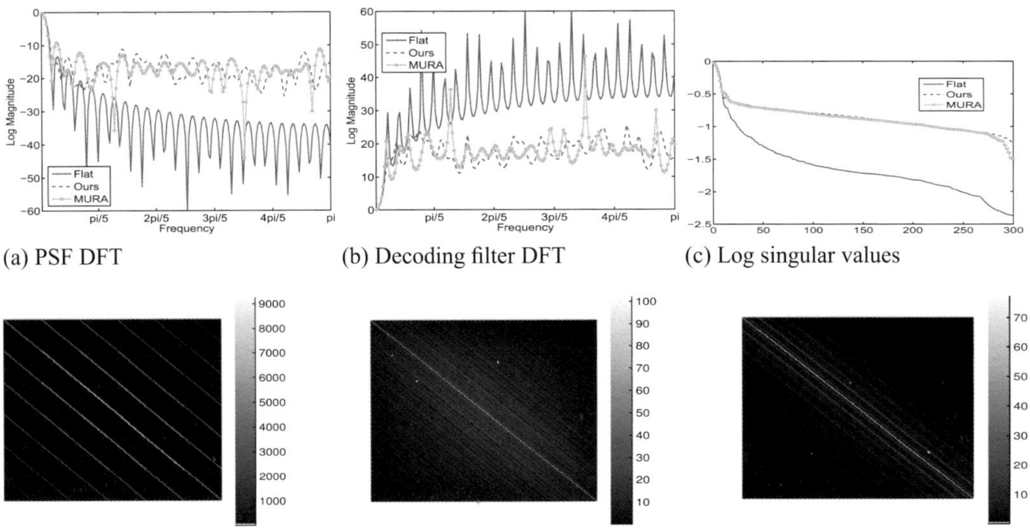

(a) PSF DFT (b) Decoding filter DFT (c) Log singular values

(d) Covariance: box (e) Covariance: MURA (f) Covariance: ours

Figure 7.3 Frequency, covariance and matrix conditionality analysis of codes. Each 52-length code is padded with 300 zeros. (a) DFT of flat blur is a sinc function and frequencies with sharp dips are lost. MURA code is better than box filter but the variance of the frequency magnitudes is high compared to our code. Our coded filter is more efficient with a nearly constant broadband response; (b) DFT of decoding filters show that box and MURA filters require significant amplification of certain high frequencies; (c) the condition number can be analyzed using singular values of the smearing matrix; (d,e,f) noise covariance matrix of A has large diagonal and sub-diagonal entries for box and MURA code, indicating ringing and banding artifacts in the deblurred results. Figure courtesy of Raskar *et al.* (2006). Reproduced with permission from ACM.

response should have low variance so that a mis-estimation of the PSF will not cause strong changes in amplification for nearby but incorrect spatial frequencies during decoding. Note that the frequency response of both the box filter sequence and the padded MURA sequence (Figure 7.3) includes deep dips or zeros, producing a high variance for both. These spikes in the frequency domain lead to the spurious amplification of frequencies during deconvolution.

Finally, one must decide the sequence length m. An ideal chop count is the one equal to the blur size k. Ideally, the camera would feature an auto-flutter mode to decide m on the fly – based on sensed optical flow, a form of motion-adaptation similar to the autofocus feature. Given our hardware constraints, we settled for a compromise value by experimentation, choosing a sequence of $m = 52$ chops with a 50% duty cycle, i.e. with 26 ones and zeros. The first and last bit of the code should be 1, which results in ${}^{50}C_{24} \approx 1.2 \times 10^{14}$ choices. Among them, there is a multitude of potential candidates with acceptable frequency magnitude profiles but different phases. We computed a near-optimal code by implementing a randomized linear search and considered approximately 3×10^6 candidate codes. We chose a code that (i) maximizes the minimum of the magnitude of the DFT values and (ii) minimizes the variance of the

DFT values. The near-optimal 52 length code we found was 1010000111000001010000 110011110111010111001001100111. The plots in Figure 7.3 demonstrate that the chosen code is a significant improvement over padded MURA code.

7.5 Linear solution for deblurring

Since the coded exposure leads to a well-conditioned linear system, one can use least-squares estimation to solve for the deblurred image $\hat{\mathbf{s}}$ (Eq. (7.6)):

$$\hat{\mathbf{s}} = A^{\dagger}\mathbf{i}, \tag{7.8}$$

where $A^{\dagger}$ is the pseudo-inverse of A. The input image can have a motion blur k different from the code size m. To handle it, we first expand/shrink the given blurred image by a factor of m/k. We then estimate the deblurred image and scale it back by k/m.

Figure 7.4 shows an example comparing our code with a traditional exposure camera. Figure 7.4(h) shows that while post-processing algorithms such as Richardson–Lucy can reduce noise amplification, they also reduce image sharpness. In contrast, the coded exposure solution can give a sharp reconstruction even without using any image priors.

7.5.1 Background estimation

We now address the problem of motion blur due to an opaque object moving in front of a stationary (non-blurred) but nonzero-valued background. This is a commonplace but difficult case because the moving object blends with the background. Thus, one also needs to estimate the background simultaneously. In the case of a nonzero background, the blurred image is given by

$$\mathbf{i} = A\mathbf{s} + A_b\mathbf{s}_b + \mathbf{n}, \tag{7.9}$$

where $\mathbf{s}$ is the moving foreground object, $\mathbf{s}_b$ is the static background, and A_b is the background attenuation matrix. A_b is a diagonal matrix whose elements attenuate the static background. A_b can be written as

$$A_b = I - \text{diag}(A \times I_{(n+k-1)\times 1}), \tag{7.10}$$

where $I_{(n+k-1)\times 1}$ is a vector of length $(n + k - 1)$ with all 1's, and diag($\mathbf{v}$) returns a square matrix by placing the vector $\mathbf{v}$ on the main diagonal. The solution for the sharp image and background $\begin{bmatrix} \mathbf{s} \\ \mathbf{s}_b \end{bmatrix}$ is obtained simultaneously as

$$\begin{bmatrix} \mathbf{s} \\ \mathbf{s}_b \end{bmatrix} = K^{\dagger}\mathbf{i}, \tag{7.11}$$

where $K = \begin{bmatrix} A & A_b \end{bmatrix}$ is the combined matrix.

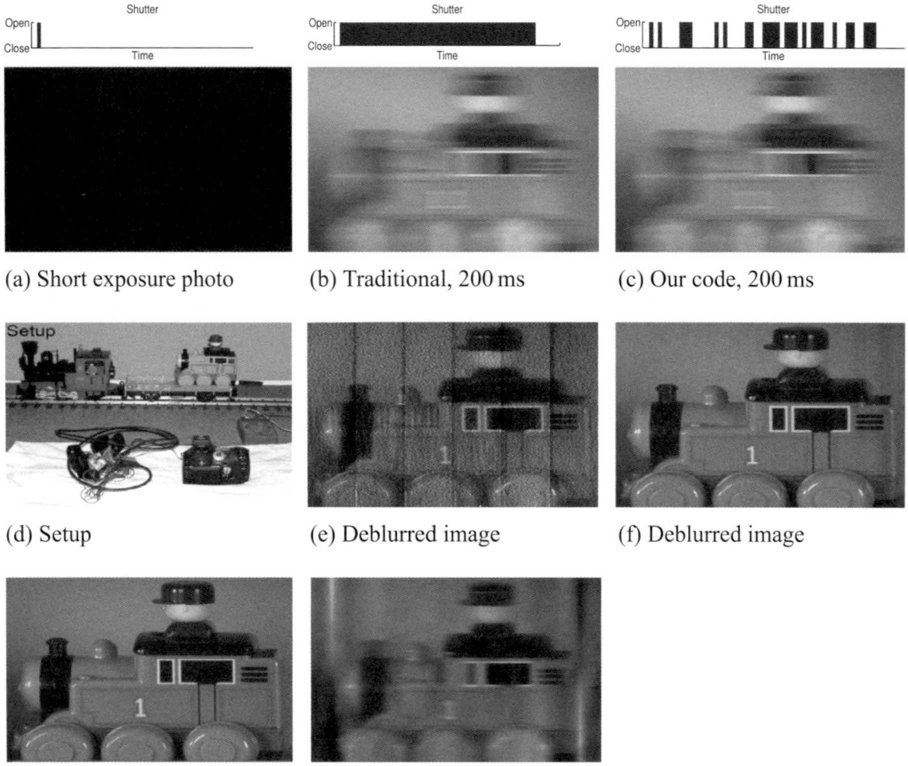

(a) Short exposure photo (b) Traditional, 200 ms (c) Our code, 200 ms

(d) Setup (e) Deblurred image (f) Deblurred image

(g) Static toy train (h) RL deblur result of (b)

Figure 7.4 Comparison between short and traditional exposures. The blur k is between 118 and 121 pixels. (a,b,c) Shutter sequence and corresponding photos; (d) experimental setup with toy train; (e,f) deblurred results using a linear solution; (g) photo of static toy train; (h) flat blurred image deblurred using Richardson–Lucy (RL) algorithm. Figure courtesy of Raskar *et al.* (2006). Reproduced with permission from ACM.

7.5.2 Motion generalization

Applying image warping can permit our technique to decode a much broader set of simple motions that project as affine transforms, in-plane rotations around a fixed center, as well as to allow movement in perspective along lines that meet at a vanishing point. While the PSF of the coded blur is nonlinear and position-dependent, several linear motions can be warped to produce an image with spatially invariant uniform-length displacement vectors aligned with image scan lines (e.g. Figures 7.5 and 7.6). As motion blur follows this same displacement vector field, the warped image provides uniform-width coded-blur regions that are now suitable for decoding. To produce the final result, we simply apply the inverse warp to return the decoded image to its original geometric form.

In the case of perspective warp, rectification can be applied after estimating the vanishing point of motion lines. Following rectification, all the warped motion lines are parallel in the camera image space. We used this technique for Figures 7.5 and 7.6.

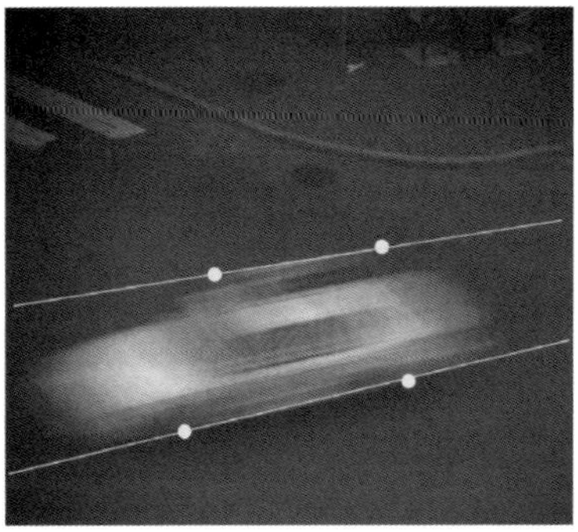

(a) Blurred image

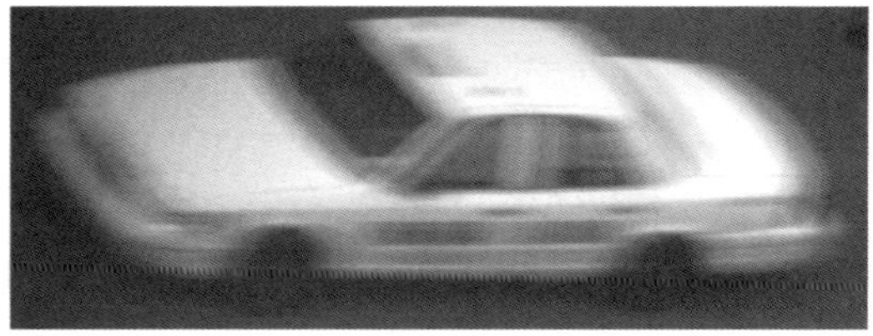

(b) Rectified crop

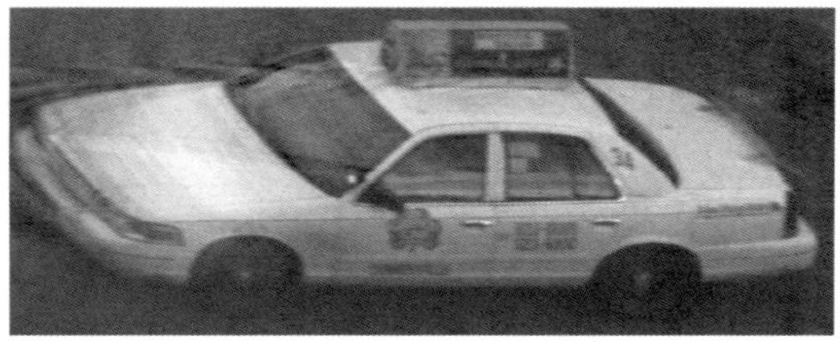

(c) Deblurred image

Figure 7.5 Simple motion generalization. (a) Photo of a fast moving vehicle with a blur of $k = 235$ pixels, which is 24% of the vehicle length in the image. We click on four points shown in gray to specify motion lines; (b) an approximate crop of the rectified motion lines; (c) despite vignetting on the left, high spatial frequencies are recovered in the deblurred result. Figure courtesy of Raskar *et al.* (2006). Reproduced with permission from ACM.

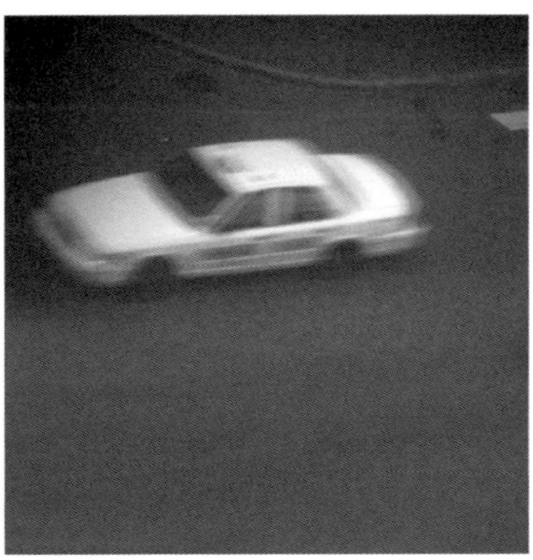

(a) Blurred image

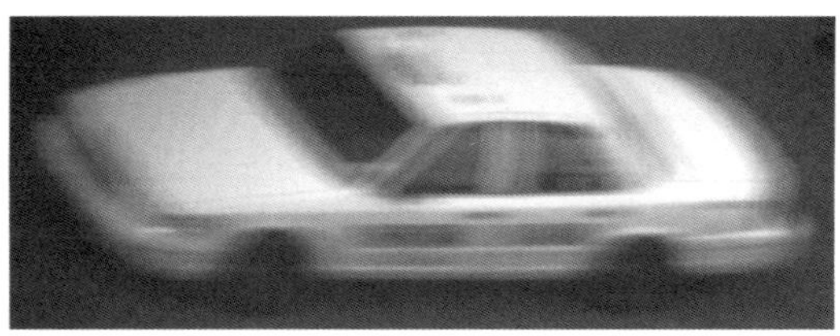

(b) Rectified crop

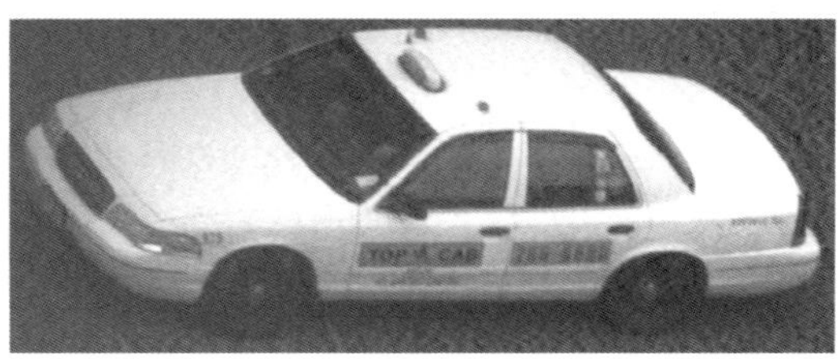

(c) Deblurred image

Figure 7.6 Coded exposure enables recovery of fine details in the deblurred image. (a) Photo of a fast moving vehicle; (b) user clicks on four points to rectify the motion lines and specifies a rough crop; (c) deblurred result. Note that all sharp features on the vehicle (such as text) have been recovered. Figure courtesy of Raskar *et al.* (2006). Reproduced with permission from ACM.

In-plane rotations (e.g. a rotating fan) create motion lines that form concentric circles around the center of rotation. These can be handled by deblurring in polar coordinates.

7.6 Resolution enhancement

Coded exposure motion deblurring can also be used to enhance the resolution (super-resolution) of a moving object (Agrawal & Raskar 2007). Both motion deblurring and image super-resolution are ill-posed problems. Using a coded-exposure camera, the combined problems of deblurring and resolution enhancement can be solved. As the blur size increases, the resolution of the moving object can be enhanced by a larger factor, albeit with a corresponding increase in the reconstruction noise.

Let r denote the resolution enhancement factor (REF) and k denote the size of blur (in pixels) in the low resolution sensor output. Let n and $n_r = n \times r$ denote the size of the object in the low resolution and desired high resolution grid, respectively. The observed image $i(x)$ will be smeared in the blur direction and will have $t = n + k - 1$ samples due to linear convolution. Combining deblurring and super-resolution, we can relate the desired high resolution image $s(x)$ to the low resolution blurred image $i(x)$ as

$$\mathbf{i} = D_r \times A_r \times \mathbf{s} + \mathbf{n}, \tag{7.12}$$

where A_r is the smearing matrix representing motion blur in the high resolution grid and D_r is the decimation matrix which transforms a high resolution image into a low resolution image. Each row of D_r has only r ones, with each column having at most a single one. Each column of A_r corresponds to a circularly shifted PSF vector of length $r \times k$ zero-padded to the length n_r, representing linear convolution.

A simple counting argument shows that the number of observations t in the low resolution motion-blurred image should be greater than the number of unknowns n_r for resolution enhancement. Thus,

$$t \geqslant n_r, \tag{7.13}$$

$$\Rightarrow (n + k - 1) \geqslant n \times r, \tag{7.14}$$

$$\Rightarrow k \geqslant n \times (r - 1) + 1. \tag{7.15}$$

Consequently, motion blur k has to be larger than the size of the object in the direction of motion by a factor of $r - 1$, where r is the factor of resolution enhancement. For example, if $r = 4$, the object should move more than three times its size in the direction of motion. Thus, as the size of the blur k increases, the REF can be increased. However, in practice, this results in more deconvolution noise and may not be desirable.

Figure 7.7 shows an example of resolution enhancement deblurring on a standard resolution chart. Camera pixel resolution is insufficient to capture individual alternating lines in a static image of a standard resolution chart (top left), since they span only $n = 18$ pixels. Upsampling by four times fails to reveal the alternating lines. Blurring the object horizontally results in a smear of $t = 74$ pixels with motion blur $k = 57$ pixels. The bottom row shows that resolution enhancement deblurring can reveal more details in comparison with standard deblurring followed by bicubic upsampling.

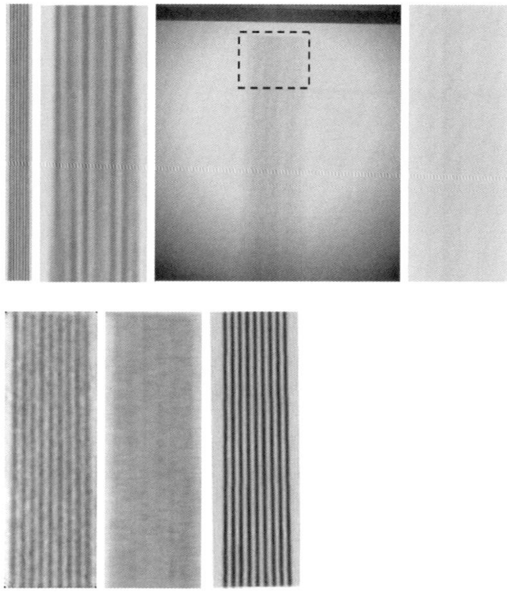

Figure 7.7 Top row, left to right: static image of the resolution chart, 4× upsampled image, captured blurred image, and cropped blurred image. Bottom row, left to right: resolution enhancement deblurring, standard deblurring followed by 4× upsampling, ground truth image. Figure courtesy of Agrawal & Raskar (2007). Reproduced with permission from IEEE.

7.7 Optimized codes for PSF estimation

So far we have discussed how to optimize the code for making the deconvolution or motion PSF invertible. However, the problem of PSF estimation still remains even if the PSF is made invertible. One needs to estimate the motion of moving objects in order to deblur them. Now we will describe codes that also help in automatic PSF estimation (Agrawal & Xu 2009).

We use the motion from blur (MFB) approach presented in Dai & Wu (2008) for PSF estimation. This approach analyzes motion blur as alpha matting and requires locally smooth foreground/background. Although this approach was demonstrated for estimating the PSF using a traditional camera, we show that the motion from blur constraint also holds for coded exposures, for those parts of the blur that correspond to *continuous* ones in the code (Figure 7.8). In the case of analyzing motion blur as alpha matting, the "foreground" corresponds to the blurred object. Since alpha matting requires locally smooth foreground/background, the optimal code for invertibility does not work well for PSF estimation because it leads to non-smooth blur. Furthermore, since the MFB algorithm relies on locally smooth alpha values to compute alpha gradients, PSF estimation becomes even more difficult. The key idea is to find an *invertible* code which also results in smooth blur for some *parts* of the blur profile to help in PSF estimation.

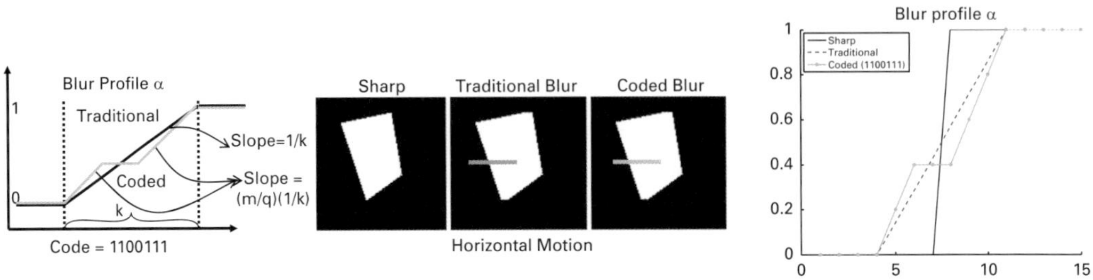

Figure 7.8 Left: motion from blur constraint holds for a coded exposure for those parts of the blur which correspond to the ones in the code. In traditional cameras, the slope of α equals $1/k$ for a blur size k. For a coded exposure, the slope increases by a factor of m/q. Middle: synthetic example showing a polygon-shaped object moving horizontally. Right: corresponding blur profiles obtained from blurred synthetic images. Figure courtesy of Agrawal & Xu (2009). Reproduced with permission from IEEE.

7.7.1 Blur estimation using alpha matting

Let $M(x)$ be a binary indicator function for the object. When the object moves in front of the background $s_b(x)$, the captured blurred image $i(x)$ is given by the sum of the blurred foreground object and partial background as

$$i(x) = s(x) * h(x) + (1 - M(x) * h(x))s_b(x). \tag{7.16}$$

This is similar to Eq. (7.9). Comparing with the familiar matting equation $I = \alpha F + (1 - \alpha)B$ (Smith & Blinn 1996), we get

$$B \sim s_b(x), \quad \alpha \sim M(x) * h(x), \quad F \sim \frac{s(x) * h(x)}{M(x) * h(x)}. \tag{7.17}$$

Note that the "foreground" for the matting algorithm is not the actual object $s(x)$, but the blurred object which depends on the PSF $h(x)$. Although matting algorithms can handle complex α (such as hair, smoke, etc.) and thus discontinuous $i(x)$, they require both the foreground and background to be locally smooth. For a traditional camera, since the PSF is a box function, it results in a smooth foreground. Motion blur estimation algorithms based on alpha matting have shown very good results on images captured using a traditional camera. However, deblurring is ill-posed due to h being low pass. Now we describe how to choose the code that results in a smooth blur as well as an invertible PSF.

7.7.2 Motion from blur

We first show that motion from blur constraint also holds for coded exposure cameras. The constraint is given by Dai & Wu (2008)

$$\nabla\alpha \cdot \mathbf{k} = \pm 1, \tag{7.18}$$

where $\mathbf{k} = [k_x, k_y]$ denotes the blur vector[1]. This constraint assumes $h(x)$ to be a box filter (traditional camera). For coded exposure, $h(x)$ is a sum of shifted box functions of varying sizes. Thus, this constraint still holds for each set of *continuous* ones in the code $c(x)$. Let $q = \sum c(x)$ be the total number of ones in the code. If the object moves with constant speed, the motion from blur constraint changes to

$$\nabla\alpha \cdot \mathbf{k} = \pm\frac{m}{q}, \quad \text{if} \quad c(x) = 1, \tag{7.19}$$

since the PSF $h(x)$ is normalized to 1. When the code is zero, no light is integrated, and hence α remains constant ($\nabla\alpha = 0$) within that time period. Only for those parts of the blur for which the code equals one does the constraint hold, as shown in Figure 7.8.

7.7.3 Code selection

Codes having the same deblurring performance could differ significantly in their resulting blur profiles. Consider two $m = 31$ codes:

$$C_1 = 1010101011100111101110101111011,$$
$$C_2 = 1111111111111000010011101000111.$$

Both codes have the same number of ones ($q = 21$), and would thus allow the same amount of light. Figure 7.9 shows the magnitude of the DFT for both codes after zero padding. The minimum DFT magnitude is the same for both codes. In fact, the increases in deconvolution noise for C_1 and C_2 are 19.7 and 20.09 dB, respectively (compared to 35.7 dB for traditional cameras). These two codes will therefore result in similar deblurring performances. However, they result in significantly different blur profiles. The number of transitions for C_1 equals 18, compared to 8 for C_2, and C_2 has a 13-long continuous string of ones. As shown in Figure 7.9, the blur profile corresponding to C_2 will be smooth at one end, with a minimum number of discontinuities compared with the blur profile corresponding to C_1. Thus, for the same deblurring performance, one could possibly choose a code which results in smooth blur for some *parts* of the entire motion blur. Since most alpha matting algorithms require local smoothness within a neighborhood (e.g. 3×3), minimizing the number of transitions in the code will reduce discontinuities in the foreground and result in better alpha map estimation. Moreover, the smoothly changing alpha values within the same region also allow better computation of gradients, facilitating PSF estimation.

Thus, the criteria for choosing the optimal code for PSF estimation and invertibility are (a) minimize deconvolution noise, (b) minimize transitions in code, and (c) maximize the run length of continuous ones. These criteria compete with each other. Note that the first and the last bit of the code has to be a 1, otherwise it will reduce to a code of smaller length. Thus, in general the search space is of order 2^{m-2} for code length m. For a small m, the search space is small and all possible codes can be tested. For a larger m, the search space is large and a sampling strategy is used by randomly sampling one million codes from the search space.

[1] Assuming constant velocity object motion in image plane.

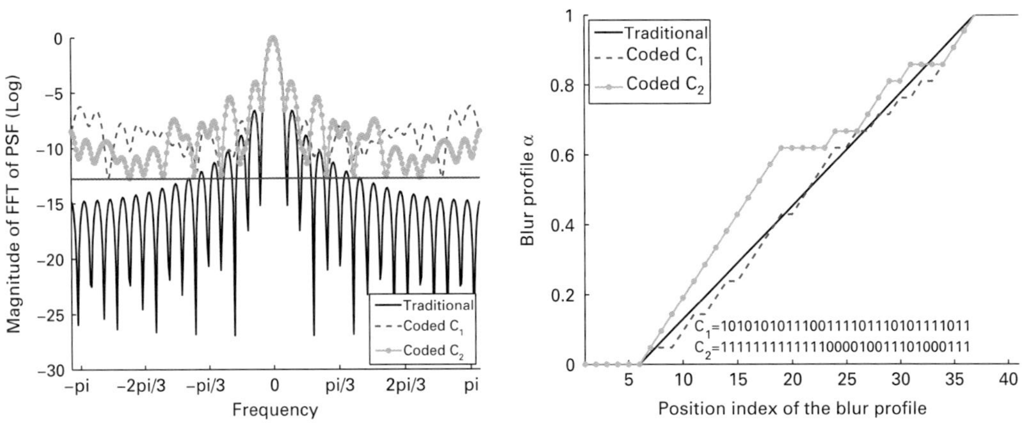

Figure 7.9 Two different codes C_1 and C_2 with the same deblurring performance but different blur profiles. Left: the DFT magnitude shows that although the minimum for C_1 and C_2 is the same (horizontal line), C_1 attenuates low frequencies much more than C_2. Right: C_2 has a small number of transitions and a long consecutive string of 1's. This results in a smooth blur profile for C_2 on one side which helps with PSF estimation. Figure courtesy of Agrawal & Xu (2009). Reproduced with permission from IEEE.

A fast procedure for code search is as follows. Fix the number of continuous ones (p) in the code. For simplicity, set the first p bits to one. The search space is reduced to 2^{m-p-2}. Among these codes, choose all the codes for which the deconvolution noise is smaller than some threshold (e.g. 20 dB) and pick the one which has the minimum number of transitions. If none of the codes satisfies the noise criteria, p is decreased by one and the search is repeated. The code C_2 as described above is found using this approach for $m = 31$ and $p = 13$ by testing only $2^{31-13-2} = 65\,536$ codes in 6.7 s on a standard PC. Thus, we see that optimized codes can be found quickly.

7.7.4 Results

Now we show real results on PSF estimation and deblurring using the codes C_1 and C_2 as described above. Notice that C_2 is optimized for both PSF estimation and invertibility, while C_1 is optimized only for invertibility. To compare the results with a traditional camera, its exposure time is reduced by m/q to ensure the *same* light level. The traditional camera image will therefore have reduced blur by the same factor.

The PSF estimation algorithm follows (Dai & Wu 2008), where alpha matting is performed first (using Levin, Lischinski & Weiss (2008)) to obtain the alpha values. However, we further improve the MFB algorithm to handle the aperture problem by using a weighted least squares (WLS) estimator. Since α-gradient values α_x and α_y together give information about the edge directions, we cluster them into eight quadrants. Specifically, we divide the samples into eight clusters depending on whether the gradients α_x, α_y are $>\tau$, $<-\tau$, or $\in[-\tau, \tau]$, where τ is a threshold (e.g. 0.02). We ignore the clusters where both α_x, α_y are $\in[-\tau, \tau]$, since those pixels do not give any useful information in

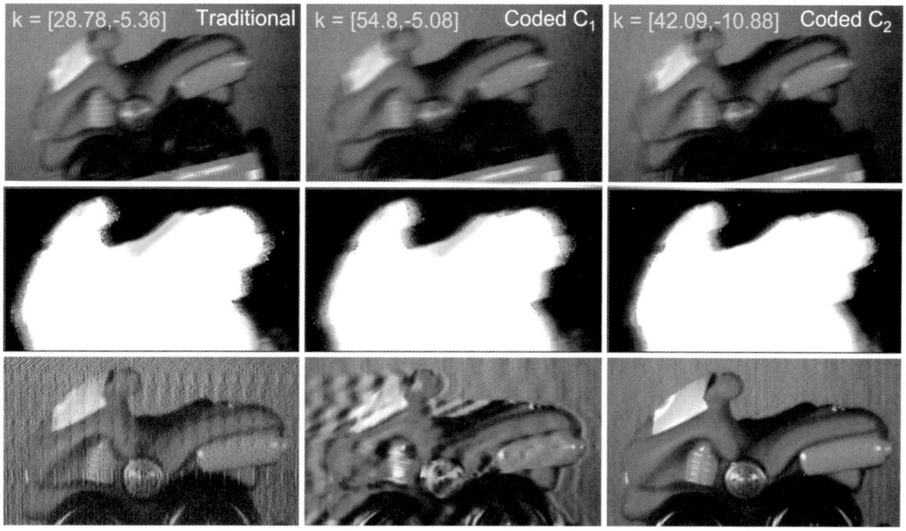

Figure 7.10 Top: blurred photos. Middle: alpha maps with inliers. Bottom: deblurred results. PSF estimation for traditional cameras is good but deblurring is poor due to non-invertible PSFs. Bad PSF estimation for code C_1 leads to poor deblurring. For C_2, the estimated PSF is good, as proved by the deblurring result. Input images are rotated using the estimated motion angle before deblurring to make the motion horizontal. For C_1, an incorrectly estimated angle cannot be used to rectify the input image. Figure courtesy of Agrawal & Xu (2009). Reproduced with permission from IEEE.

the presence of noise. Then a WLS estimation is performed, where the weights are the inverse of the cluster sizes. This ensures that edges having different directions get equal weights in blur estimation, so that the estimation is not biased towards a particular edge direction.

Figure 7.10 shows real results on a toy motorcycle, where the motion is non-horizontal in the image plane. The captured blurred photos and deblurred results are shown in the top and bottom rows, respectively, for traditional and coded exposure cameras using C_1 and C_2 codes. Note that the estimated PSF using C_2 is close to ground truth, as shown by the good deblurring result. PSF estimation for traditional cameras is also good but deblurring is bad due to the PSF being non-invertible. Figure 7.10 (middle row) also shows inliers (a different color for each cluster) obtained from the MFB algorithm. For traditional cameras, inliers span *all* parts of the blur as expected while, for coded blur, the α-motion blur constraint only holds for those parts of the blur that correspond to 1's in the code. Note that for C_2, most of the inliers are present in one end of the blur, corresponding to the long string of 1's in C_2. However, for C_1, inliers are scattered all over the blur which shows that the alpha estimation and MFB algorithm were not successful. Figure 7.11 also shows the ground truth photo and the deblurring result for C_1 if the PSF estimated using C_2 is used. This clearly demonstrates that the deblurring performances for C_1 and C_2 are similar. However, C_2 assists in PSF estimation, while C_1 does not.

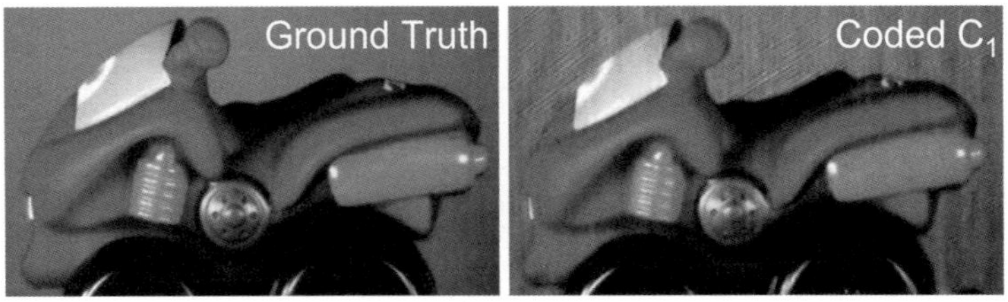

Figure 7.11 Left: ground truth sharp image. Right: deblurring result for C_1, using the motion PSF estimated from C_2, shows that the deblurring performances are similar for C_1 and C_2, but PSF estimation fails using C_1 (see Figure 7.10, middle image in the bottom row). Figure courtesy of Agrawal & Xu (2009). Reproduced with permission from IEEE.

7.8 Implementation

The coded exposure camera can be built using a standard off-the-shelf SLR camera as shown in Figure 7.12. We used an 8 MP Canon Pro 1 digital still camera and interfaced it to a PIC microcontroller connected to the camera's hot shoe. The PIC controls a *Displaytech* ferroelectric shutter placed over the camera lens, and drives the shutter to follow the binary coded sequence, when triggered by the hotshoe signals. The ferroelectric shutter's on/off contrast is high (1000 : 1) and switching time is less than 100 µs. However, the ferroelectric shutter costs $\approx$\$500 and requires an external microcontroller for control. In addition, the external shutter leads to vignetting in images, it has limited contrast (opaque versus transparent), and loses light even when it is transparent due to its polarization layers.

Coded exposure can also be implemented on consumer grade machine vision cameras by *on-chip* fluttered integration with zero additional cost, and so avoid the above issues. In fact, it can be achieved with any camera that supports IEEE DCAM Trigger mode 5. This trigger mode supports multiple pulse-width triggers with a *single* readout. We used the Dragonfly2 camera from Point Grey (Point Grey 2006) to obtain the results shown in Figure 7.10. The camera is triggered using the parallel port of a PC. Each bit of the code corresponds to 1 ms of the exposure time in the implementation. To implement a particular code, the camera is triggered at $0 \rightarrow 1$ transition and held until the next $1 \rightarrow 0$ transition. For example, for the code 11101000011, three triggers will be sent at 0, 4, and 9 ms and held for a duration of 3, 1, and 2 ms, respectively. Note that the number of triggers is not equal to the number of ones in the code; for each continuous set of ones, one trigger is sent.

7.9 Analysis

7.9.1 Noise analysis

We now compare coded exposure deblurring with traditional exposure deblurring in terms of noise amplification using linear system analysis. Noise amplification in deblurring depends on the condition number of the smear matrix A. Assuming zero mean IID

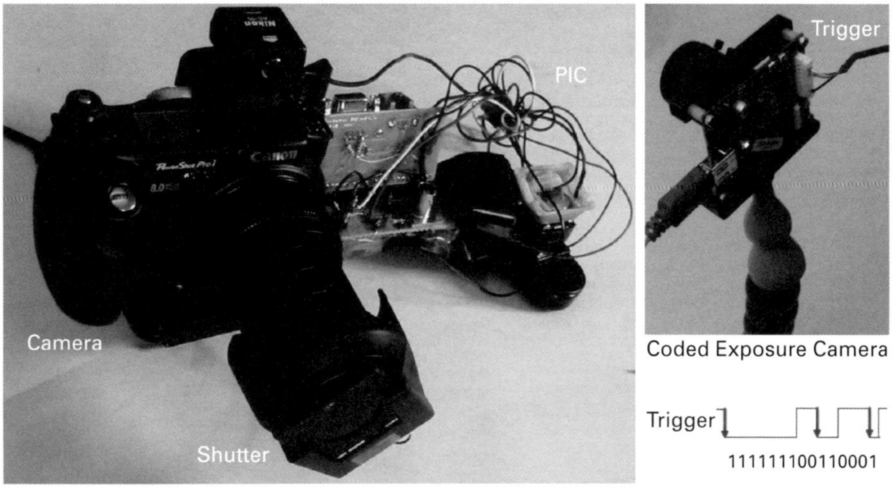

Figure 7.12 Left: prototype coded-exposure camera with a ferroelectric liquid crystal shutter. Right: coded exposure implementation on an off-the-shelf machine vision camera using an external hardware trigger. Figures courtesy of Raskar *et al.* (2006) & Agrawal & Xu (2009), respectively. Reproduced with permission from IEEE/ACM.

Gaussian noise with variance σ^2, the covariance matrix of the estimation error, $\hat{\mathbf{s}} - \mathbf{s}$, is equal to

$$\Sigma = (A^T A)^{-1} A^T \sigma^2 A (A^T A)^{-T} = \sigma^2 (A^T A)^{-1}. \tag{7.20}$$

Figure 7.3 shows the absolute of the covariances matrices (assuming $\sigma^2 = 1$). For traditional exposure (flat blur), Σ has large off-diagonal entries, indicating that noise in any one pixel severely affects several other pixels in the deblurred image. For the 52 chop sequence that we have used, the effect of the off-diagonal entries in the corresponding Σ matrix is significantly reduced.

The root mean square error (RMSE) of the estimation error is given by

$$\text{RMSE} = \sigma \sqrt{\frac{\text{trace}(A^T A)^{-1}}{n}}. \tag{7.21}$$

Thus, the noise amplification factor f is given by

$$f = \sqrt{\frac{\text{trace}(A^T A)^{-1}}{n}}. \tag{7.22}$$

Note that the noise amplification factor depends only on the smear matrix A, and hence the PSF. For a traditional camera, assuming an object length of 300 pixels and a blur of 52 pixels, the noise amplification factor $f = 38.45$ dB. In contrast, using the 52 length sequence for the coded exposure camera, $f = 18.70$ dB, giving a noise improvement of 19.75 dB. However, since the coded exposure camera is on approximately half the time, it loses half the light compared to a traditional camera. To compare the signal-to-noise ratio (SNR), the light loss should be taken into account as described in Agrawal & Raskar (2009).

Using coded exposure, one can deblur the image to the extent of the motion within a single chop. Let us compare this to an image captured with an exposure of a single chop, which is equal to T/m seconds. As the cumulative exposure time for coded exposure is roughly $T/2$, SNR is potentially better by a factor of $m/2$ in the blurred region. A key advantage with respect to a short exposure photo is that in the areas without motion blur (which do not need to be deblurred), coded exposure cameras can record a sharp image with reduced noise.

7.9.2 Resolution analysis

Now we analyze the choice of the code length for coded exposure. A long code (large m) subdivides the exposure time finely and allows decoding of a large amount of blur, but it proves ineffective for small amounts of blur. Conversely, a short code ($k > m$) has a longer duration per chop and blur within a single chop cannot be resolved. We would like to keep the ratio k/m close to 1 pixel per chop to achieve best possible sampling of the blur. Figure 7.13 shows the captured images and the corresponding deblurred results for the test pattern with varying spatial frequencies for blur sizes 44, 78, and 193. Note that when the blur is 193 pixels, high frequencies (fourth row from bottom of test pattern) are not recovered. Thus, as the blur size k differs from chop count m, the decoding fails to resolve very fine frequencies. However, in practice, we do not need to know the blur size to determine the code length. Using the 52 chop code, we were able to handle motion blur ranging from $k = 27$ to $k = 300$ pixels.

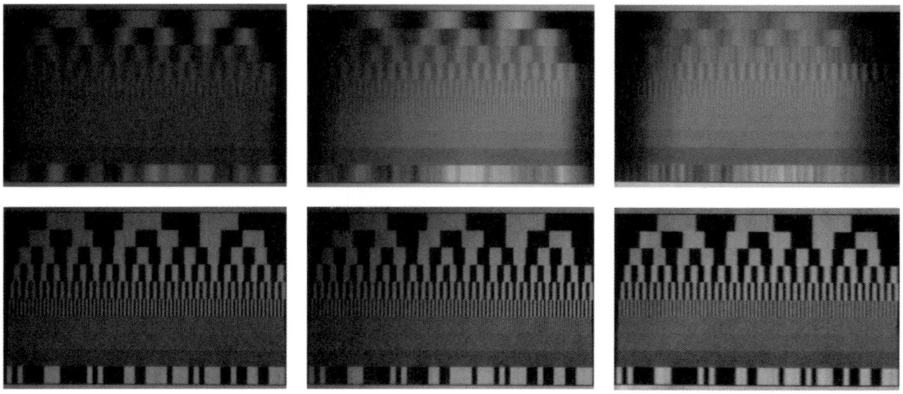

Figure 7.13 For a given code size m, the resolution of the deblurred image is related to blur size k. Top row: blurred images corresponding to $k = 44, 78$, and 193 respectively using a code length of $m = 52$. Bottom row: corresponding deblurred images. Note that at $k = 193$, high frequencies (fourth row from bottom of chart) are not recovered, but they are recovered when $k = 44$. Figure courtesy of Raskar *et al.* (2006). Reproduced with permission from ACM.

7.10 Summary

In this chapter, we explained the concept of coded exposure photography for motion deblurring. Motion blur leads to the loss of high spatial frequencies in the captured photo. While traditional image processing algorithms attempt to recover the lost information utilizing image priors, coded exposure modifies the imaging process itself to avoid the loss of high frequency information in the first place. This allows even linear algorithms to recover a sharp image without any image priors. However, we believe that combining image-based priors with smart capture time decisions will further improve the quality of the deconvolved image.

We showed how to design codes that make the PSF invertible. However, such codes result in non-smooth blur and could make PSF estimation worse. We then described how to design codes that also assist in PSF estimation while being invertible. We also showed that the resolution of a motion-blurred object can be increased while simultaneously deblurring, in certain cases.

Current Point Grey cameras allow on-chip fluttering which is the same for all pixels. In the future, red, green, and blue pixels might each use a different binary code that exploits the Bayer grid (color sensor interleaving) for finer spatio-temporal resolution. We have explored constant duration codes, but variable width codes may prove beneficial in videos to adapt to intra-frame motion. The coded exposure photography may also inspire a manual or automatic motion knob on cameras. The camera can have an auto-flutter facility, similar to autofocus systems, wherein the camera electronics can determine the best code sequence length and duration on the fly. An ultrasound sensor or an auxiliary low resolution camera (Ben-Ezra & Nayar 2004) can trigger fluttering by detecting and measuring object motion. As the coded sequencing preserves the look and feel of the conventional flat motion-blurred image, we feel that the camera manufacturers could add fluttered integration without annoying uninterested consumers.

The trend in computational photography is to capture coded information smartly, which allows greater post-capture image processing. We believe that coded exposure offers a promising avenue for computational methods to improve motion photography.

References

Agrawal, A. & Raskar, R. (2007). Resolving objects at higher resolution from a single motion-blurred image. In *IEEE Conference on Computer Vision and Pattern Recognition*, pp. 1–8.

Agrawal, A. & Raskar, R. (2009). Optimal single image capture for motion deblurring. In *IEEE Conference on Computer Vision and Pattern Recognition*, pp. 2560–7.

Agrawal, A. & Xu, Y. (2009). Coded exposure deblurring: optimized codes for PSF estimation and invertibility. In *IEEE Conference on Computer Vision and Pattern Recognition*, pp. 2066–73.

Ben-Ezra, M. & Nayar, S. (2004). Motion-based Motion Deblurring. *IEEE Transactions on Pattern Analysis and Machine Intelligence*, **26**(6), 689–98.

Canon (2006). What is optical shift image stabilizer? http://www.canon.com/bctv/faq/optis.html.

Dai, S. & Wu, Y. (2008). Motion from blur. In *IEEE Conference on Computer Vision and Pattern Recognition*, pp. 1–8.

Edgerton, H. (1951–63). Rapatronic Photographs. http://simplethinking.com/home/rapatronic_photographs.htm.

Gottesman, S. R. & Fenimore, E. E. (1989). New family of binary arrays for coded aperture imaging. *Applied Optics*, **28**(20), 4344–52.

Jain, A. (1988). *Fundamentals of Digital Image Processing*, 1st edn. Prentice-Hall.

Levin, A., Lischinski, D. & Weiss, Y. (2008). A closed-form solution to natural image matting. *IEEE Transactions on Pattern Analysis and Machine Intelligence*, **30**(2), 228–42.

Liu, X. & Gamal, A. (2001). Simultaneous image formation and motion blur restoration via multiple capture. In *Proceedings of the International Conference on Acoustics, Speech, and Signal Processing*, Vol. 3, pp. 1841–4.

Lucy, L. (1974). An iterative technique for the rectification of observed distributions. *Journal of Astronomy*, **79**, 745–54.

Nikon (2005). Precise camera-shake compensation at every angle. www.nikon.co.jp.

Point Grey (2006). Article 239: External trigger modes supported by Point Grey cameras. http://www.ptgrey.com.

Raskar, R., Agrawal, A. & Tumblin, J. (2006). Coded exposure photography: motion deblurring using fluttered shutter. *ACM Transactions on Graphics*, **25**(3), 795–804.

Richardson, W. (1972). Bayesian-based iterative method of image restoration. *Journal of Optical Society of America*, **62**(1), 55–9.

Smith, A. R. & Blinn, J. F. (1996). Blue screen matting. In *ACM Special Interest Group on Graphics and Interactive Techniques*, pp. 259–68.

Tikhonov, A. N. & Arsenin, V. I. (1977). *Solutions of ill-posed problems [Metody resheniia nekorrektnykh zadach]*. New York: Halsted Press.

8 Richardson–Lucy deblurring for scenes under a projective motion path

Yu-Wing Tai and Michael S. Brown

8.1 Introduction

Motion blur from camera egomotion is an artifact in photography caused by the relative motion between the camera and an imaged scene during exposure. Assuming a static and distant scene, and ignoring the effects of defocus and lens aberration, each point in the blurred image can be described as the convolution of the un-blurred image by a point spread function (PSF) that describes the relative motion trajectory at that point's position. The aim of image deblurring is to reverse this convolution process to recover the clear image of the scene from the captured blurry image as shown in Figure 8.1.

A common assumption in existing motion deblurring algorithms is that the motion PSF is spatially invariant. This implies that all pixels are convolved with the same motion blur kernel. However, as discussed by Levin, Weiss, Durand & Freeman (2009) the global PSF assumption is often invalid. In their experiments, images taken with camera shake exhibited notable amounts of rotation that attributed to spatially-varying motion blur within the image. Figure 8.2 shows a photograph that illustrates this effect. As a result, Levin *et al.* (2009) advocated the need for a better motion blur model as well as image priors to help regularize the solution space when performing deblurring. This chapter addresses the former issue by introducing a new and compact motion blur model that is able to describe spatially-varying motion blur caused by a camera undergoing egomotion.

We refer to this blur model as the *projective motion blur model* as it represents the degraded image as an integration of the clear scene under a sequence of planar projective transforms (i.e. homographies). Figure 8.3 shows a diagram of this representation. One key benefit of this model is that it is better suited to represent camera egomotion than the conventional kernel-based PSF parameterization that would require the image to be segmented into uniform blur regions, or worse, a separate blur kernel per pixel. However, because our approach is not based on convolution with an explicit PSF, it has no apparent frequency-domain equivalent. One of the key contributions of this chapter is to show how our blur model can be used to extend the conventional pixel-domain *Richardson–Lucy* (RL) deblurring algorithm. We refer to this modified RL algorithm as the *projective motion Richardson–Lucy* algorithm. Similar to the conventional RL deblurring, regularization based on various priors can be incorporated in our algorithm.

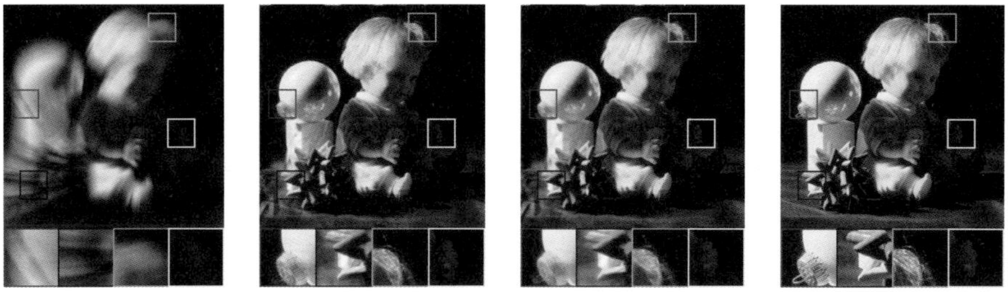

(a) RMS: 47.27 (b) RMS: 10.51 (c) RMS: 10.35 (d) Ground truth

Figure 8.1 (a) An image degraded by spatially-varying motion blur due to camera egomotion; (b) the result from our basic algorithm; (c) our result with added regularization; (d) ground truth image. The RMS errors are also shown below each image. © IEEE 2011.

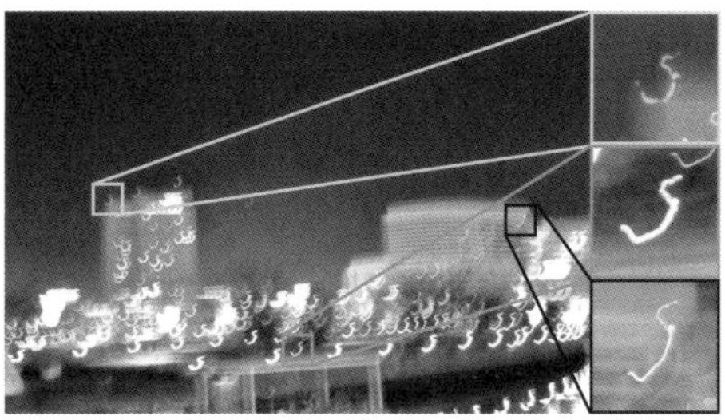

Figure 8.2 This example demonstrates the spatially-varying nature of camera shake. The motion paths of the saturated point light sources (shown zoomed-in) represent the PSF at various locations in the image. It is clear that the PSFs are not uniform in appearance. © IEEE 2011.

We note that the idea of using a sequence of homographies to model camera-shake blur was independently proposed by Whyte, Sivic, Zisserman & Ponce (2010) and Joshi, Kang, Zitnick & Szeliski (2010). Our chapter is focused on developing the projective motion blur model and the associated RL algorithm. As such, we assume that the motion path is given. The works in Whyte *et al.* (2010) and Joshi *et al.* (2010) address more clearly the problem of estimating the unknown motion. As part of our chapter, we also describe methods useful for estimating the projective motion path in Section 8.5. As with other camera-shake deblurring approaches, we assume that the scene is distant and void of moving objects.

The remainder of this chapter is organized as follows: Section 8.2 discusses related work; Section 8.3 details our motion blur model; Section 8.4 derives the projective motion Richardson–Lucy deconvolution algorithm; Section 8.5 discusses potential methods that can be used to estimate the projective motion path; Section 8.6 provides analysis of the convergence properties of our algorithm, its sensitivity to noise, and comparisons with other approaches. A discussion and summary of this work is presented in Section 8.7.

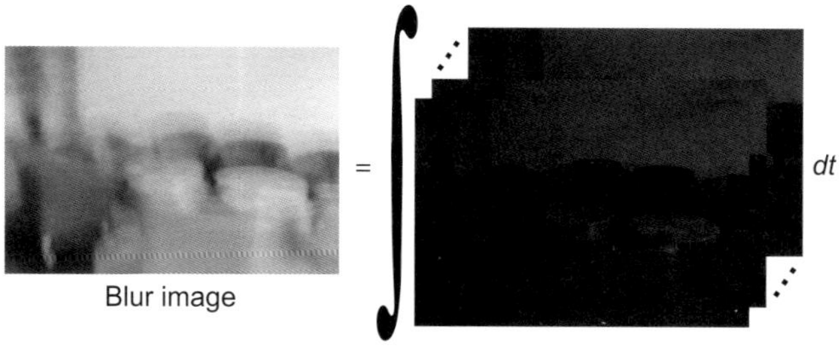

Figure 8.3 Our input image is considered the integration of an image scene under projective motion. © IEEE 2011.

8.2 Related work

Existing work targeting image blur due to camera egomotion has assumed a global PSF for the entire image. When the blur PSF is known, or can be estimated, well-known deblurring algorithms such as Richardson–Lucy (Richardson 1972, Lucy 1974) and Wiener filter (Wiener & Norbert 1949) can be applied to deblur the image. Due to poor kernel estimation, or convolution with PSFs that result in unrecoverable frequencies, these conventional deblurring algorithms can introduce undesirable artifacts in the deblurred result such as "ringing" and amplification of image noise.

Consequently, research addressing image deblurring, including camera shake and other types of blur, typically target either blur kernel estimation or ways to regularize the final result, or both. For example, Dey *et al.* (Dey, Blanc-Fraud, Zimmer, Kam, Roux, Olivo-Marin & Zerubia 2004) and Chan and Wong (Chan & Wong 1998) utilized total variation regularization to help ameliorate ringing and noise artifacts. Fergus *et al.* (Fergus, Singh, Hertzmann, Roweis & Freeman 2006) demonstrated how to use a variational Bayesian approach combined with gradient-domain statistics to estimate a more accurate PSF. Raskar *et al.* (Raskar, Agrawal & Tumblin 2006, Agrawal & Raskar 2007) coded the exposure to make the PSF more suitable for deconvolution. Jia (Jia 2007) demonstrated how to use an object's alpha matte to better compute the PSF. Levin *et al.* (Levin, Fergus, Durand & Freeman 2007) introduced a gradient sparsity prior to regularize results for images exhibiting defocus blur. This prior is also applicable to motion-blurred images. Yuan *et al.* (Yuan, Sun, Quan & Shum 2008) proposed a multi-scale approach to progressively recover blurred details while Shan *et al.* (Shan, Jia & Agarwala 2008) introduced regularization based on high order partial derivatives to reduce image artifacts. Cho and Lee (Cho & Lee 2009) introduced a fast almost real-time deconvolution algorithm utilizing sharp edge prediction. Krishnan and Fergus (Krishnan & Fergus 2009) proposed hyper-Laplacians priors for fast deconvolution. Xu and Jia (Xu & Jia 2010) separated the kernel estimation and deconvolution into two phases and applied L1-norm regularization in the frequency domain for fast kernel estimation and deconvolution. Cho *et al.* (Cho, Wang & Lee 2011) used the EM approach to handle outliers in non-blind deconvolution. Tai and Lin (Tai & Lin 2012) used motion-aware filtering to remove image noise before deconvolution, which was shown to be effective.

These previous approaches all work under the uniform PSF assumption. As mentioned in Section 8.1, camera egomotion causes a spatially-varying motion blur and this aberration cannot be modeled accurately with a uniform PSF. Prior work has recognized the need to handle non-uniform motion blur for camera egomotion, moving objects, and defocus blur. For example, early work by Sawchuk (1974) addressed motion blur from a moving camera by first using a log-polar transform to transform the image such that the blur could be expressed as a spatially-invariant PSF. The range of motion that could be addressed was limited to rotation and translation. When addressing moving objects, the input image can be segmented into multiple regions each with a constant PSF as demonstrated by Levin (Levin 2006), Bardsley *et al.* (Bardsley, Jefferies, Nagy & Plemmons 2006), Cho *et al.* (Cho, Matsushita & Lee 2007) and Li *et al.* (Li, Yu & Chai 2008). Such segmented regions, however, should be small to make the constant PSF assumption valid for the spatially-varying motion blur in camera-shake motion. For example, Tai *et al.* (Tai, Du, Brown & Lin 2008, Tai, Du, Brown & Lin 2010) extended the hybrid camera framework used by Ben-Ezra and Nayar (Ben-Ezra & Nayar 2004) to estimate a PSF per pixel using an auxiliary video camera. This need for a per-pixel PSF revealed the futility of relying on the conventional kernel-based PSF model for spatially-varying blur due to egomotion.

The impetus of this work is to introduce a better blur model for camera egomotion. The utility of this projective motion path model has already been demonstrated by Tai *et al.* (Tai, Kong, Lin & Shin 2010) for deblurring moving objects and Li *et al.* (Li, Kang, Joshi, Seitz & Huttenlocher 2010) for generating sharp panoramas from motion-blurred image sequences. As previously mentioned, Whyte *et al.* (2010) and Joshi *et al.* (2010) independently proposed a similar blur formulation and correction algorithms. Whyte *et al.* (2010) showed the camera shake was mostly related to rotational motion about the camera's optical axis and developed an algorithm to determine the motion path under this assumption. Joshi *et al.* (2010) exploited motion inertia to estimate the camera motion. The motion blur model proposed by these works is identical to the one we present. Recent work has leveraged this new blur model to estimate the spatially-varying blur kernel using filter flow (Hirsch, Schuler, Harmeling & Scholkopf 2011) or multiple images (Cho, Cho, Tai & Lee 2012).

8.3 The projective motion blur model

In this section, we describe our blur model based on a planar projective motion path. This model will be used in the following section to derive the deblurring algorithm.

In photography, the pixel intensity of an image is determined by the amount of light received by the imaging sensor over the exposure time:

$$I(x) = \int_0^T \delta I(x,t)\,dt \approx \sum_{i=1}^{N} \Delta I(x,t_i), \tag{8.1}$$

where $I(x)$ is the image recorded after exposure; $\delta I(x,t)$ and its discrete equivalent $\Delta I(x,t_i)$ refer to the image captured by the sensor within an infinitesimal time interval

dt at time instance t; $[0, T]$ is the total exposure time and x is a 3×1 vector indicating the homogeneous pixel coordinates. In our model, we assume N (the discrete sampling rate over exposure time) is large enough so that the difference between continuous integration and discrete integration is negligible.

When there is no relative motion between the camera and the scene, assuming the effects of sensor noise are small, $\Delta I(x, t_1) \cong \Delta I(x, t_2) \cong \cdots \cong \Delta I(x, t_N)$ and $I(x) \cong N\Delta I(x, t_0) \triangleq I_0(x)$ is a clear image. When there is relative motion, $I(x)$ is the summation of multiple *unaligned* images $\Delta I(x, t_i)$. For a static distant scene, the relative motion causes a planar projective transform in the image plane, i.e. $\Delta I(x, t_i) = \Delta I(h_i x, t_{i-1})$. Here, h_i is a homography[1] defined by a 3×3 non-singular matrix up to a scalar. Suppose all h_i are known, we can then express $\Delta I(x, t_i)$ by $I_0(x)$ using the following formulation:

$$\Delta I(x, t_i) = \Delta I\left(\prod_{j=1}^{i} h_j x, t_0\right) = \frac{1}{N} I_0(H_i x), \tag{8.2}$$

where $H_i = \prod_{j=1}^{i} h_j$ is also a homography. Hence, we obtain our projective motion blur model as:

$$B(y) = \sum_{i=1}^{N} \Delta I(x, t_i) = \frac{1}{N} \sum_{i=1}^{N} I_0(H_i x), \tag{8.3}$$

where $B(y)$ is the motion-blurred image, and $I_0(x)$ is the clear image we want to estimate. According to our model, the blur image is the average of multiple clear images, each of which is a planar projective transformation of the clear image $I_0(x)$.

Figure 8.4 illustrates the relationship between our blur model and the conventional representation. The conventional spatially-invariant PSF representation is a special case of our model for which every h_i is a translation. Note that for in-plane translational motion, the conventional kernel-based model provides a more compact representation of the motion blur than our model. However, in the cases of other motions, e.g. in-plane/out-of-plane rotation, our projective motion model is more compact and intuitive.

8.4 Projective motion Richardson–Lucy

In this section, we describe how to modify the Richardson–Lucy algorithm to incorporate our blur model. To do so, we first give a brief review of the Richardson–Lucy algorithm (Richardson 1972, Lucy 1974) and then derive our algorithm in a similar manner. For simplicity, the term I is used instead of I_0 to represent the clear image to be estimated.

[1] We use a homography for its ability to model all planar transformations. More restrictive transformations, e.g. rotation, translation, can be used instead when prior knowledge of the camera's motion path is known.

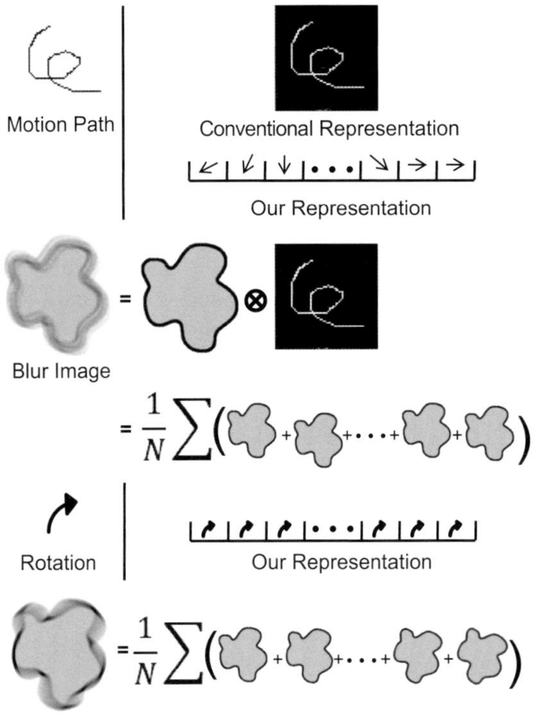

Figure 8.4 This figure compares our blur model and the conventional model. Given the motion path (PSF), a conventional model uses a rectangular kernel (analogous to an image patch) to represent the PSF. In comparison, our model uses a sequence of transformation matrices. For rotational motion, our representation encodes the rotation via a sequence of homographies naturally, while the conventional approach would need to store pixelwise PSFs. © IEEE 2011.

8.4.1 Richardson–Lucy deconvolution algorithm

The derivation in this section of the Richardson–Lucy deconvolution algorithm (Richardson 1972, Lucy 1974) is based on the work by Shepp and Vardi (Shepp & Vardi 1982). In particular, the derivation from Shepp and Vardi shows that the RL algorithm can be considered a maximum likelihood solution using the Poisson distribution to model the likelihood probability $P(B, k|I)$:

$$P(B, k|I) = \prod_{x \in I} \frac{g(x)^{B(x)} e^{-g(x)}}{B(x)!}, \tag{8.4}$$

$$g(x) = \sum_{y \in k} I(y) k(x - y), \tag{8.5}$$

where B is the observed motion-blurred image, k is the motion PSF, i.e. $\sum_{y \in k} k(y) = 1$, I is the clear image we want to estimate, and $g(x)$ is a convolution process for a pixel located at x. Equation (8.4) assumes that the likelihood probability is conditionally independent for each x. Since $\sum_{y \in k} k(y) = 1$ and $\sum_{x \in B} B(x) = \sum_{x \in I} I(x)$, the overall intensity is preserved.

In Shepp & Vardi (1982), Shepp and Vardi show that Eq. (8.4) is a concave function by showing the matrix of second derivatives of Eq. (8.4) is negative semi-definite. In order to optimize Eq. (8.4), it follows from Theorem 2.19(e) of Vardi (1969) that the sufficient conditions for I to be a maximizer of Eq. (8.4) are the Kuhn–Tucker conditions where all x satisfy:

$$I(x)\frac{\partial}{\partial I(x)}\ln\left(\prod_{x\in I}\frac{g(x)^{B(x)}e^{-g(x)}}{B(x)!}\right) = 0, \tag{8.6}$$

and

$$\frac{\partial}{\partial I(x)}\ln\left(\prod_{x\in I}\frac{g(x)^{B(x)}e^{-g(x)}}{B(x)!}\right) \leqslant 0, \quad \text{if } I(x) = 0. \tag{8.7}$$

To obtain an iterative update rule for the RL algorithm, we use the first condition in Eq. (8.6)[2], for all $x \in I$:

$$\begin{aligned} I(x)\frac{\partial}{\partial I(x)}\ln\left(\prod_{x\in I}\frac{g(x)^{B(x)}e^{-g(x)}}{B(x)!}\right) &= 0,\\ I(x)\sum_{x\in I}\frac{\partial}{\partial I(x)}(B(x)\ln(g(x)) - g(x) - \ln(B(x)!)) &= 0,\\ I(x)\sum_{x\in I}\frac{B(x)}{g(x)}\frac{\partial}{\partial I(x)}g(x) - I(x)\sum_{x\in I}\frac{\partial}{\partial I(x)}g(x) &= 0,\\ I(x)\sum_{y\in k}\frac{B(y)}{g(y)}k(y-x) - I(x)\sum_{y\in k}k(y-x) &= 0. \end{aligned}$$

Since $\sum_{y\in k} k(y) = 1$, we have $\sum_{y\in k} k(y-x) = 1$. After adding the iteration index, t, we get:

$$I^{t+1}(x) = I^t(x)\sum_{y\in k}\frac{B(y)}{\sum_{z\in k} I^t(z)k(y-z)}k(y-x). \tag{8.8}$$

Utilizing the convolution operation for the whole image, we obtain the RL algorithm:

$$I^{t+1} = I^t \times \tilde{k} \otimes \frac{B}{k \otimes I^t}, \tag{8.9}$$

where $\tilde{k}$ is the transpose of k that flips the shape of k upside down and left to right, $\otimes$ is the convolution operation and $\times$ is a pixel-wise multiplication operation. To understand Eq. (8.9), we consider that $B'^t = k \otimes I^t$ is the prediction of a blurred image according to the current estimation of clear image I^t and the given point spread function k. Thus, B/B'^t are the residual errors (by pixel-wise division) between the real blurred image B and the predicted blurred image B'^t. The correlation operation ($\tilde{k}\otimes$) integrates the residual errors distributed according to $\tilde{k}$. The update rule in Eq. (8.9) essentially

[2] The second condition in Eq. (8.7) is used to relate the RL algorithm with the EM algorithm (Dempster, Laird & Rubin 1977) for the convergence proof. For further details, we refer readers to Shepp & Vardi (1982).

computes a clear image I^∞ that would generate the blurred image B, given a known point spread function k. Typically, the algorithm starts with an initial guess of $I^0 = B$.

8.4.2 Projective motion Richardson–Lucy algorithm

With the basic Richardson–Lucy algorithm, we can derive our projective motion Richardson–Lucy algorithm. From the projective motion blur model defined in Eq. (8.3), we can integrate the motion path at each pixel location y and define a spatially-varying motion PSF k_y:

$$B(y) = \frac{1}{N}\sum_{i=1}^{N} I(H_i x) = \sum_{x \in k_y} I(x) k_y(x), \tag{8.10}$$

and

$$k_y(x) = \begin{cases} \frac{1}{N}, & \text{if } x = H_i^{-1} y \\ 0, & \text{otherwise} \end{cases}, \tag{8.11}$$

where $\sum_{x \in k_y} k_y(x) = \sum_{i=1}^{N} \frac{1}{N} = 1$. Because $x = H_i^{-1} y$ does not correspond to a discrete integer pixel coordinate, bicubic interpolation is used to estimate the pixel values for the points x.

Substituting and replacing Eq. (8.5) with Eq. (8.10) for the RL algorithm, we get:

$$I^{t+1}(x) = I^t(x) \sum_{y \in k_x} \frac{B(y)}{\sum_{z \in k_y} I^t(z) k_y(z)} \tilde{k}_x(y), \tag{8.12}$$

which is the general form for spatially-varying motion deblurring using the RL algorithm. The motion path in $\tilde{k}_x$ is the reverse direction of the motion path in k_x.

Since the motion path in k_x, according to Eq. (8.10), can be described by a sequence of homographies, $H_1 \ldots H_N$, we can also group the motion path of $\tilde{k}_x$. Grouping the motion path of $\tilde{k}_x$ forms a new sequence of homographies which is the original homography sequence but with each matrix inverted and applied in reverse order, i.e. $H_N^{-1} \ldots H_1^{-1}$. For each point along the motion path in $H_1 \ldots H_N$, $H_N^{-1} \ldots H_1^{-1}$ reverse the transformation and integrate the errors along the motion path. Thus, we obtain the iterative update rule for the projective motion blur model:

$$I^{t+1} = I^t \times \frac{1}{N}\sum_{i=1}^{N} E^t(H_i^{-1} x), \tag{8.13}$$

where $E^t(x) = \frac{B(x)}{\frac{1}{N}\sum_{i=1}^{N} I^t(H_i x)}$ is the residual error between the real blurred image B and the predicted blurred image $B'^t = \frac{1}{N}\sum_{i=1}^{N} I^t(H_i x)$. Note that although we computed the per-pixel motion PSF during the derivation, Eq. (8.13) processes the image as a whole and does not need to reconstruct the per-pixel motion PSF explicitly. This is similar to the global convolution and correlation process in the conventional RL algorithm.

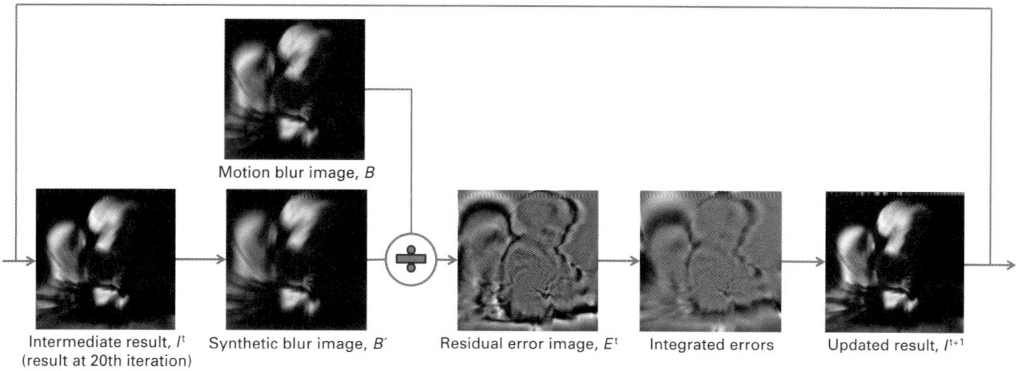

Figure 8.5 Overview of the projective motion RL algorithm. Given the current estimation I^t, we compute a synthetic blur image B' according to the given motion in terms of H_i. The residual error image $E^t = B/B'$ is computed by pixel-wise division. The residual errors are then integrated according to H_i^{-1} to produce an updated estimation I^{t+1}. The I^{t+1} is then used as the initial guess for the next iteration. This process is repeated until convergence or after a fixed number of iterations. In our implementation, the number of iterations is fixed to 500. © IEEE 2011.

In essence, our approach replaces the convolution and correlation operators in the conventional RL algorithm with a sequence of forward projective motions and their inverses via the homographies. Figure 8.5 illustrates our projective motion RL algorithm.

8.4.3 Gaussian noise

If we assume that the image statistics follow a Gaussian distribution instead of a Poisson distribution, we can model the likelihood probability $P(B, k|I)$ as follows:

$$\begin{aligned}
&\arg\max_I P(B,k|I),\\
&= \arg\max_I \prod_{x\in I} \exp\left(-\frac{\|g(x)-B(x)\|^2}{\sigma_g^2}\right),\\
&= \arg\min_I \sum_{x\in I} \|g(x)-B(x)\|^2, \qquad (8.14)
\end{aligned}$$

where $g(x) = \sum_{y\in k} I(y)k(x-y)$ is defined in Eq. (8.5), and σ_g^2 is the variance of a Gaussian distribution. To solve Eq. (8.14), we can derive an additive update rule based on the gradient descent method as follows:

$$\begin{aligned}
I^{t+1}(x) &= I^t(x) + \sum_{y\in k}\left(B(y) - \sum_{z\in k} I^t(z)k(y-z)\right)k(y-x),\\
I^{t+1} &= I^t + \tilde{k}\otimes(B - k\otimes I^t). \qquad (8.15)
\end{aligned}$$

Similarly, for our projective motion blur model, we obtain:

$$I^{t+1} = I^t + \frac{1}{N}\sum_{i=1}^{N} E'^t\left(H_i^{-1}x\right), \qquad (8.16)$$

where $E'^t(x) = B(x) - \frac{1}{N}\sum_{i=1}^{N} I^t(H_i x)$ is the residual error obtained by pixel-wise subtraction.

If we introduce regularization as a prior model $P(I)$, we now have to maximize a posterior probability $P(I|B,k)$. After some mathematical rearrangement, we can obtain the following energy function for minimization:

$$\begin{aligned} &\arg\max_I P(I|B,k), \\ &= \arg\max_I P(B,k|I)P(I), \\ &= \arg\min_I \sum_{x\in I} \|g(x) - B(x)\|^2 + \lambda R(I), \end{aligned} \tag{8.17}$$

where $\|g(x) - B(x)\|^2$ is the data term, and $R(I) = -\frac{\sigma_g^2}{\lambda}\log(P(I))$ is the regularization term.

Computing the derivative of Eq. (8.17), we can obtain another set of iterative update rules based on the gradient descent method:

$$I^{t+1} = I^t + \tilde{k} \otimes (B - k \otimes I^t) + \lambda \nabla R(I^t), \tag{8.18}$$

and

$$I^{t+1} = I^t + \frac{1}{N}\sum_{i=1}^{N} E'^t(H_i^{-1}x) + \lambda \nabla R(I^t), \tag{8.19}$$

for the conventional motion blur model and our projective motion blur model, respectively.

8.5 Motion estimation

Although the focus of this chapter is the derivation of our projective motion path blur model and its associated deblurring algorithm, for completeness we describe some promising methods to compute the projective motion within the exposure time. General algorithms for motion estimation are an area for future research.

Auxiliary hardware or imaging modifications

A direct method to estimate camera motion in terms of homographies during the exposure is to use a hybrid camera (Ben-Ezra & Nayar 2004, Tai, Du, Brown & Lin 2010) that captures an auxiliary high frame-rate low resolution video. From the auxiliary video, the motion trajectory at each pixel (i.e. optical flow) can be computed as shown in Tai, Du, Brown & Lin (2010) and a sequence of homographies can be fitted to each frame. Another promising hardware coupling is the use of accelerometer and inertial sensors for estimating the camera's motion as demonstrated in Hong, Rahmati, Wang & Zhong (2008) and Joshi *et al.* (2010). Recently, Tai *et al.* (Tai, Kong, Lin & Shin 2010) showed how a coded exposure could be used to capture an image containing a motion-blurred moving object. Through the analysis of motion discontinuities protected by the coded exposure, homographies describing the motion path could be computed.

Uniform motion assumption

If the motion blur is caused by a constant motion path over the exposure time, then the path can be described as $h_1 = h_2 = \cdots = h_N$. According to the definition of $H_i = \prod_{j=1}^{i} h_j$, we find

$$h_i = \sqrt[N]{H_N}, \qquad 1 \leqslant i \leqslant N. \tag{8.20}$$

Thus, we can obtain h_i by computing the Nth matrix root (Bini, Higham & Meini 2005) of H_N, and the estimation of the series of h_i is reduced to the estimation of H_N. The easiest technique to estimate this uniform motion is described by Shan *et al.* (Shan, Xiong & Jia 2007) which relies on the user to supply image correspondences to establish the transformation. Another highly promising technique is that proposed by Dai and Wu (Dai & Wu 2008) that uses the blurred objects' alpha matte to estimate H_N. In our experiments, we use a former approach in Shan *et al.* (2007). Recently, Li *et al.* (2010) proposed a method to estimate a global homography between successful video images to perform mosaicing. The frames could also be deblurred using a motion path model similar to that proposed in this chapter.

Recently, Whyte *et al.* (2010) used a variational approach to estimate the weight of quantized rotation parameters without the need for auxiliary hardware or a uniform assumption. Their estimation of the motion PSF, however, is not sufficiently accurate, leading to deconvolution artifacts in the deblurred result. This work concurrently showed a method to estimate the spatially-varying PSF using blurred and noisy image pairs.

8.6 Experiment results

In this section, we empirically examine our algorithm's convergence properties by plotting the RMS error against the number of iterations. Robustness is analyzed by comparing the results with different amounts of additive noise. We also evaluate the quality of our projective motion RL algorithm with and without regularization by creating a set of synthetic test cases. Our results are also compared with the spatially-invariant deblurring method, e.g. Levin *et al.* (2007), with synthetic test cases that resemble real motion-blurred images caused by camera shake. Finally, we show results on real images for which the projective motion paths were estimated using the methods described in Section 8.5.

8.6.1 Convergence analysis

While the conventional RL algorithm guarantees convergence, in this section we empirically examine the convergence of our projective motion RL algorithm. At each iteration, we compute the log ratio $\log(I^{t+1}/I^t)$ and the RMS error of the current result against the ground truth image. We run our algorithm for a total of 5000 iterations for each case. The convergence rates of our basic projective motion RL regularized algorithm with fixed regularization weight ($\lambda = 0.5$ for an intensity range between 0 and 255) and the

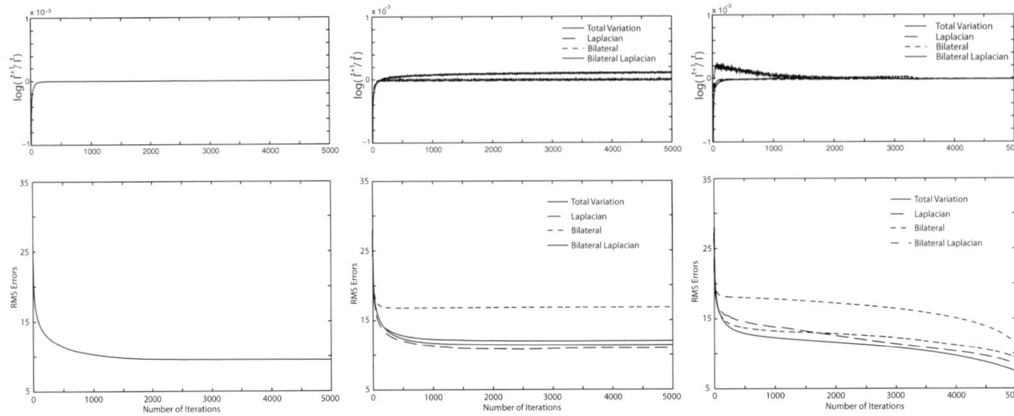

Basic algorithm | With regularization (λ = 0.5) | With regularization (decreasing λ)

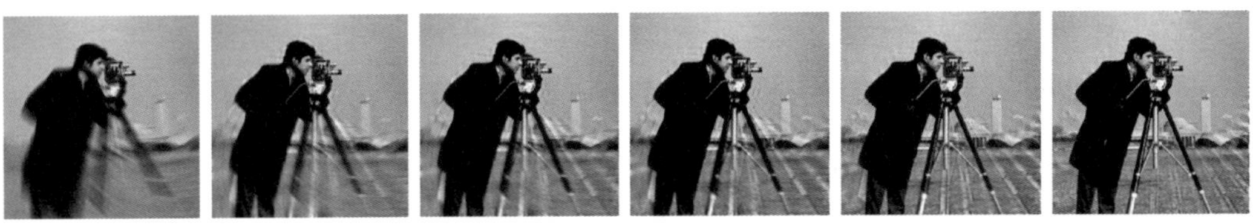

RMS: 35.89 Input blur image | RMS: 18.59 20th iteration | RMS: 16.34 50th iteration | RMS: 14.69 100th iteration | RMS: 11.36 500th iteration | RMS: 9.46 5000th iteration

Figure 8.6 Convergence rates of our projective motion RL algorithm. The first row shows the plot of I^{t+1}/I^t, the second row shows the plot of RMS errors against the number of iterations for our basic algorithm, regularized algorithm with fixed weight ($\lambda = 0.5$), and regularized algorithm with decreasing weight (linearly from 1 to 0) as the number of iterations increases. Note that the motion blur is different for the two test cases. The third row shows the input motion blur image and intermediate results at the 20th, 50th, 100th, 500th, and 5000th iterations for the basic algorithm.

regularized algorithm with decreasing regularization weight (linearly from 1 to 0) are compared. The RMS error is plotted with respect to the ground truth image. Figure 8.6 shows the graphs of RMS errors versus the number of iterations.

Typically, our method converges within 300 to 400 iterations for both the basic algorithm and the regularized algorithm with fixed regularization weight. As the number of iterations increases, the difference of RMS errors between successive iterations decreases; however, after 500 iterations the visual improvement in the deblurred result is unnoticeable as shown in some intermediate results in Figure 8.6. We found that the algorithm with regularization produced RMS errors that were higher than those produced using the basic algorithm. The main reason is that the test cases demonstrated here are noise-free, and the regularization tends to smooth out high frequency details resulting in higher RMS errors. However, as we will show in the next subsection, when noise is present in the image, incorporating regularization becomes effective. Using the decreasing regularization weight scheme we still observe a decrease in RMS errors, even at 5000 iterations. This approach also achieves RMS errors lower than both the basic algorithm and the regularized algorithm using a fixed regularization weight. This convergence study confirms the effectiveness of this scheme observed by Shan *et al.* (2008), and Yuan *et al.* (2008).

8.6.2 Noise analysis

To test for the robustness of our algorithm, we added different amounts of zero-mean Gaussian noise to the synthetic blur images. To simulate sensor noise, the added image noise does not undergo the convolution process, and is independent of the motion blur effect.

Figures 8.7 and 8.8 show our deblurring results with different amounts of Gaussian noise[3] added. We show results of our algorithm with and without regularization. For this experiment, we use the Poisson noise model with total variation regularization[4]. The test pattern is a resolution chart. As expected, our deblurring results without regularization amplify image noise just like other deblurring algorithms. The quality of our deblurring algorithm degrades as the amount of noise increases. In addition, larger motions tend to produce noisier results. In such cases, the added noise in the image around the center of rotation becomes less apparent than that in image regions with larger motion blur (i.e. the image boundaries). In the presence of image noise, the regularization term becomes important to improve the visual quality of the results. Better results can be achieved by increasing the regularization weight as the noise level increases. The difference between the regularized and un-regularized results are significant both in terms of visual quality and RMS errors. However, when the amount of image noise added is very large, e.g. $\sigma^2 \geqslant 20$, the regularization term cannot suppress image noise effectively.

8.6.3 Qualitative and quantitative analysis

Figure 8.1 has already shown a synthetic example of spatially-varying motion blur with known h_i. To further evaluate our algorithm quantitatively, we have created a test set consisting of fifteen test cases and five test images: *Mandrill*, *Lena*, *Cameraman*, *Fruits*, and *PAMI*. The *Mandrill* example contains significant high frequency details in the hair regions. The *Lena* example contains high frequency details, smooth regions, and also step edges. The *Cameraman* image is a monotone image but with noise added independently to each RGB channel. The *Fruits* example also contains high frequency details and smooth regions. Lastly, the *PAMI* example is a text-based binary image (i.e. black and white). For each test case, we add additional Gaussian noise ($\sigma^2 = 2$) to simulate camera sensor noise. The parameter settings are the same for all test cases and the values we used are the same as in the implementation of Levin *et al.* (2007).

Our results for the RMS of the input motion-blurred image, the RMS of the deblurred image using our basic projective motion RL algorithm, and the RMS of the deblurred

[3] The noise variance σ^2 of Gaussian noise added are with respect to an intensity range between 0 and 255.

[4] For details of regularization, please refer to Tai, Tan & Brown (2011).

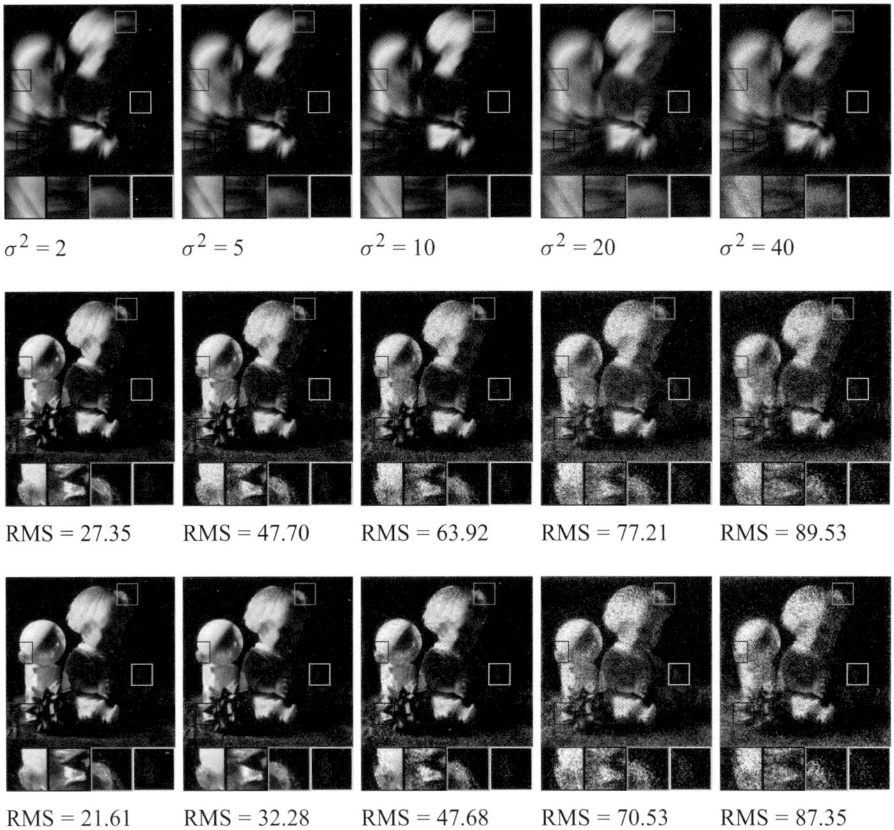

Figure 8.7 Evaluation on the robustness of our algorithm by adding different amounts of noise to blurred images. Top row: noisy blurred image, the amplitude of noise is determined by the noise variance (σ^2). Second row: deblurring results with our basic projective motion RL algorithm. Third row: deblurring results with total variation regularization. In the presence of image noise, our deblurring algorithm amplified the image noise in deblurred results. The effects of regularization hence become significant in suppressing amplified image noise. © IEEE 2011.

image with regularization (total variation) are shown in Figure 8.9. Our projective motion RL is effective, especially when regularization is used with our suggested implementation scheme. These test cases also demonstrate that our approach is effective in recovering spatially-varying motion blur that satisfied our projective motion blur assumption. We note that in some test cases, e.g. examples from the *Fruits* test case, the RMS errors of the deblurred results are larger than the RMS errors of the input motion-blurred images. This is due to the effects of amplified noise; after noise suppression with regularization, the reverse is true.

8.6.4 Comparisons with spatially-invariant method

To evaluate our projective motion blur model for the motion blur effects caused by camera-shake, another test set was created to simulate blur from camera-shake motion.

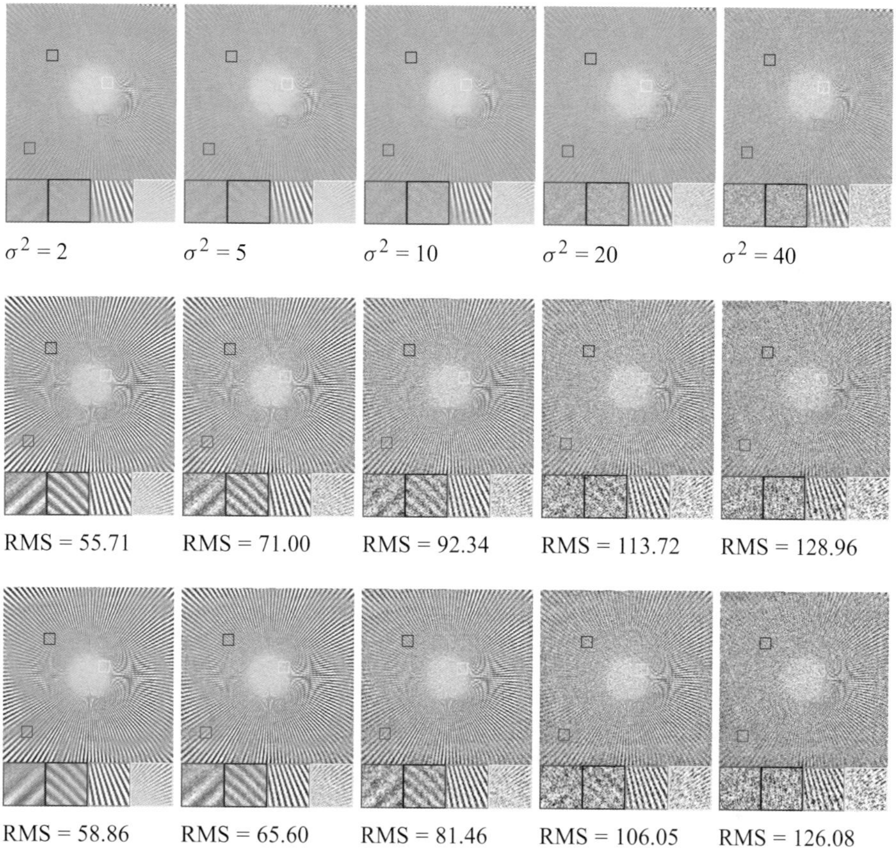

Figure 8.8 Evaluation on the robustness of our algorithm by adding different amounts of noise in blurred images. Top row: noisy blurred image. Second row: deblurring results with our basic projective motion RL algorithm. Third row: deblurring results with total variation regularization. © IEEE 2011.

Our results are compared with results obtained using a spatially-invariant blur model based on the deblurring algorithm in Levin *et al.* (2007).

This test set consists of eight test cases and five images: *Doll*, *Einstein*, *Lotus*, *Tower*, and *Palace*. For each test case, we first use homographies to approximate the shape and intensity variation of "real" motion blur kernels. The real kernels are from the ground truth kernels in the data provided by Levin *et al.* (2009). Figure 8.10 shows their ground truth kernels and our approximated kernels using homographies to describe the projective motion path. Since Levin *et al.* locked the z-axis rotation handle of the tripod when they captured the images, their data do not contain any rotation. However, real camera shake usually contains a small amount of rotation. We added one degree of rotation in total to the overall motion path to simulate the effects of in-plane rotation of camera shake as shown in Figure 8.11. Although the motion blur of this test set contains rotation, the effect of rotation is almost invisible from the motion-blurred

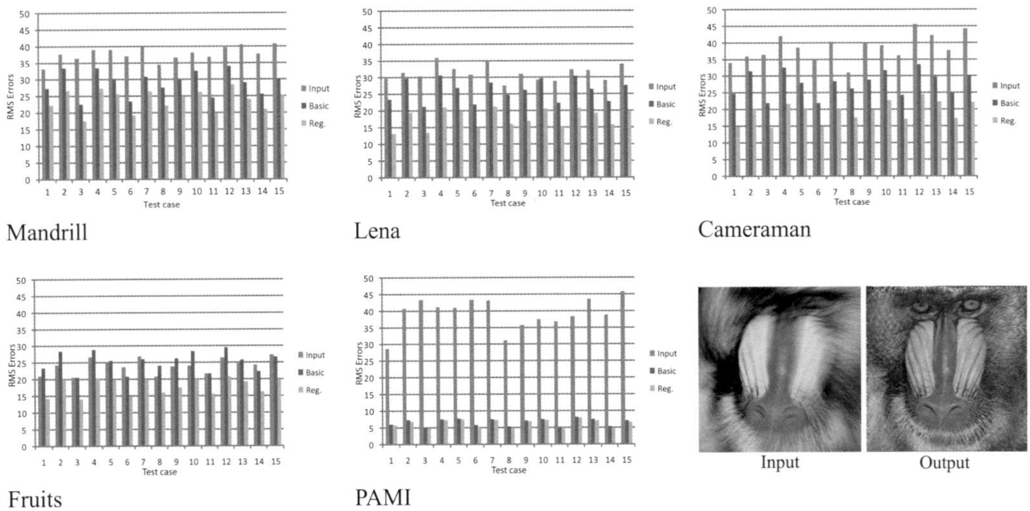

Figure 8.9 RMS pixel errors for different examples and test cases in our synthetic trials. We compare the RMS of the input blur image, the deblurred image using basic projective motion RL, and the deblurred image using projective motion RL with (total variation) regularization. © IEEE 2011.

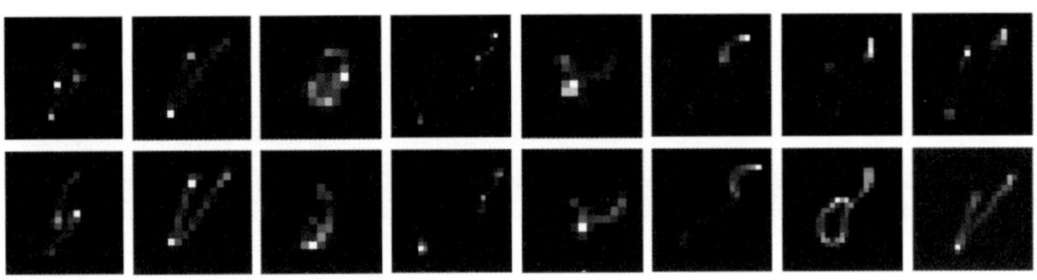

Figure 8.10 Top row: ground truth PSF from Levin *et al.* (2009). Bottom row: our approximated PSF using projective motion paths. © IEEE 2011.

images themselves since the translational motion dominates the effects of motion blur. However, when we apply a spatially-invariant motion deblurring algorithm (Levin *et al.* 2007) to these images, the effects of rotation are obvious. Similar to the previous test set, we also added Gaussian noise ($\sigma^2 = 2$) to simulate camera sensor noise.

Figure 8.12 shows the RMS errors of our results (with TV regularization) compared with the results from Levin *et al.* (2007) using nine different PSFs sampled at different locations of the image. We use the source code provided by Levin *et al.* (2007) with parameter $\lambda = 0.001$ to obtain the results for comparison. Our results consistently produce smaller RMS errors and fewer visual artifacts when compared to the results from Levin *et al.* (2007). These test cases also show the insufficiency of conventional spatially-invariant motion to model camera shake. Levin *et al.* (2007) obtain good results around local regions where the PSF is sampled; however, visual artifacts become

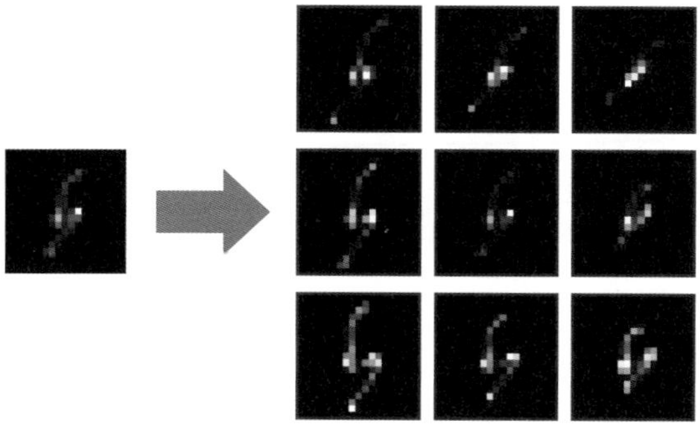

Figure 8.11 Our simulated spatially-varying motion blur PSF for camera shake. From the approximated kernels in Figure 8.10, we included 1 degree of rotation (in total) in the motion path. This was done by first estimating the PSF using homographies (Figure 8.10) and then introducing a rotation in the motion path of a single degree distributed over the whole range of motion. To demonstrate the effect of this rotation, we reconstructed the PSFs at nine different positions in the image. These reconstructed PSFs will be used by a spatially-invariant motion deblurring algorithm (Levin *et al.* 2007) for comparison. With only 1 degree of rotation we can see a significant change to the camera-shake PSFs in different regions of the image. Rotational motion for real camera shake (e.g. as shown in Figure 8.2) would be larger than shown here. © IEEE 2011.

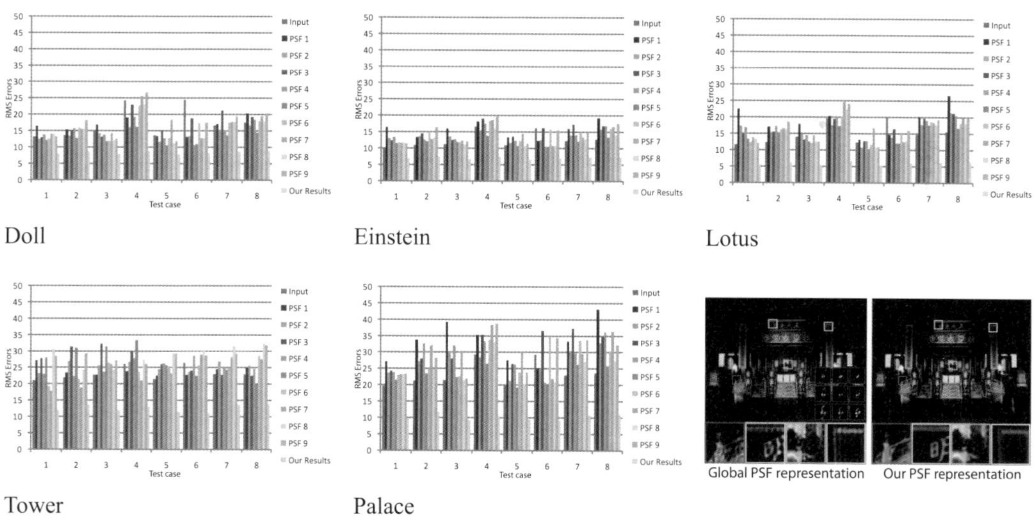

Figure 8.12 We compare our results with the spatially-invariant motion deblurring algorithm in Levin *et al.* (2007). The RMS pixel errors for different examples and test cases are shown. For the results of Levin *et al.* (2007), we sampled the PSF at nine different locations in the images and therefore obtained nine different PSFs for deblurring. Our results are compared to the results from Levin *et al.* (2007) using all nine PSFs. Our approach consistently produces smaller RMS errors in all examples for all test cases. © IEEE 2011.

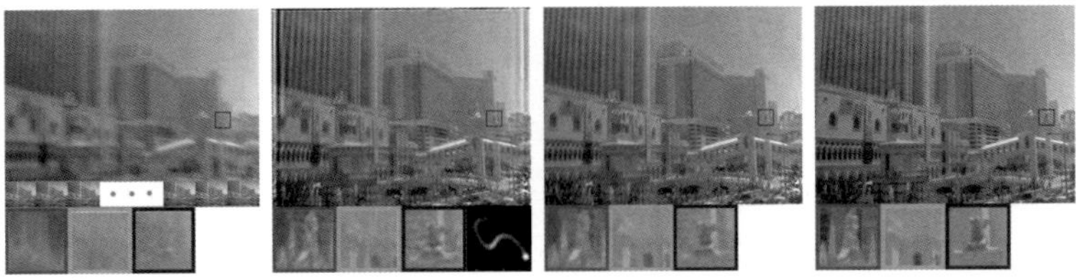

(a) (b) (c) (d)

Figure 8.13 Image deblurring using globally invariant kernels. (a) Input from a hybrid camera where the high frame-rate low resolution images are also shown; (b) result generated by Ben-Ezra & Nayar (2003) (standard RL algorithm from Matlab); (c) result from our projective motion RL with regularization; (d) the ground truth sharp image. Close-up views and the estimated global blur kernels are also shown. © IEEE 2011.

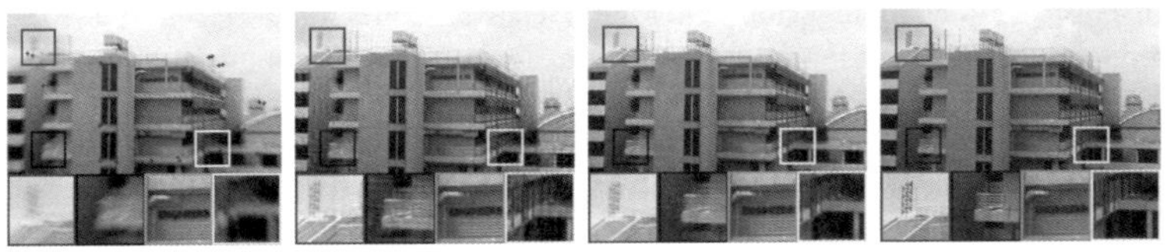

(a) (b) (c) (d)

Figure 8.14 Example of zoomed-in motion. (a) Blurred input image and user markups for motion estimation; (b) our result from using the basic projective motion RL algorithm; (c) our result including regularization; (d) ground truth image. © IEEE 2011.

increasingly noticeable as we move further away from the sampled point of the PSF used for deconvolution.

8.6.5 Real examples

Figure 8.13 shows an example of global motion blur obtained from our previous work using a hybrid camera system (Tai, Du, Brown & Lin 2010). To obtain the motion path homographies, we use the motion vectors in the low resolution high frame-rate camera as a set of point correspondences and fit a global homography per low resolution frame. We also show the effectiveness of our regularization compared with previous results. Our approach achieves comparable results with Tai, Du, Brown & Lin (2010), however we do not use the low resolution images for regularization as in Tai, Du, Brown & Lin (2010).

Figures 8.14 and 8.15 show more real examples with zoomed-in motion[5] and rotational motion, respectively. These input images were obtained with a long exposure time and a low ISO setting. The motion-blurred matrix H_N is obtained by fitting the transformation matrix with user markup as shown in Figures 8.14(a) and 8.15(a), respectively. Each h_i is computed by assuming the motion is uniform. We show the deblurred results from our algorithm without and with regularization in (b) and (c). The

[5] The approach in Shan *et al.* (2007) only discusses how to estimate rotational motion, but the estimation for zoom motion can be derived using a similar method.

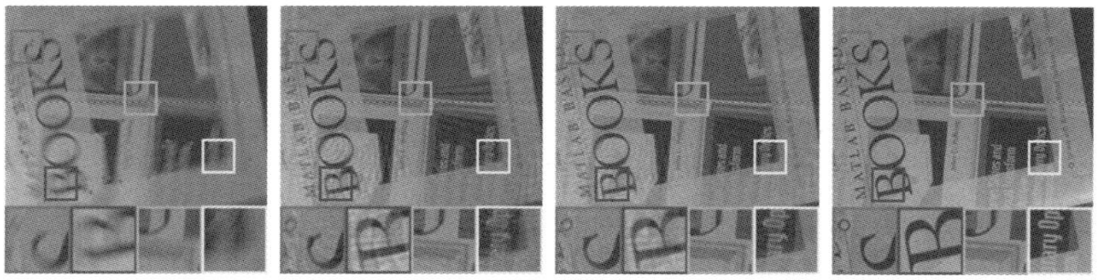

(a) (b) (c) (d)

Figure 8.15 Example of rotational motion. (a) Blurred input image and user markups for motion estimation; (b) our result from using the basic projective motion RL algorithm; (c) our result including regularization; (d) ground truth image. © IEEE 2011.

ground truth image is shown in (d) for comparison. We note that our real examples contain more visual artifacts than the synthetic examples. This is due to estimation errors in h_i. The effects of image noise in our real examples are not as significant as in our synthetic test case due to the long exposure time. We also note that the motions in our real examples are not as large as in our synthetic example.

8.7 Discussion and conclusion

This chapter has introduced two contributions for addressing image blur due to camera egomotion. The first is a formulation of the motion blur as an integration of the scene that has undergone a motion path described by a sequence of homographies. The advantage of this motion blur model is that it can parameterize camera egomotion better than conventional approaches which rely on a uniform blur kernel. In addition, this "kernel-free" formulation more closely models the physical phenomena causing the blur. Our second contribution is an extension to the RL deblurring algorithm to incorporate our motion blur model in a correction algorithm. We have outlined the basic algorithm and provided details on incorporating state-of-the-art regularization. Experimental results have demonstrated the effectiveness of our approach on a variety of examples.

We will now discuss issues and limitations of this blur model, as well as convergence, running time, and future work.

8.7.1 Conventional PSF representation versus projective motion blur model

While this chapter advocates a new blur model for camera egomotion to replace the conventional kernel-based approach, we note that the conventional representation has several advantages. First, the kernel-based PSF provides an easy to understand and intuitive representation of "point spread" about a point. Second, the kernel-based model can include other global blurring effects (e.g. defocus) and motion blur in a unified representation, while our representation only targets motion blur from camera motion. Third, by assuming the motion blur is the same globally, image deconvolution can be done in the frequency domain, while our projective motion RL algorithm can only be done

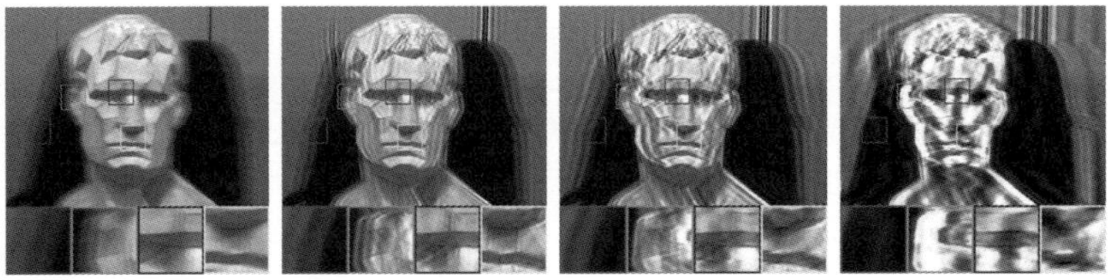

(a) (b) (c) (d)

Figure 8.16 Case of non-projective motion (images from Tai, Kong, Lin & Shin (2010)). (a) Input; (b–d) deblurring results with different motion PSFs estimated from different areas of the face. © IEEE 2011.

in the spatial domain. As demonstrated several times in this chapter, the conventional representation, however, is not suitable for dealing with spatially-varying motion blur. For such cases, our projective motion RL formulation becomes advantageous.

8.7.2 Limitations

Our projective motion RL algorithm has several limitations similar to other deblurring algorithms. A fundamental limitation to our algorithm is that the high frequency details that have been lost during the motion blur process cannot be recovered. Our algorithm can only recover the hidden details that remain inside the motion blur images. Another limitation is that our approach does not deal with moving or deformable objects or scenes with occlusion and/or disocclusion, or with significant depth variation. Figure 8.16, for instance, shows an object with out-of-plane rotational motion and a large variation in relative depth captured with a coded exposure camera. Here, we assume the motion to be strictly horizontal, and estimate the motion PSF using only local observations around the mouth (b), ear (c), and chair (d). Our deblurring results accurately recovered scene details only for the local regions used for PSF estimation; other regions are distorted by the incorrect PSF. Better results could potentially have been obtained by first segmenting the image into different regions, each of which satisfies the projective motion model, and then applying our motion deblurring algorithm on each region separately. The problem with occlusions, disocclusions, and depth variations is common to existing deblurring techniques, since the exhibited image blur cannot be adequately represented by a PSF in such cases. Other limitations of our approach include the problem of pixel color saturations and severe image noise, as demonstrated in our experiments.

8.7.3 Running time analysis

Our current implementation takes about 15 minutes to run 500 iterations on an image size 512×512 with an Intel(R) CPU L2400 @ 1.66 GHz and 512 MB of RAM. The running time of our algorithm depends on several factors, including image size $|I|$, number of discrete sampling N (number of homographies), and the number of iterations T. Hence, the running time of our algorithm is $O(|I|NT)$. Comparing our running time to approaches which reconstruct pre-pixel PSF for deblurring (e.g. Tai, Du, Brown & Lin (2010)), their running time is $O(|I|MT)$ where M is the window size of the square blur kernel used to represent the PSF. Our approach has a significant advantage with $N \ll M$. In our experiments, we found that $N = 50$ is sufficient to model very large variations over the projective motion path. However, a conventional PSF kernel size M can be as large as $31 \times 31 = 961$ for small to mid-range motion. We also note that the majority of our running time is spent in the bicubic interpolation process – necessary when applying the homographies. Significant speed-ups could undoubtedly be obtained with better implementation exploiting a GPU to perform the image warping.

Acknowledgements

This chapter is based on the work that appeared in Tai *et al.* (2011). We gratefully acknowledge the IEEE for their permission to reproduce portions here.

References

Agrawal, A. & Raskar, R. (2007). Resolving objects at higher resolution from a single motion-blurred image. In *IEEE Conference on Computer Vision and Pattern Recognition*, pp. 1–8.

Bardsley, J., Jefferies, S., Nagy, J. & Plemmons, R. (2006). Blind iterative restoration of images with spatially-varying blur. In *Optics Express*, pp. 1767–82.

Ben-Ezra, M. & Nayar, S. (2003). Motion deblurring using hybrid imaging. In *IEEE Conference on Computer Vision and Pattern Recognition*, vol. I, pp. 657–64.

Ben-Ezra, M. & Nayar, S. (2004). Motion-based motion deblurring. *IEEE Transactions on Pattern Analysis and Machine Intelligence*, **26**(6), 689–98.

Bini, D. A., Higham, N. J. & Meini, B. (2005). Algorithms for the matrix *p*th root. *Numerical Algorithms*, **39**(4), 349–78.

Chan, T. F. & Wong, C.-K. (1998). Total variation blind deconvolution. *IEEE Transactions on Image Processing*, **7**(3), 370–5.

Cho, S., Cho, H., Tai, Y.-W. & Lee, S. (2012). Registration based non-uniform motion deblurring. *Computer Graphics Forum (Special Issue on Pacific Graphics)*, **31**(7), 2183–92.

Cho, S. & Lee, S. (2009). Fast motion deblurring. *ACM Transactions on Graphics*, **28**(5), 145:1–8.

Cho, S., Matsushita, Y. & Lee, S. (2007). Removing non-uniform motion blur from images. In *IEEE International Conference on Computer Vision*, pp. 1–8.

Cho, S., Wang, J. & Lee., S. (2011). Handling outliers in non-blind image deconvolution. In *IEEE International Conference on Computer Vision*, pp. 495–502.

Dai, S. & Wu, Y. (2008). Motion from blur. In *IEEE Conference on Computer Vision and Pattern Recognition*, pp. 1–8.

Dempster, A. D., Laird, N. M. & Rubin, D. B. (1977). Maximum likelihood from incomplete data via the EM algorithm. *Journal of the Royal Statistical Society*, **39**, 1–38.

Dey, N., Blanc-Fraud, L., Zimmer, C., Kam, Z., Roux, P., Olivo-Marin, J. & Zerubia, J. (2004). A deconvolution method for confocal microscopy with total variation regularization. In *IEEE International Symposium on Biomedical Imaging: Nano to Macro*, pp. 1223–6.

Fergus, R., Singh, B., Hertzmann, A., Roweis, S. T. & Freeman, W. T. (2006). Removing camera shake from a single photograph. *ACM Transactions on Graphics*, **25**(3), 787–94.

Hirsch, M., Schuler, C., Harmeling, S. & Scholkopf, B. (2011). Fast removal of non-uniform camera shake. In *IEEE International Conference on Computer Vision*, pp. 463–70.

Hong, G. M., Rahmati, A., Wang, Y. & Zhong, L. (2008). Sensecoding: Accelerometer-assisted motion estimation for efficient video encoding. In *Proceedings of the 16th ACM International Conference on Multimedia*, pp. 749–52.

Jia, J. (2007). Single image motion deblurring using transparency. In *IEEE Conference on Computer Vision and Pattern Recognition*, pp. 1–8.

Joshi, N., Kang, S. B., Zitnick, L. & Szeliski, R. (2010). Image deblurring with inertial measurement sensors. *ACM Transactions on Graphics*, **29**(4), 30:1–9.

Krishnan, D. & Fergus, R. (2009). Fast image deconvolution using hyper-Laplacian priors. In *Neural Information Processing Systems Conference*, pp. 1033–41.

Levin, A. (2006). Blind motion deblurring using image statistics. In *Neural Information Processing Systems Conference*, pp. 841–8.

Levin, A., Fergus, R., Durand, F. & Freeman, W. T. (2007). Image and depth from a conventional camera with a coded aperture. *ACM Transactions on Graphics*, **26**(3), 70:1–9.

Levin, A., Weiss, Y., Durand, F. & Freeman, W. (2009). Understanding and evaluating blind deconvolution algorithms. In *IEEE Conference on Computer Vision and Pattern Recognition*, pp. 1964–71.

Li, F., Yu, J. & Chai, J. (2008). A hybrid camera for motion deblurring and depth map super-resolution. In *IEEE Conference on Computer Vision and Pattern Recognition*, pp. 1–8.

Li, Y., Kang, S. B., Joshi, N., Seitz, S. & Huttenlocher, D. (2010). Generating sharp panoramas from motion-blurred videos. In *IEEE Conference on Computer Vision and Pattern Recognition*, pp. 2424–31.

Lucy, L. (1974). An iterative technique for the rectification of observed distributions. *The Astronomical Journal*, **79**, 745–54.

Raskar, R., Agrawal, A. & Tumblin, J. (2006). Coded exposure photography: motion deblurring using fluttered shutter. *ACM Transactions on Graphics*, **25**(3), 795–804.

Richardson, W. (1972). Bayesian-based iterative method of image restoration. *Journal of the Optical Society of America*, **62**(1), pp. 55–9.

Sawchuk, A. A. (1974). Space-variant image restoration by coordinate transformations. *Journal of the Optical Society of America*, **64**(2), 138–44.

Shan, Q., Jia, J. & Agarwala, A. (2008). High-quality motion deblurring from a single image. *ACM Transactions on Graphics*, **27**(3), 73:1–10.

Shan, Q., Xiong, W. & Jia, J. (2007). Rotational motion deblurring of a rigid object from a single image. In *IEEE International Conference on Computer Vision*, pp. 1–8.

Shepp, L. A. & Vardi, Y. (1982). Maximum likelihood reconstruction for emission tomography. *IEEE Transactions on Medical Imaging*, **1**(2), 113–22.

Tai, Y., Du, H., Brown, M. & Lin, S. (2008). Image/video deblurring using a hybrid camera. In *IEEE Conference on Computer Vision and Pattern Recognition*, pp. 1–8.

Tai, Y., Du, H., Brown, M. & Lin, S. (2010). Correction of spatially varying image and video blur using a hybrid camera. *IEEE Transactions on Pattern Analysis and Machine Intelligence*, **32**(6), 1012–28.

Tai, Y., Kong, N., Lin, S. & Shin, S. (2010). Coded exposure imaging for projective motion deblurring. In *IEEE Conference on Computer Vision and Pattern Recognition*, pp. 2408–15.

Tai, Y.-W. & Lin, S. (2012). Motion-aware noise filtering for deblurring of noisy and blurry images. In *IEEE Conference on Computer Vision and Pattern Recognition*, pp. 17–24.

Tai, Y.-W., Tan, P. & Brown, M. (2011). Richardson–Lucy deblurring for scenes under projective motion path. *IEEE Transactions on Pattern Analysis and Machine Intelligence*, **33**(8), 1603–18.

Vardi, Y. (1969). *Nonlinear Programming*. Englewood Cliffs, NJ: Prentice-Hall.

Whyte, O., Sivic, J., Zisserman, A. & Ponce, J. (2010). Non-uniform deblurring for shaken images. In *IEEE Conference on Computer Vision and Pattern Recognition*, pp. 491–8.

Wiener & Norbert (1949). *Extrapolation, Interpolation, and Smoothing of Stationary Time Series*. New York: Wiley.

Xu, L. & Jia., J. (2010). Two-phase kernel estimation for robust motion deblurring. In *European Conference on Computer Vision*, pp. 157–70.

Yuan, L., Sun, J., Quan, L. & Shum, H.-Y. (2008). Progressive inter-scale and intra-scale non-blind image deconvolution. *ACM Transactions on Graphics*, **27**(3), 74:1–10.

9 HDR imaging in the presence of motion blur

C. S. Vijay, C. Paramanand, and A. N. Rajagopalan

9.1 Introduction

Digital cameras convert incident light energy into electrical signals and present them as an image after altering the signals through different processes which include sensor correction, noise reduction, scaling, gamma correction, image enhancement, color space conversion, frame-rate change, compression, and storage/transmission (Nakamura 2005). Although today's camera sensors have high quantum efficiency and high signal-to-noise ratios, they inherently have an upper limit (full well capacity) for accumulation of light energy. Also, the sensor's least acquisition capacity depends on its pre-set sensitivity. The total variation in the magnitude of irradiance incident at a camera is called the dynamic range (DR) and is defined as DR = (maximum signal value)/(minimum signal value). Most digital cameras available in the market today are unable to account for the entire DR due to hardware limitations. Scenes with high dynamic range (HDR) either appear dark or become saturated. The solution for overcoming this limitation and estimating the original data is referred to as high dynamic range imaging (HDRI) (Debevec & Malik 1997, Mertens, Kautz & Van Reeth 2007, Nayar & Mitsunaga 2000).

Over the years, several algorithmic approaches have been investigated for estimation of scene irradiance (see, for example, Debevec & Malik (1997), Mann & Picard (1995), Mitsunaga & Nayar (1999)). The basic idea in these approaches is to capture multiple images of a scene with different exposure settings and algorithmically extract HDR information from these observations. By varying the exposure settings, one can control the amount of energy received by the sensors to overcome sensor bounds/limits. The camera exposure is defined as $\left(\pi d^2/4\right).\Delta t$, where d is the aperture diameter and Δt is the exposure time (Mitsunaga & Nayar 1999). Hence the exposure settings can be varied either by changing the aperture diameter or by changing the exposure duration. Since changing the aperture affects the depth of field, varying exposure times/shutter speeds is preferable. The exposure times needed for a real scene can vary from values as small as $\frac{1}{500}$ s to as high as 3 s. An important and practical issue associated with long exposure times in cameras is that of image blurring, especially in HDR imaging where one has to capture images using long exposure times. While a camera can be held still for the shorter exposures, it would be extremely difficult for a photographer to hold a camera steady for longer periods while taking multiple shots of a static scene. Any incidental shake of the camera will result in motion blur. Although one can resort to tripods to avoid motion blur, carrying such sturdy accessories can be cumbersome. Worse still, tripods are off-limits in locations such as heritage sites.

The problem of restoring motion-blurred images has been widely studied in the literature. A lot of approaches exist that simultaneously solve for the latent image and the blur kernel (blind deconvolution) (Cai, Ji, Liu & Shen 2012, Shan, Jia, Agarwala *et al.* 2008, Fergus, Singh, Hertzmann, Roweis & Freeman 2006, Levin, Weiss, Durand & Freeman 2011). To reduce the ill-posedness of the motion deblurring problem, techniques exist that use more than one observation of the scene (Šroubek & Flusser 2005, Yuan, Sun, Quan & Shum 2007). In all these restoration techniques, blurring is assumed to be space invariant. However, camera shake typically results in space-variant blur due to camera rotation during image capture (Levin, Weiss, Durand & Freeman 2009, Whyte, Sivic, Zisserman & Ponce 2010). In fact, the influence of rotation on blurring is more significant than that of translation. Thus, in real scenarios, the space-invariant convolution model is inadequate to describe the blurring process.

Recent techniques have addressed the problem of non-uniform (i.e. space-variant) motion deblurring (Gupta, Joshi, Lawrence Zitnick, Cohen & Curless 2010, Tai, Tan & Brown 2011, Whyte *et al.* 2010) by modeling the blurred observation as an average of projectively transformed instances of the original image. For this model, Tai *et al.* (2011) developed a deblurring scheme based on the Richardson–Lucy algorithm. However, they do not address the problem of estimating the blurring function. Gupta *et al.* (2010) describe motion blur in terms of a motion density function (MDF) which denotes the camera motion. Their estimates of MDF and deblurred image are based on an existing space-invariant deblurring technique (Shan *et al.* 2008). Whyte *et al.* (2010) use a similar model but consider the camera motion to be composed of rotations about all three axes. Using the blind deblurring framework of Fergus *et al.* (2006), they estimate the blur kernel. None of the above non-uniform deblurring works addresses the variation of blur due to parallax effects. For the case of translational camera motion, Sorel & Flusser (2008) proposed a technique to restore space-variantly blurred images by jointly solving for the latent image as well as the scene depth map. Blur variation due to parallax has also been exploited for estimating the depth map of a scene (Paramanand & Rajagopalan 2012). All the algorithms mentioned above are applied on intensity images and the scenes are assumed to be well lit. The effect of saturation and such outliers on deblurring has been discussed in Cho, Wang & Lee (2011) and Whyte, Sivic & Zisserman (2011). In a very recent work (Kim, Tai, Kim, Brown & Matsushita 2012), the effect of a camera's nonlinear response on deblurring has been explored.

While the standard deblurring works do not consider the high dynamic nature of a scene, the problem of jointly estimating the HDR scene irradiance, the mapping between image intensity and irradiance, and blur kernels from motion blurred observations was first addressed in a recent work by Lu, Huang, Wu, Cheng & Chuang (2009). However, they consider a space-invariant blur model which restricts the camera motion to in-plane translations. Although this is a good step forward, it is limited in scope as the effect of rotation on blur is greater than that of translation (Whyte *et al.* 2010) and hence cannot be ignored.

In this chapter, we discuss a method to obtain the HDR image of a scene from a set of differently exposed observations which are likely to have been affected by non-uniform motion blur. While capturing multiple images with a handheld camera, when

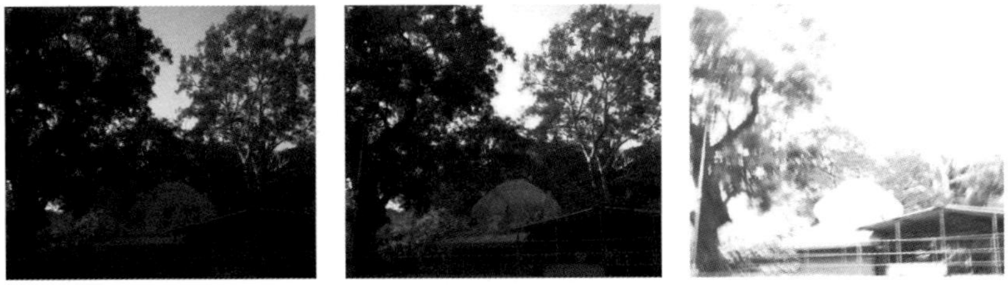

Figure 9.1 Multiple images where longer exposures are motion-blurred. © IEEE 2013.

the exposure time is short, there is less possibility of an observation getting blurred. However, when the exposure duration is long, it is highly probable that the observation gets blurred. Consequently, we consider a scenario wherein the input image set consists of blurred longer exposures and non-blurred shorter exposures. The scene is assumed to be far enough from the camera that blur variations due to parallax effects are negligible. Although the images that are captured with longer exposure times are blurred, they contain information not present in the less exposed images. Figure 9.1 shows one such image set. One simplistic solution would be to restore the blurred observations and then reconstruct the HDR image using existing methods. However, such an approach is not effective because blind deblurring does not always guarantee good quality output. In our approach, we estimate the latent scene irradiance from the given set of observations. In our framework, we judiciously exploit the tight coupling that inherently exists between the latent scene irradiance and each of the differently exposed and blurred observations. Our deblurring step subsumes all the blurred observations, thereby leading to a robust estimate. To model space-variant blur, we adopt the notion of transformation spread function (TSF) (Paramanand & Rajagopalan 2010, Vijay, Paramanand & Rajagopalan 2012, Vijay, Paramanand, Rajagopalan & Chellappa 2013), also known as the motion density function (Gupta *et al.* 2010). In the TSF model, a blurred observation is considered to be a weighted average of differently transformed versions of the latent image. We first estimate the TSFs of each of the blurred images using locally estimated point spread functions (PSFs). We follow such an approach because we found that this leads to a more accurate and straightforward solution rather than attempting to estimate the TSF directly as in Whyte *et al.* (2010). To estimate the local PSFs, we use patches from the observations and consider the blur to be space invariant within a patch. After estimating the TSFs, we formulate a cost function to arrive at the scene irradiance that best explains the given input observations. We impose total variation prior on the irradiance for regularization. Thus, we simultaneously achieve the twin objectives of deblurring and HDR imaging.

9.2 Existing approaches to HDRI

In this section, we briefly introduce different HDRI approaches proposed over the years. These include hardware-based HDRI techniques as well as algorithmic attempts. While the aim is to provide an overview, those with direct relevance to our work are subsequently elaborated.

9.2.1 Spatially-varying pixel exposures

A hardware-based approach for HDR imaging was proposed in Brajovic & Kanade (1996). Their design involved an atypical image sensor in which each sensor element included a measuring component to find the time it takes to attain full potential well capacity (saturation). Since the response of all the sensors is assumed to be identical, the measured time in a pixel is proportional to scene irradiance. The HDR irradiance image is derived using these time durations. The approach, although attractive, is restricted by resolution and fabrication costs. Besides, incorporation of such a sensor in a handheld camera is susceptible to motion blur.

Nayar & Mitsunaga (2000) introduced the idea of varying pixel sensitivities for HDR imaging. They identified different ways to spatially vary the exposures, including: (i) fixing different sensor sensitivities for adjacent pixels during fabrication; (ii) applying an optical mask to the sensor array; (iii) using different integration times for adjacent pixels; and (iv) embedding different apertures (micro-lenses) for the potential wells of the pixels. The energy recorded at these pixels depends on both sensitivity and scene irradiance. Hence, this approach results in different measurements in adjacent pixels but represents similar irradiance values, assuming similar irradiance values are incident on adjacent pixels. This spatio-exposure sampling is exploited to compute a high dynamic range image of the scene.

9.2.2 Multiple exposures (irradiance)

The use of multiple exposures is the most widely known and followed approach for HDR imaging. In digital imaging, a wide range of irradiance quantities are mapped to a handful of intensity values; these intensity values are dependent on exposure settings. Varying the exposure settings remaps the scene irradiance to a different set of intensities. Thus, different mappings of the scene irradiance can be exploited to derive the original HDR image. Algorithms based on this approach were initially proposed in Mann & Picard (1995), Debevec & Malik (1997), and Mitsunaga & Nayar (1999). It is well known that the incident irradiance is developed, refined and digitized to obtain an intensity image. These processes can be equivalently represented by a nonlinear function known as the camera response function (CRF), which is a mapping between the total accumulated light energy in the sensors and the intensity image. Given the exposure duration, the mapping to the irradiance values from the intensity image can be viewed as the inverse of the CRF. Mann & Picard (1995) assume the CRF to be a gamma curve. The associated parameters of the expression are estimated using the input images. However, in this chapter, we discuss the irradiance recovery method proposed by Debevec & Malik (1997) since not all CRFs can be defined by gamma curves. In Debevec & Malik (1997), a non-parametric and generic form is considered for the CRF by assuming a smooth function. Mitsunaga & Nayar (1999) proposed that the

CRF could be modeled as a polynomial function and showed that the CRF can be estimated without accurate knowledge of the exposure durations. Since modern cameras can be expected to provide exposure durations accurately, the method in Debevec & Malik (1997) is widely followed. The final HDR irradiance image is then calculated as a weighted average of irradiance maps corresponding to all the input images. The HDR result is stored in radiance formats such as .hdr, OpenEXR, or radiance RGBE. In order to be able to display these images on regular screens, they are tone-mapped to low dynamic range (LDR) intensity images. We give more details in Section 9.3.

9.2.3 Multiple exposures (direct)

Attempts have also been made to obtain HDR tone-mapped equivalent images directly from differently exposed intensity images (Mertens *et al.* 2007, Raman & Chaudhuri 2009, Raskar, Ilie & Yu 2005, Ward 2003). These approaches generate an HDR equivalent (lookalike) LDR image by compositing differently exposed images. Note that compositing directly on image intensity does not lead to the generation of HDR irradiance images. Composition is carried out by finding a weighted average of the input images. The resultant LDR image can be encoded in common 8 bits/channel. These methods are easy to implement, do not require tone mapping, and are faster.

Several multi-exposure fusion methods exist in the literature. The weights of the exposures are determined using features derived from the images. Mertens *et al.* (2007) weigh the exposures based on local contrast, saturation and well-exposedness of a pixel. Ward (2003) chooses the weights based on the median threshold bitmap (MTB) of each of the exposures. Fattal, Lischinski & Werman (2002) and Raskar *et al.* (2005) propose fusion based on the gradient images of different frames. The result in all these techniques is a well-composed LDR image. All the works, except for Ward (2003), assume that the frames in the dataset are registered and that the scene is static.

To give some insight into feature and weight selection, in Mertens *et al.* (2007), the contrast measure (C) of an image is obtained by finding the absolute value of the Laplacian filter output when applied on the grayscale version of the image. The color saturation (S) at each pixel is derived from the standard deviation of values corresponding to the RGB channels. The well-exposedness (G) of a pixel is calculated from a Gaussian curve which is defined in terms of image intensity values. While the mid-intensity corresponds to the mean of the Gaussian function, the variance is defined by the user. Once the parameters C, S, and G are obtained for a pixel (u, v) in the jth image, the weights are derived as

$$W_{u,v,j} = (C_{u,v,j})^{\omega C} + (S_{u,v,j})^{\omega S} + (G_{u,v,j})^{\omega G} \tag{9.1}$$

where ω_C, ω_S, and ω_G are user-assigned values representing the prominence of different measures (typically close to unity). The fused output from N_E observations or exposures (denoted by Z) is obtained as

$$I_{u,v} = \frac{\sum_{j=1}^{N_E} W_{u,v,j} Z_{u,v,j}}{\sum_{j=1}^{N_E} W_{u,v,j}}. \tag{9.2}$$

The weight maps corresponding to input images act as corresponding alpha masks. For seamless blending, fusion is implemented at different resolutions using image pyramids of the masks and input frames.

Although the idea of combining information from multiple images is straightforward, there are restrictions too (in either the irradiance domain or the intensity domain). When there are misalignments between different observations due to relative motion during image capture, or when some of the images suffer from motion blur, the resultant HDR image will be erroneous. While misalignments can be somewhat accounted for by performing registration (Ward 2003), motion blur is not straightforward to handle.

9.3 CRF, irradiance estimation, and tone-mapping

Given several digitized intensity images captured with different exposure times, each of the images can be associated with the same scene irradiance (assuming no relative motion). Initially, the inverse CRF must be estimated from the differently exposed images. The second step is to estimate the scene irradiance by averaging the irradiance images obtained by mapping each of the input intensity images to the irradiance domain (using the estimated inverse CRF). The scene irradiance must be finally tone-mapped for display purposes. We now discuss each of these steps in detail.

9.3.1 Estimation of inverse CRF and irradiance

The relation between the different intensity images and the irradiance image can be written as $Z_{ij} = f(E_i \Delta t_j)$ where Z denotes the observed intensity image, E is the irradiance image, Δt_j is the exposure duration of the jth input image, and i is the spatial index (Debevec & Malik 1997). The CRF f in the above equation is assumed to be smooth, monotonic, and invertible. After taking the natural logarithm, the imaging relation can be written as $\ln f^{-1}\left(Z_{ij}\right) = \ln(E_i) + \ln\left(\Delta t_j\right)$. Let us denote the log-inverse of the CRF i.e. $\ln f^{-1}$ by g.

The objective of HDR imaging is to find E given N_E different input observations $Z_1, Z_2, \ldots, Z_{N_E}$ and corresponding exposure durations Δt_j for the jth observation. If g is known, then the irradiance image can be calculated as

$$E = \frac{\exp\left(g(Z_j)\right)}{\Delta t_j}. \tag{9.3}$$

Considering $Z_{\min}$ and $Z_{\max}$ as the minimum and maximum intensity values of the images, Debevec & Malik (1997) calculate $(Z_{\max} - Z_{\min} + 1)$ values of the log-inverse CRF g and the values of E at N_q pixel locations by minimizing the following cost function

$$\Theta(E, g) = \sum_{i=1}^{N_q} \sum_{j=1}^{N_E} \left(\left[g(Z_{ij}) - \ln E_i - \ln \Delta t_j\right]\right)^2 + \lambda \sum_{z=Z_{\min}+1}^{Z_{\max}-1} \left[g''(z)\right]^2 \tag{9.4}$$

where λ is the smoothness parameter chosen appropriately for the amount of noise expected in Z, and g'' represents the second derivative of g. The choice of N_q is discussed subsequently.

The following points are to be noted (Debevec & Malik 1997):

1. Since the above objective function can result in a scaled version of E_i and g, i.e. a scaled $g + \epsilon$ will be balanced by $((\ln E_i) + \epsilon)$ in the cost function, an additional constraint of $g(Z_{\text{mid}}) = 0$ is added, with $Z_{\text{mid}} = (Z_{\min} + Z_{\max})/2$.
2. Since the CRF is expected to be highly nonlinear, especially at very high and very low intensities, weights are introduced to enhance the contribution of mid-intensities in the cost function. A 256-point (for every intensity) hat function w which gives highest weight to mid-intensity is used to derive the weight maps. The modified cost function is given by

$$\Theta(E, g) = \sum_{i=1}^{N_q} \sum_{j=1}^{N_E} \left(w(Z_{ij}) \left[g(Z_{ij}) - \ln E_i - \ln \Delta t_j \right] \right)^2 + \lambda \sum_{z=Z_{\min}+1}^{Z_{\max}-1} \left[w(z) g''(z) \right]^2 . \tag{9.5}$$

3. This is an overdetermined problem and it is not necessary to use all available pixels of the input image in estimation. If N_E different images are available, the minimum number of pixels N_q required for estimation has to be greater than $(Z_{\max} - Z_{\min})/(N_E - 1)$. These N_q pixels are chosen from regions in the input frame having low variations in intensity and with a reasonably even distribution between $Z_{\min}$ and $Z_{\max}$.

Having recovered the log-inverse CRF g, any of the intensity images Z_j can be mapped to a corresponding irradiance image E. However, there will be inconsistencies among these irradiance images due to the nonlinearity of the CRF (ideally, they should all map to the same irradiance image). The differences between irradiance maps are greater at pixels with very low or high intensities, i.e. the sensors which were either saturated or insensitive. Hence, the final scene irradiance is obtained as a weighted average to render robustness to the solution and is calculated as

$$\ln E_i = \frac{\sum_{j=1}^{N_E} w(Z_{ij}) \left(g(Z_{ij}) - \ln \Delta t_j \right)}{\sum_{j=1}^{N_E} w(Z_{ij})}, \qquad 1 \leqslant i \leqslant N_T \tag{9.6}$$

where E_i is the irradiance value corresponding to pixel location i and N_T is the total number of pixels in E. Using multiple images to calculate E has the advantage of both noise and artifact suppression. The estimated HDR irradiance image cannot be directly displayed and must be tone-mapped before it can be viewed on general 8-bit display systems.

9.3.2 Tone-mapping

The HDR irradiance image can only be displayed on special-purpose display systems which are not yet commercially prevalent (Seetzen, Heidrich, Stuerzlinger, Ward,

Whitehead, Trentacoste, Ghosh & Vorozcovs 2004). General display systems cannot handle the floating point, high-bit resolution of irradiance values. Hence, the HDR images need to be tone-mapped to low dynamic range (LDR) images. Several algorithms have been proposed to carry out this conversion effectively (Bae, Paris & Durand 2006, Drago, Myszkowski, Annen & Chiba 2003, Durand & Dorsey 2002, Fattal *et al.* 2002, Reinhard, Stark, Shirley & Ferwerda 2002). The objective of tone mapping algorithms is to convincingly represent the HDR images on LDR displays, while preserving visually important features (Reinhard, Heidrich, Pattanaik, Debevec, Ward & Myszkowski 2010).

In the popular algorithm of Reinhard *et al.* (2002), a user-specified middle-gray value a is accepted for the displayed image on a scale from 0 to 1. The world luminance of the HDR image is then linearly mapped as $L(u,v) = (a/\bar{L}_W)L_W(u,v)$, where L_W is the input world luminance derived from the scene irradiance map, and $\bar{L}_W$ is the log-average luminance. The final display luminance is obtained as $L_d(u,v) = [L(u,v)]/[1+L(u,v)]$. This operator ensures compression of high intensities and delivers luminance in the display range. We too use this method to tone-map our results. While the algorithm in Reinhard *et al.* (2002) attempts to faithfully reproduce the HDR image, techniques such as Durand & Dorsey (2002) can provide artistic mappings of the scene irradiance.

9.4 HDR imaging under uniform blurring

A technique for obtaining HDR images from differently exposed, blurred input images was first proposed by Lu *et al.* (2009). The input consists of blurred intensity images Z_j, where j is the temporal index taking the values $1, 2, \ldots, N_E$. As discussed in Section 9.3.1, given the inverse CRF f^{-1}, each intensity image can be mapped to the corresponding irradiance image through the inverse camera response function as $B_j = [f^{-1}(Z_j)]/\Delta t_j$. Note that f^{-1} is the inverse CRF and not the log-inverse CRF, and B_j represents the vectorized blurred irradiance image. Lu *et al.* (2009) formulate the problem of simultaneously estimating the irradiance image, the camera response function, and blur kernels using alternating minimization. Based on the assumption that the convolution of the latent irradiance with the kernel is equal to the irradiance-domain equivalent of the blurred input image, Lu *et al.* (2009) obtain a cost function of the form

$$O_1(E, H_j) = \sum_{j=1}^{N_E} \left\| H_j E - B_j \right\|^2 + \lambda_1^2 \left\| H_j \right\|^2 + \lambda_2^2 \left\| E \right\|^2 \tag{9.7}$$

where E is the vectorized latent irradiance image, H_j represents the blurring associated with the jth blurred input image, and the terms λ_1 and λ_2 denote the Tikhonov regularization parameters. This cost is minimized using the Landweber method (Engl, Hanke & Neubauer 1996) alternately for E and the N_E different blur kernels H_j. The inverse

CRF f^{-1} is estimated by inserting E and H estimated in the previous iteration into a cost function which is derived from Eq. (9.5) as

$$O_2(f^{-1}) = \sum_{i=1}^{N_q} \sum_{j=1}^{N_E} \left(w(Z_{ij}) \left[f^{-1}(Z_{ij}) - (H_j E)\Delta t_j \right] \right)^2 + \lambda \sum_{z=Z_{\min}+1}^{Z_{\max}-1} \left[w(z) \left(f^{-1} \right)''(z) \right]^2 \tag{9.8}$$

where w, Δt_j, and N_q are as defined in Section 9.3.1. Even though this is a good step forward, it falls short of being able to handle real-world situations. A major drawback of this approach is the assumption of a uniform blur model, which is not adequate for practical scenarios.

9.5 HDRI for non-uniform blurring

While capturing images with handheld cameras, the possibility of incurring motion blur depends on exposure time. For short exposure times, it is reasonable to assume that no blurring will occur since the displacement of the camera will be negligible during that period. With a handheld camera, images captured beyond a certain exposure duration typically get blurred. The un-blurred shorter exposures provide focused scene irradiance information only in certain regions. While the longer exposures contain information that is not available in the shorter exposures, that information cannot be utilized directly because it is blurred. We now discuss and develop a technique to estimate the latent un-blurred HDR irradiance from differently exposed and possibly blurred images. Since we allow for camera rotations, the blur kernel can vary at every image point. Consequently, a single blur kernel is not adequate to model blurring in this situation. We adopt the TSF model (Paramanand & Rajagopalan 2010) for handling non-uniform motion blur in the HDR imaging problem. The details of the TSF model, estimation of the TSF from locally determined blur kernels (PSFs), PSF estimation, and latent irradiance estimation will now be described.

9.5.1 Image accumulation

As per our earlier discussion in Section 9.3.1, the pixel intensity value in an image is a monotonic but nonlinear function of the irradiance and exposure period (Δt). The accumulated light energy in time Δt is given by

$$X = \int_{t=0}^{\Delta t} E \, dt. \tag{9.9}$$

For a static scene and for a still camera, $X = E\Delta t$ and the relation between irradiance and final intensity may be written as $Z = f(E\Delta t)$. Given the intensity image and CRF, the estimate of irradiance can then be obtained as $E = f^{-1}(Z)/\Delta t$ where f^{-1} is the inverse of the smooth and monotonous CRF.

For a static scene with constant depth (i.e. a fronto-parallel planar scene or a distant scene with negligible depth changes), when there is camera shake, at each instant of time during exposure, the 2D irradiance image is the projection of the 3D scene. The image intensity corresponding to accumulated energy over the exposure duration Δt is given by

$$Z = f\left(\int_{t=0}^{\Delta t} \mathbf{\Gamma}_t (E)\, dt\right) \tag{9.10}$$

where Z is the final intensity image and $\mathbf{\Gamma}_t$ refers to the transformation operation (homography) at time instant t. Note that $\mathbf{\Gamma}_t$ reflects the camera motion at time t, i.e. it yields the path of camera motion. When integrating over time, the temporal information is lost. But this is not an issue since the temporal information is not relevant to our problem. Rather than averaging over time, we adopt the TSF model (Paramanand & Rajagopalan 2010) which averages the warped images over transformations to describe the blurred image.

9.5.2 TSF and its estimation

The TSF can be understood as follows. For a flat static scene, the blurred image resulting from camera shake can be described as a weighted sum of warped versions of the original image, the TSF itself being a function that assigns non-negative weights to the homographies corresponding to each warp. In other words, the TSF is a mapping from a set of homographies to non-negative scalar values in the range [0,1].

We extend the notion of TSFs to the irradiance domain by observing that the irradiance images received at different instants (within the exposure duration) are differently warped versions of the latent irradiance image. The homographies are due to the relative motion between camera and scene. For a scene with constant depth d_o, the 2D images projected at different instances of time (due to camera motion) can be related by a homography (Hartley & Zisserman 2000). Let $\mathcal{H}$ denote the transformation space (i.e. the space of homographies) where a transformation is a point in $\mathcal{H}$ that describes the homography between incident images through the relation

$$P_\nu \left(R + \frac{1}{d_o} T[0\ 0\ 1]\right) P_\nu^{-1}, \tag{9.11}$$

where P_ν is a 3×3 camera matrix, R is a 3×3 rotation matrix, and T is a 3×1 translation vector $[T_X\ T_Y\ T_Z]^T$. The matrix P_ν is assumed to be of the form $P_\nu = \begin{bmatrix} \nu & 0 & 0 \\ 0 & \nu & 0 \\ 0 & 0 & 1 \end{bmatrix}$ where ν denotes the focal length. The rotation matrix R can be expressed in terms of a matrix exponential as $R = e^{\Phi}$ where

$$\Phi = \begin{bmatrix} 0 & -\theta_z & \theta_y \\ \theta_z & 0 & -\theta_x \\ -\theta_y & \theta_x & 0 \end{bmatrix}.$$

The terms θ_x, θ_y, and θ_z represent the angles of rotation about the three standard axes (Whyte *et al.* 2010). Based on the TSF model (Whyte *et al.* 2010, Paramanand &

Rajagopalan 2010, Gupta *et al.* 2010), the blurred intensity image Z in Eq. (9.10) can be equivalently expressed as

$$Z = f\left(\left(\sum_{k=1}^{N_{\mathbf{H}}} \left[\bar{h}_{\mathrm{TSF}}\left(\boldsymbol{\Upsilon}_k\right)\right] \boldsymbol{\Upsilon}_k(E)\right) \Delta t\right) \tag{9.12}$$

where $\boldsymbol{\Upsilon}_k$ is a point in $\mathcal{H}$ and is given by Eq. (9.11), $\bar{h}_{\mathrm{TSF}}$ is the TSF which maps to a non-negative value that denotes the fraction of the total exposure duration spent by the camera in transformation $\boldsymbol{\Upsilon}_k$, while $N_{\mathbf{H}}$ is the total number of homographies in $\mathcal{H}$ (after discretization). We assume that the bounds on $\mathcal{H}$ (as considered in the problem) and the sampling rate provide a sufficient number of transformations ($N_{\mathbf{H}}$) to approximately represent the blurred image. Thus, the total energy accumulated can be viewed as the sum of differently weighted and warped versions of the latent irradiance image projected over the exposure duration.

A camera has 6 degrees of freedom, viz. 3 translations (t_x, t_y, and t_z) and 3 rotations (yaw θ_y, pitch θ_x, and roll θ_z). However, when the scene is largely flat and static (as assumed in this work), the blur induced by (small) out-of-plane rotations is akin to that of translation. Also, while capturing images using handheld cameras, the variation in t_z can be treated as negligible (Gupta *et al.* 2010). Hence, the motion can be described using only in-plane translations and rotation i.e. t_x, t_y, and θ_z (Gupta *et al.* 2010, Whyte, Sivic, Zisserman & Ponce 2012). These three dimensions constitute our 3D transformation space $\mathcal{H}$. As mentioned earlier, each point in this space corresponds to a different transformation or homography. The discrete relation in the argument of f in Eq. (9.12) can be written as a linear equation $Z = f([\mathbf{K} \cdot E] \cdot \Delta t)$, where $\mathbf{K}$ is a large sparse matrix with nonzero elements derived from TSF coefficients $\bar{h}_{\mathrm{TSF}}(\boldsymbol{\Upsilon}_k)$ and bilinearly interpolated weights representing non-integer pixel locations resulting from $\boldsymbol{\Upsilon}_k$ (Whyte *et al.* 2010). Every discrete TSF $\bar{h}_{\mathrm{TSF}}$ can be associated with a corresponding matrix $\mathbf{K}$.

Each of the given blurred observations is related to the latent scene irradiance by a different TSF and each of these TSFs must be estimated in order to solve for the unblurred scene irradiance. In Whyte *et al.* (2010) and Gupta *et al.* (2010), a blind estimation method is used for finding the TSF. We believe that such an approach is not suitable for finding multiple TSFs simultaneously since it does not fully utilize the information available in different observations. The procedure that we employ to estimate the TSF is described next.

For a given blurred observation, the local PSFs at each pixel location can be related to the TSF as

$$h\left(u, v; m, n\right) = \sum_{k=1}^{N_{\mathbf{H}}} \bar{h}_{\mathrm{TSF}}\left(\boldsymbol{\Upsilon}_k\right) \delta_d\left(m - \left(u^k_{\boldsymbol{\Upsilon}} - u\right), n - \left(v^k_{\boldsymbol{\Upsilon}} - v\right)\right) \tag{9.13}$$

where $\left(u^k_{\boldsymbol{\Upsilon}}, v^k_{\boldsymbol{\Upsilon}}\right)$ denotes the position to which (u, v) goes when transformation $\boldsymbol{\Upsilon}_k^{-1}$ is applied, h is the PSF at (u, v), and δ_d denotes the 2D Kronecker delta function. From Eq. (9.13), we note that each local PSF can be considered as a weighted sum of the components of the TSF of the image, i.e. each PSF (vectorized) can be related to the

TSF as $h_{\mathbf{p}_l} = M_l \bar{h}_{\text{TSF}}$ where $l = 1 \ldots N_{\mathbf{p}}$, and M_l is a matrix whose entries are determined by the location $\mathbf{p}_l$ of the blur kernel and interpolation coefficients. When $N_{\mathbf{p}}$ such PSFs are given, stacking these PSF relations leads to the linear relation $\hat{h} = \mathbf{M}\bar{h}_{\text{TSF}}$ where $\mathbf{M}$ is a large sparse matrix derived from M_ls and $\hat{h}$ is a vector consisting of local PSFs.

Unlike optical defocus which has a large spatial support, it is important to note that camera shake results in motion that can be represented with only a small subset of transformations. Hence, a sparsity prior can be fittingly enforced on the TSF $\bar{h}_{\text{TSF}}$. Several other works dealing with camera motion blur also use this constraint (Fergus *et al.* 2006, Whyte *et al.* 2010, Yuan *et al.* 2007). The TSF can then be derived by minimizing the cost function

$$\Theta(\bar{h}_{\text{TSF}}) = \left\| \hat{h} - \mathbf{M}\bar{h}_{\text{TSF}} \right\|_2^2 + \lambda_s \left\| \bar{h}_{\text{TSF}} \right\|_1 . \tag{9.14}$$

While the least-squares term in Eq. (9.14) ensures consistency with the observed blur kernels, the sparsity constraint is imposed by the l_1-norm term. To minimize the cost function, we use the toolbox available in Liu, Ji & Ye (2009). We discuss the estimation of local PSFs in the following subsection.

We conducted an experiment to evaluate the effectiveness of the TSF estimation from locally determined PSFs. Although the experiment was carried out using images in the intensity domain, the approach is valid even on irradiance images. The transformation space was defined as follows: t_x and t_y ranged between -11 and 11 in steps of 1 pixel, θ took values between $0°$ and $2.5°$ in steps of $0.25°$. Along the θ axis, the TSF had nonzero values only between $1°$ and $1.75°$, as plotted in the fourth row of Figure 9.2. On blurring the reference image (Figure 9.2(a)) with the TSF, we obtained the observation shown in Figure 9.2(b). Using patches cropped from four different locations (marked in Figures 9.2(a,b)), we estimated the blur kernels. The PSFs were of size 33×33 and these are shown in the second row of Figure 9.2. With these blur kernels, the TSF was estimated using the proposed method. The components of the estimated TSF for angles between $1°$ and $1.75°$ are shown in the fifth row of Figure 9.2. From the fourth and fifth rows of Figure 9.2, we can see that our estimate is quite close to the ground truth value. Furthermore, the blur kernels generated using the estimated TSF (third row of Figure 9.2) closely match the observed blur kernels (second row of Figure 9.2). The rms error between the observation of Figure 9.2(b) and the image obtained by blurring the reference image with the estimated TSF was found to be only 1.1% of the maximum image intensity. This serves to demonstrate that the TSF was estimated quite accurately from locally derived PSFs.

9.5.3 PSF estimation

Given N_E observations, those which were captured below a certain exposure time are assumed not to have undergone blurring. In order to find the local PSFs of a blurred frame Z_s, local patches from Z_s and a non-blurred shorter exposure (say) Z_r are cropped in such a way that the pixels contain intensities in the mid-range so as to avoid the clipping effects. If there are multiple non-blurred observations, then Z_r corresponds to the one with the longest exposure duration. Note that the intensities of images Z_r, Z_s

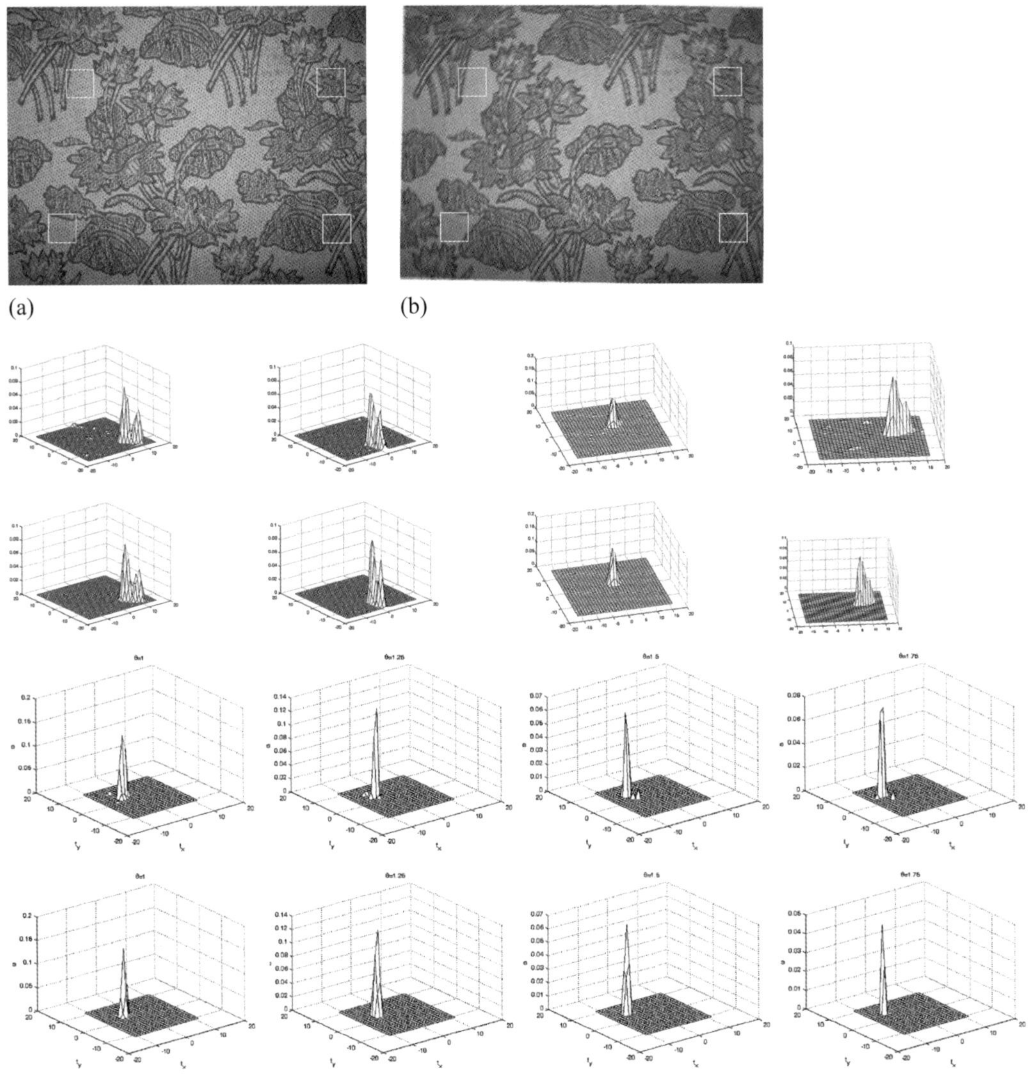

Figure 9.2 (a) Original image; (b) blurred observation. Second row: blur kernels determined from patches. Third row: blur kernels obtained by the estimated TSF. Fourth row: true TSF. Fifth row: estimated TSF.

are not directly comparable. Therefore, the estimation operation is carried out in the irradiance domain using the knowledge of the inverse CRF. Within a blurred frame, we consider $N_{\mathbf{p}}$ different image patches at image locations $\mathbf{p}_1, \mathbf{p}_2, \ldots, \mathbf{p}_{N_{\mathbf{p}}}$. The blur is assumed to be locally space invariant. For each blurred patch in Z_s, it is reasonable to assume that $B^s_{\mathbf{p}_l} = E^r_{\mathbf{p}_l} * h^s_{\mathbf{p}_l}$ where $B^s_{\mathbf{p}_l}$ is the irradiance patch in Z_s at location $\mathbf{p}_l$, $E^r_{\mathbf{p}_l}$ is the corresponding latent irradiance image patch obtained from Z_r (an un-blurred frame), $h^s_{\mathbf{p}_l}$ is the blur kernel at location $\mathbf{p}_l$ of the blurred frame Z_s, and $*$ denotes convolution. We apply a mask m to limit any inconsistencies that might arise due to the truncating nature

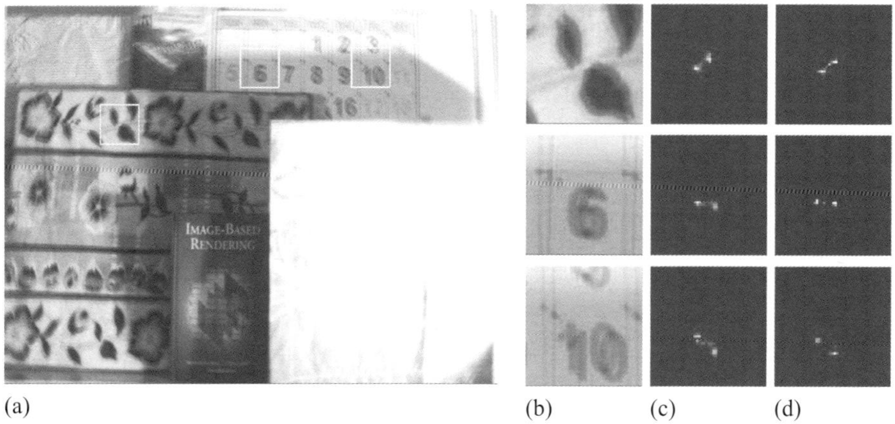

Figure 9.3 Example of PSF estimation. (a) Selected patches are marked in the blurred image; (b) closeups of the cropped patches; (c) the corresponding true PSFs; and (d) the estimated PSFs for each patch. © IEEE 2013.

of the CRF. Each input image has a corresponding mask which is derived based on its intensity values. A single equivalent mask m is applied here to restrict the unreliable pixels in both image patches. The details of mask generation are elaborated later. Each of the blur kernels $h^s_{\mathbf{p}_l}$ is individually estimated by minimizing

$$\Theta(h^s_{\mathbf{p}_l}) = \left\| \left(m \cdot B^s_{\mathbf{p}_l} - m \cdot \left(h^s_{\mathbf{p}_l} * E^r_{\mathbf{p}_l} \right) \right) \right\|^2 \tag{9.15}$$

using the conjugate gradient method. The same procedure is followed for estimating the local PSFs of all the blurred frames. Note that the entire estimation scheme is carried out in the irradiance domain. Figure 9.3(a) shows a synthetically blurred image. The image patches (at locations marked in Figure 9.3(a)), the true PSFs, and estimated PSFs corresponding to these patches are shown in Figures 9.3(b–d), respectively. Note that the PSFs have been estimated accurately by solving Eq. (9.15). The variations in the PSFs across spatial locations clearly reveal the space-variant nature of the underlying blur.

9.5.4 Irradiance image recovery

The objective of our work is to recover the deblurred HDR latent irradiance image. Given N_E differently exposed intensity images $Z_1, Z_2 \ldots Z_{N_E}$, each input image Z_j is associated with its corresponding TSF $\bar{h}^j_{\text{TSF}}$, for $j = 1, 2, \ldots, N_E$. The TSFs of the non-blurred observations consist of a unit weight at the transformation corresponding to zero translation and rotation. Small misalignments across frames are accounted for inherently in the TSF model, i.e., when an observation is transformed (but not blurred) with respect to the reference image, its TSF will have a unity entry at a position corresponding to its transformation. When there are large shifts across frames, although they can still be represented by TSFs, for the ease of computation, one can align them using any established image registration technique. The irradiance image map B_j corresponding

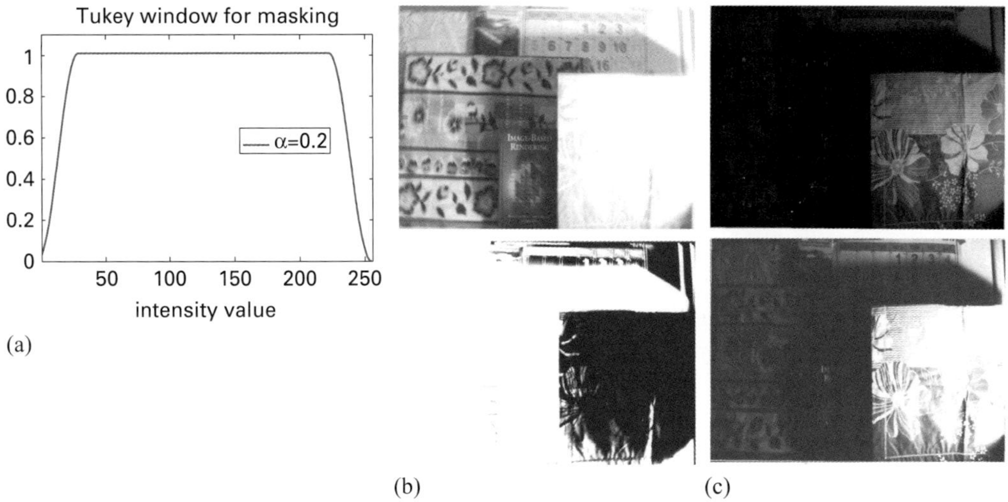

Figure 9.4 (a) Plot of Tukey window function; (b,c) input observations corresponding to different exposure times and their respective masks. © IEEE 2013.

to the observation Z_j is given by $B_j = f^{-1}(Z_j)/\Delta t_j$, where Δt_j is the exposure time of the observation Z_j. Note that the inverse CRF $\left(f^{-1}\right)$ is independent of the scene. Hence, it can be determined offline. In our algorithm, we assume that the knowledge of the inverse CRF $\left(f^{-1}\right)$ is available. For a particular camera setting, by using the standard algorithm in Debevec & Malik (1997), one can determine the inverse CRF.

In order to achieve deblurring and to estimate the latent HDR irradiance image E, the cost function can be formulated as

$$\sum_{j=1}^{N_E} \|\mathbf{K}_j E - B_j\|^2 \tag{9.16}$$

where the term $\mathbf{K}_j E$ represents the blurring of E by the jth TSF. We replace the summation in Eq. (9.16) by concatenating all the blurred irradiance images (B_js) into a single vector $\tilde{B}$ and the corresponding $\mathbf{K}_j$s into a single matrix $\tilde{\mathbf{K}}$ to arrive at $\|\tilde{\mathbf{K}}E - \tilde{B}\|^2$. Considering the nonlinearity of the CRF, we introduce a mask in Eq. (9.16) which is derived based on the grayscale image equivalent of the observations Z_j. When mapped to an irradiance domain, a pixel which is saturated in different exposures or is close to zero leads to different irradiance values, whereas image intensities that are neither too low nor too high get mapped to comparable irradiances. Hence, we propose to use masks to remove inconsistencies in the cost function caused by the CRF. The mask is generated from a 256-point (corresponding to each intensity value) Tukey window (Tukey 1967) which has a smooth transition at both ends when going from 0 to 1. The smoothness is controlled by a parameter α. The Tukey window for $\alpha = 0.2$ is shown in Figure 9.4(a). In Figures 9.4(b,c), we show the masks obtained for images captured at exposure times $\frac{1}{30}$ s and 1 s, respectively. Observe that both saturated and low intensity pixels are assigned negligible weights. In order to maintain color balance, the same mask is applied to all three color channels. On incorporating the effect of masking, we get a modified cost function

$$\|\bar{m} \cdot [\tilde{\mathbf{K}}E] - \bar{m} \cdot \tilde{B}\|^2 \tag{9.17}$$

where $\bar{m}$ is a large diagonal matrix containing masks corresponding to each of the observations.

For regularization, we employ the total variation (TV) prior which incorporates smoothness as well as preserves edges. The latent irradiance image E is obtained by minimizing the following cost function

$$O(E) = \|\bar{m}\tilde{\mathbf{K}}E - \bar{m}\tilde{B}\|^2 + \lambda_{\mathrm{TV}} \int_{\Omega} |\nabla E| \, dx. \tag{9.18}$$

To incorporate the TV prior, we employ the half-quadratic scheme (following the method in Šroubek & Flusser (2005)). The cost function is iteratively minimized by the conjugate gradient method. The derivative of the cost is given by

$$\frac{\partial O}{\partial E} = \bar{m}.\tilde{\mathbf{K}}^T \left(\bar{m}. \left(\tilde{\mathbf{K}}E - \tilde{B}\right)\right) - \lambda_{\mathrm{TV}} \left(\mathrm{div}\left(\frac{\nabla E}{|\nabla E|}\right)\right) \tag{9.19}$$

where $\tilde{\mathbf{K}}^T$ represents the adjoint of the blurring operation and corresponds to inverse warping (Tai *et al.* 2011). The term $\tilde{\mathbf{K}}E$ is obtained by blurring E with each $\mathbf{K}_j$ present in $\tilde{\mathbf{K}}$. It must be noted that in our implementation, we do not generate the matrices $\bar{m}$ or $\tilde{\mathbf{K}}$. Instead of evaluating the term $\tilde{\mathbf{K}}E$ by matrix multiplication, we directly blur the irradiance image E using the TSFs (through warping and averaging). Also, we directly mask the pixel locations instead of generating the matrix $\bar{m}$.

To summarize, the steps involved in the proposed method for estimating the latent scene irradiance from a multi-exposure dataset consisting of blurred images are given below:

1. Calculate (offline) the log-inverse CRF (Section 9.3.1), and map each input image to the irradiance domain.
2. Estimate the PSFs using the highest exposed non-blurred irradiance image and the blurred image (Section 9.5.3).
3. Compute the TSF associated with each blurred image using the locally derived PSFs.
4. Reconstruct the scene irradiance (Section 9.5.4).
5. Tone-map for display purposes.

9.6 Experimental results

Initially, we show the results of our scheme on an image set in which we synthetically blurred the longer exposures. The first row of Figure 9.5 shows a set of images captured at different exposures with a camera placed on a tripod. Two longer exposures that were captured at $\frac{1}{2}$ s and 2 s were blurred using known TSFs. The TSFs were chosen in such a way that they replicated blurring caused by a real camera shake. The support for the

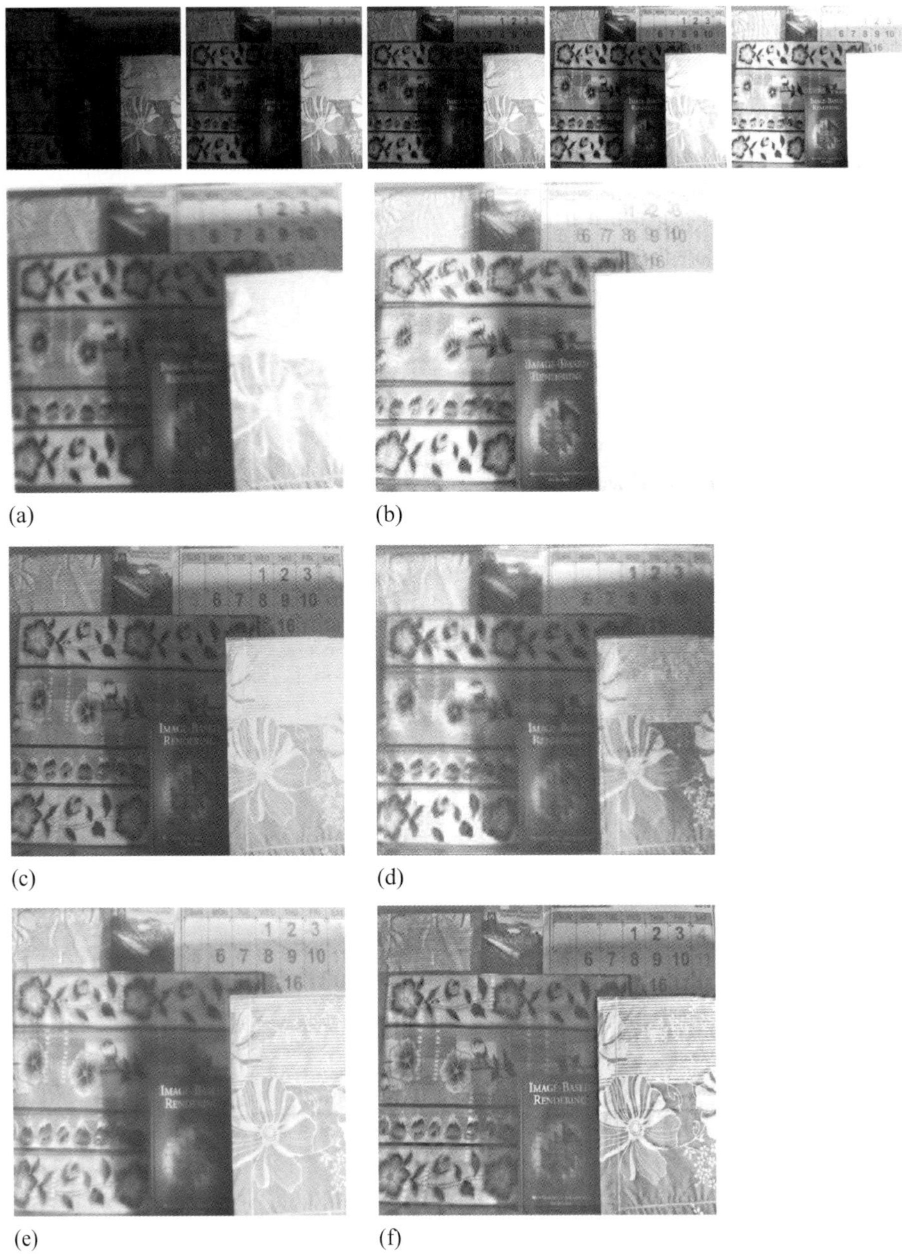

Figure 9.5 First row: different un-blurred exposures captured from a still camera. (a,b) Synthetically blurred longer exposures; (c) tone-mapped HDR image derived from non-blurred images; (d) HDR obtained from separately deblurred images; (e) result obtained from Lu *et al.* (2009); (f) result of our method.

TSFs was defined as follows: t_x and t_y took integer values in the range $[-12, 12]$, and θ_z ranged between $[-1.5°, 1.5°]$ in steps of 0.25°. On blurring, the observations were obtained as shown in Figures 9.5(a,b). These images, along with the un-blurred shorter exposures, were used as input images. In the input image set, some of the regions that are not visible in the shorter exposures are visible only in the blurred longer exposures. When the original un-blurred set (first row of Figure 9.5) was used to determine the irradiance using the method in Debevec & Malik (1997), the tone-mapped HDR image was obtained as shown in Figure 9.5(c). Since this result was obtained in an ideal scenario, it can be considered as a ground truth reference. When we performed irradiance estimation by using the deblurred estimates of blurred longer exposures (using the technique in Whyte *et al.* (2010)) along with the un-blurred shorter exposures, the HDR image was estimated as shown in Figure 9.5(d). It is quite apparent that the resultant image in Figure 9.5(d) has a significant residual blur. Figure 9.5(e) shows the result obtained by the method proposed in Lu *et al.* (2009) for our non-uniformly blurred data. Although the image has been deblurred in certain regions such as the illuminated parts and the book title, residual blur can be seen in the outer region where the effect of rotation is greater. This is not surprising since the method in Lu *et al.* (2009) is not equipped to handle non-uniform blur. The result of our method is shown in Figure 9.5(f), wherein we can see that all the regions of the scene are well illuminated, as well as sharp. Our result is comparable to the ground truth shown in Figure 9.5(c).

Figure 9.6 shows the result of our method on two different sets of real data. The first set shown in the first row of Figure 9.6 has images of size 640×480 captured at exposure times $\frac{1}{15}, \frac{1}{8}, \frac{1}{2}$, and 2 s. Among the observations, the two longest exposures were motion-blurred due to camera shake. We observe that the motion-blurred frames contain information about the lower portion of the scene which is not visible in the less-exposed frames. The second set shown in the second row of Figure 9.6 shows three images whose exposure durations were $\frac{1}{8}, \frac{1}{3}$, and 1 s, respectively. In this set, the observations corresponding to $\frac{1}{3}$ and 1 s are blurred.

The estimated HDR images (after tone-mapping) using our method are shown in Figures 9.6(a,b), wherein we observe that the images are well lit and there are no saturated regions. Also, the blurred information present in the longer exposures has been restored while retaining the information from the shorter exposures. We show closeup patches from the longest exposed images and the estimated HDR image (side-by-side) in the last two rows of Figure 9.6. From the zoomed-in patches, it is clear that smearing in the longest exposure is not present in the resultant image. In the patches corresponding to the second set, we observe that the result is shifted with respect to the input blur patches. This shift is due to the compensation for the misalignment between input patches. The results further reinforce the fact that our method satisfactorily achieves the twin objectives of HDR imaging and deblurring.

Next we considered an outdoor scene. The three observations (shown earlier in Figure 9.1) were captured with exposure times $\frac{1}{100}, \frac{1}{50}$, and $\frac{1}{2}$ s, respectively, from a handheld camera. The longest exposed image captured at $\frac{1}{2}$ s was found to be blurred. By applying the proposed method on this dataset, the deblurred HDR result was obtained as shown in Figure 9.7(a). We observe that the resultant image is deblurred and also well

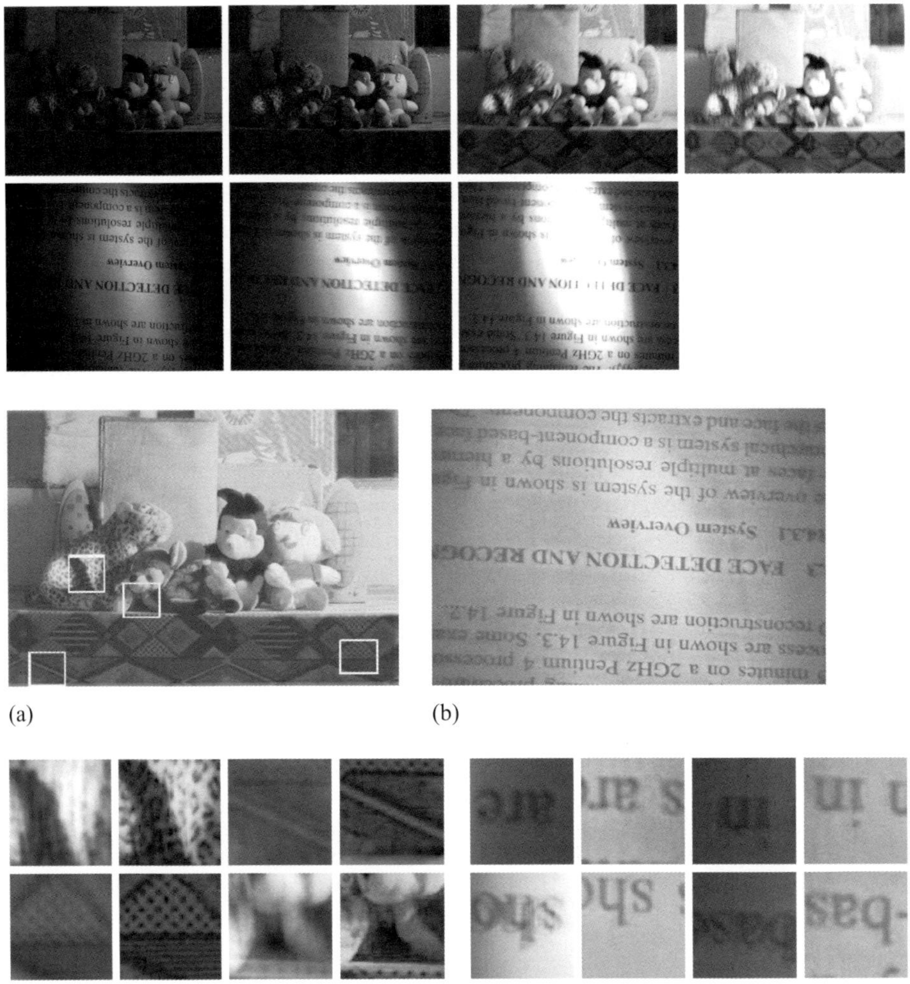

Figure 9.6 Top two rows: images corresponding to two different scenes. (a,b) Estimated HDR images using our scheme. Bottom two rows: closeups of blurred data (available only in the longer exposures) and the corresponding patches in the output images. © Springer 2012.

illuminated. We show the zoomed-in patches from the observations and the resultant images in the two columns of Figure 9.7(b). While patches in the top three rows are from the three observations, the bottom row contains patches from the estimated HDR image. We observe that the railings are visible only in the longest exposure of $\frac{1}{2}$ s, but are blurred. However, in our final result, the railings appear deblurred. Also, the darker background region behind the railing is visible in the result. In the second closeup which corresponds to a brightly illuminated region, information is available only in the shorter non-blurred exposures but is completely saturated in the longer exposures. The closeups from the estimated result highlight the effectiveness of our algorithm.

When the convolution model of Lu *et al.* (2009) was applied, we obtain the result as shown in Figure 9.7(c). Although this result is correctly lit in all regions, deblurring is

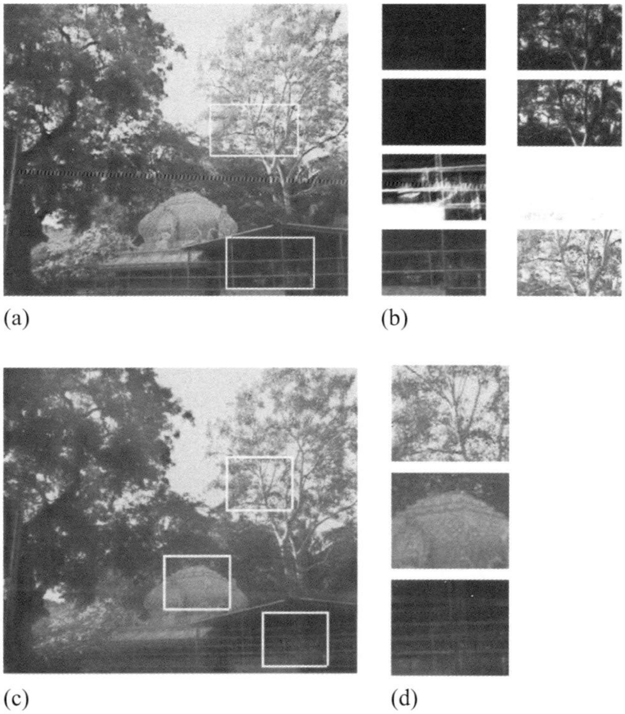

Figure 9.7 Real HDR imaging experiment with blurred longer exposures captured using a handheld camera. (a) The final deblurred HDR result; (b) closeup patches from the input observations are shown in the top three rows while the corresponding output patches from our method are shown in the bottom row; (c) result obtained using the convolution model of Lu *et al.* (2009); (d) zoomed-in patches from (c). © IEEE 2013.

not completely achieved. The image appears sharp in the center and in regions where the scene was brightly illuminated (as can be seen in the top two patches of Figure 9.7(d)). However, the third patch in Figure 9.7(d) shows residual blur near the railings. This result again demonstrates the effectiveness of our method for space-variant blurring situations.

9.7 Conclusions and discussions

We discussed an approach to estimate the high dynamic range irradiance of a static scene from a set of observations captured from a handheld camera. We considered that among the observations, the longer exposures underwent non-uniform motion blur due to camera shake. A two-step procedure was proposed in which we first obtained the TSF of each blurred image, then estimated the latent scene irradiance. We exploited the linear relationship between the TSF of an observation with its local PSFs and developed an elegant method for TSF estimation. The scene irradiance was then estimated by minimizing a suitably derived cost involving all the frames.

There are many directions in which to proceed further. In our method, we assumed that only the longer exposures were blurred. However, our framework can be extended to the scenario wherein *all* the observations are blurred. While capturing images of a 3D scene, blur will generally vary due to parallax effects. Estimation of latent scene irradiance under such conditions would be quite challenging. Also, the problem of HDR imaging of a dynamic scene affected by camera shake is potentially an interesting one to tackle.

Acknowledgments

This chapter is based on our work in Vijay *et al.* (2012) and Vijay *et al.* (2013). We are grateful to IEEE and Springer for permitting us to reproduce portions from them.

References

Bae, S., Paris, S. & Durand, F. (2006). Two-scale tone management for photographic look. *ACM Transactions on Graphics*, **25**(3), 637–45.

Brajovic, V. & Kanade, T. (1996). A sorting image sensor: an example of massively parallel intensity-to-time processing for low-latency computational sensors. In *IEEE Proceedings of International Conference on Robotics and Automation*, vol. 2, pp. 1638–43.

Cai, J., Ji, H., Liu, C. & Shen, Z. (2012). Framelet-based blind motion deblurring from a single image. *IEEE Transactions on Image Processing*, **21**(2), 562–72.

Cho, S., Wang, J. & Lee, S. (2011). Handling outliers in non-blind image deconvolution. In *IEEE International Conference on Computer Vision*, pp. 495–502.

Debevec, P. E. & Malik, J. (1997). Recovering high dynamic range radiance maps from photographs. In *ACM Special Interest Group on Graphics and Interactive Techniques*, pp. 369–78.

Drago, F., Myszkowski, K., Annen, T. & Chiba, N. (2003). Adaptive logarithmic mapping for displaying high contrast scenes. *Computer Graphics Forum*, **22**(3), 419–26.

Durand, F. & Dorsey, J. (2002). Fast bilateral filtering for the display of high-dynamic-range images. *ACM Transactions on Graphics*, **21**(3), 257–66.

Engl, H., Hanke, M. & Neubauer, A. (1996). *Regularization of Inverse Problems*. Springer.

Fattal, R., Lischinski, D. & Werman, M. (2002). Gradient domain high dynamic range compression. *ACM Transactions on Graphics*, **21**(3), 249–56.

Fergus, R., Singh, B., Hertzmann, A., Roweis, S. & Freeman, W. (2006). Removing camera shake from a single photograph. *ACM Transactions on Graphics*, **25**(3), 787–94.

Gupta, A., Joshi, N., Lawrence Zitnick, C., Cohen, M. & Curless, B. (2010). Single image deblurring using motion density functions. In *Proceedings of European Conference on Computer Vision*. Springer, pp. 171–84.

Hartley, R. & Zisserman, A. (2000). *Multiple View Geometry in Computer Vision*, vol. 2. Cambridge University Press.

Kim, S., Tai, Y., Kim, S., Brown, M. & Matsushita, Y. (2012). Nonlinear camera response functions and image deblurring. In *IEEE Proceedings of Computer Vision and Pattern Recognition*, pp. 25–32.

Levin, A., Weiss, Y., Durand, F. & Freeman, W. (2011). Understanding blind deconvolution algorithms. *IEEE Transactions on Pattern Analysis and Machine Intelligence*, **33**(12), 2354–67.

Levin, A., Weiss, Y., Durand, F. & Freeman, W. T. (2009). Understanding and evaluating blind deconvolution algorithms. In *IEEE Proceedings of Computer Vision and Pattern Recognition*, pp. 1964–71.

Liu, J., Ji, S. & Ye, J. (2009). Sparse learning with efficient projections (SLEP). http://www.public.asu.edu/~jye02/Software/SLEP.

Lu, P., Huang, T., Wu, M., Cheng, Y. & Chuang, Y. (2009). High dynamic range image reconstruction from hand-held cameras. In *IEEE Proceedings of Computer Vision and Pattern Recognition*, pp. 509–16.

Mann, S. & Picard., R. W. (1995). On being 'undigital' with digital cameras: extending dynamic range by combining differently exposed pictures. In *Proceedings of 46th Annual Conference of Imaging, Science and Technology*, pp. 422–8.

Mertens, T., Kautz, J. & Van Reeth, F. (2007). Exposure fusion. In *15th IEEE Pacific Conference on Computer Graphics and Applications*, pp. 382–90.

Mitsunaga, T. & Nayar, S. K. (1999). Radiometric self calibration. In *Proceedings of Computer Vision and Pattern Recognition*, pp. 374–80.

Nakamura, J. (2005). *Image Sensors and Signal Processing for Digital Still Cameras*. CRC Press.

Nayar, S. & Mitsunaga, T. (2000). High dynamic range imaging: spatially varying pixel exposures. In *IEEE Proceedings of Computer Vision and Pattern Recognition*, vol. 1, pp. 472–9.

Paramanand, C. & Rajagopalan, A. (2012). Depth from motion and optical blur with an unscented Kalman filter. *IEEE Transactions on Image Processing*, **21**(5), 2798–811.

Paramanand, C. & Rajagopalan, A. N. (2010). Inferring image transformation and structure from motion-blurred images. In *Proceedings of the British Machine Vision Conference*, 73:1–12.

Raman, S. & Chaudhuri, S. (2009). Bilateral filter based compositing for variable exposure photography. In *Eurographics*, pp. 1–4.

Raskar, R., Ilie, A. & Yu, J. (2005). Image fusion for context enhancement and video surrealism. In *ACM Special Interest Group on Graphics and Interactive Techniques Courses*, p. 4.

Reinhard, E., Heidrich, W., Pattanaik, S., Debevec, P., Ward, G. & Myszkowski, K. (2010). *High Dynamic Range Imaging: Acquisition, Display, and Image-based Lighting*. Morgan Kaufmann.

Reinhard, E., Stark, M., Shirley, P. & Ferwerda, J. (2002). Photographic tone reproduction for digital images. *ACM Transactions on Graphics*, **21**(3), 267–76.

Seetzen, H., Heidrich, W., Stuerzlinger, W., Ward, G., Whitehead, L., Trentacoste, M., Ghosh, A. & Vorozcovs, A. (2004). High dynamic range display systems. *ACM Transactions on Graphics*, **23**(3), 760–8.

Shan, Q., Jia, J. & Agarwala, A. (2008). High-quality motion deblurring from a single image. *ACM Transactions on Graphics*, **27**(3), 73–81.

Sorel, M. & Flusser, J. (2008). Space-variant restoration of images degraded by camera motion blur, *IEEE Transactions on Image Processing*, **17**(2), 105–16.

Šroubek, F. & Flusser, J. (2005). Multichannel blind deconvolution of spatially misaligned images. *IEEE Transactions on Image Processing*, **14**(7), 874–83.

Tai, Y., Tan, P. & Brown, M. (2011). Richardson–Lucy deblurring for scenes under a projective motion path. *IEEE Transactions on Pattern Analysis and Machine Intelligence*, **33**(8), 1603–18.

Tukey, J. (1967). An introduction to the calculations of numerical spectrum analysis. *Spectral Analysis of Time Series*, pp. 25–46.

Vijay, C., Paramanand, C. & Rajagopalan, A. N. (2012). HDR imaging under non-uniform blurring. In *European Conference on Computer Vision, Workshops and Demonstrations*, Springer, pp. 451–60.

Vijay, C. S., Paramanand, C., Rajagopalan, A. N. & Chellappa, R. (2013). Non-uniform deblurring in HDR image reconstruction. *IEEE Transactions on Image Processing*, **22**(10), 3739–50.

Ward, G. (2003). Fast, robust image registration for compositing high dynamic range photographs from hand-held exposures. *Journal of Graphics Tools*, **8**(2), 17–30.

Whyte, O., Sivic, J. & Zisserman, A. (2011). Deblurring shaken and partially saturated images. In *IEEE International Conference on Computer Vision Workshops*, pp. 745–52.

Whyte, O., Sivic, J., Zisserman, A. & Ponce, J. (2010). Non-uniform deblurring for shaken images. In *IEEE Proceedings of Computer Vision and Pattern Recognition*, pp. 491–8.

Whyte, O., Sivic, J., Zisserman, A. & Ponce, J. (2012). Non-uniform deblurring for shaken images. *International Journal of Computer Vision*, **98**(2), 168–86.

Yuan, L., Sun, J., Quan, L. & Shum, H. (2007). Image deblurring with blurred/noisy image pairs. *ACM Transactions on Graphics*, **26**(3), 1–8.

10 Compressive video sensing to tackle motion blur

Ashok Veeraraghavan and Dikpal Reddy

10.1 Introduction

Spatial resolution of imaging devices is steadily increasing: mobile phone cameras have 5–10 MP while point-and-shoot cameras have 12–18 MP. As the spatial resolution of imaging devices increases, the effect of either camera or subject motion on the captured images is magnified, resulting in the acquisition of heavily blurred images. Since the motion-blur kernel is a low pass kernel, the high frequencies in the image are heavily attenuated. Deblurring the effects of such low pass blur kernels results in the introduction of significant high frequency artifacts. When the size of the blur kernel is small, these artifacts can be mitigated by the use of appropriate image regularizers that allow for the hallucination of the high frequency detail. Unfortunately there are several scenarios in which such a technique, relying primarily on image regularization, cannot be directly applied. They fall into the following three main categories.

1. Large motion

The resolutions of image sensors are rapidly increasing, while the hands of photographers are not becoming any steadier. This results in ever larger sizes of blur kernels. While image regularization allows us to handle blur kernels of a moderate size (say about 10–15 pixels), larger blur kernels test the ability of even modern deblurring algorithms, especially in regions of high texture. Such large motion blur kernels necessitate active approaches that shape the blur kernel to be invertible in order to handle these robustly.

2. Object motion

Object motion causes blur that is spatially variant. In particular, the blur kernel depends on the local velocity of the objects. To apply traditional image deblurring techniques we require object segmentation which is a hard problem in itself. We can avoid explicit object segmentation by solving a video estimation problem with the help of video priors.

3. Complex camera motion

Complex camera motions such as those due to an unsteady hand during long exposures result in a spatially variant blur. Deblurring such images requires complex motion models. We pose deblurring as a video estimation problem, thereby allowing us to accommodate complex camera motions.

10.1.1 Video compressive sensing to handle complex motion

Traditional methods for image deblurring make strict assumptions about the camera or scene motion and when these assumptions hold true, the observed blurry image can be represented as a spatially-invariant convolution of the sharp image with a blur kernel. Unfortunately, there are several practical scenarios, as mentioned above, when this simplistic assumption breaks down. Traditional image deblurring techniques suffer significant performance degradation when this occurs. An alternative viewpoint of image deblurring that sidesteps the spatial invariance assumption is to reformulate the image deblurring problem as a video estimation problem. Over the last few years, multiple methods have been developed that follow this line of enquiry, resulting in the ability to acquire sharp high temporal resolution videos from one or more acquired blurred images (Takeda & Milanfar 2011, Gupta, Agrawal, Veeraraghavan & Narasimhan 2010, Bub, Tecza, Helmes, Lee & Kohl 2010, Gu, Hitomi, Mitsunaga & Nayar 2010, Sankaranarayanan, Turaga, Baraniuk & Chellappa 2010, Veeraraghavan, Reddy & Raskar 2011, Marcia & Willett 2008, Hitomi, Gu, Gupta, Mitsunaga & Nayar 2011). Two common characteristics of these methods are as follows.

1. **Coded multiplexing**
An imaging architecture that allows for coded multiplexed measurements to be obtained on the sensor. Typically, codes are carefully written to optimize reconstruction performance. While the specifics of the imaging architecture vary, these different choices provide a rich tradeoff between complexity of the architecture and optimality of the multiplexing measurements that can be acquired.

2. **Statistical priors on video**
Since multiple frames from a video get multiplexed onto a single acquired image, the resulting system of equations is under-determined. These methods assume some statistical redundancy in the form of video priors to help solve this under-determined system of equations. The rich choice of reconstruction algorithms allows for a tradeoff between computational complexity and reconstruction performance.

In this chapter, we describe one particular imaging architecture and reconstruction algorithm that we proposed originally in 2011 (Reddy, Veeraraghavan & Chellappa 2011). However, other video compressive sensing techniques (Gupta *et al.* 2010, Bub *et al.* 2010, Gu *et al.* 2010, Sankaranarayanan *et al.* 2010, Veeraraghavan *et al.* 2011, Marcia & Willett 2008, Hitomi *et al.* 2011) tackling image deblurring inherit similar advantages and result in similar performance characteristics.

10.2 Related work

High-speed and high efficiency sensors

Sensors with high quantum efficiency and very low thermal noise (usually achieved by active cooling) are necessary if we need to completely avoid motion blur at acquisition time for fast moving subjects. Moreover, when acquiring videos, apart from light sensitivity, high communication bandwidth is also necessary to acquire images at fast frame rates. Such high-speed cameras are expensive due to the requirement of high light sensitivity and large communication bandwidth. These cameras (*www.photron.com*) usually have limited on-board memory with a dedicated bus connecting the sensor. The acquisition time is limited by the on-board memory. For example, the $300K high-speed FastCam SA5 video camera can capture, at most, 3 seconds of video at 7 500 fps and 1 MP.

Single-multi-image spatial super-resolution

The size of the motion blur kernel is dependent on the resolution of the sensor – more pixels within the same imaging area implies larger motion blur in pixels. This suggests that acquiring a slightly lower resolution image followed by image super-resolution may be a reasonable choice for certain applications, given the maturity of image super-resolution algorithms. Unfortunately, there are fundamental limits to such methods (Lin & Shum 2004, Baker & Kanade 2000). Recent methods tackle this problem either by using matching high–low resolution image pairs (Freeman, Jones & Pasztor 2002) or by modeling images as compressible in appropriate transform basis (Sun, Xu & Shum 2008). These methods have enabled image super-resolution even from a single image (Fattal 2007, Glasner, Bagon & Irani 2010).

Temporal super-resolution

Spatial super-resolution from multiple images (Borman & Stevenson 1998, Park, Park & Kang 2003) is a classical problem which has been studied extensively. Shechtman *et al.* (Shechtman, Caspi & Irani 2005) perform spatio-temporal super-resolution by using multiple cameras with staggered exposures. Similarly, Wilburn *et al.* (Wilburn, Joshi, Vaish, Talvala, Antunez, Barth, Adams, Horowitz & Levoy 2005) use a dense 30 fps camera array to generate a 1000 fps video. Recently Agrawal *et al.* (Agrawal, Gupta, Veeraraghavan & Narasimhan 2010) showed that combining this idea with per camera flutter shutter (FS) (Raskar, Agrawal & Tumblin 2006) significantly improves the performance of such staggered multi-camera high-speed acquisition systems. Ben-Ezra (Ben-Ezra & Nayar 2004) built a hybrid camera where motion is measured using an additional higher frame-rate sensor and then used to estimate the point spread function for deblurring.

Traditional motion deblurring

When a fast phenomenon is acquired via a low frame-rate camera, one can obtain either noisy and aliased sharp images using short exposures, or blurred images using long exposures. Motion deblurring has made great progress by incorporating spatial regularization terms within the deconvolution framework (Shan, Jia, Agarwala *et al.* 2008, Fergus, Singh, Hertzmann, Roweis & Freeman 2006). Novel hardware architectures (Raskar *et al.* 2006, Levin, Sand, Cho, Durand & Freeman 2008) have also been designed to improve deconvolution. These techniques require the knowledge of motion magnitude and/or direction and cannot handle general scenes exhibiting complex motion.

Compressive sensing (CS) of videos

One of the non-traditional techniques for tackling motion blur is to treat the motion blur problem as a compressive video reconstruction problem. In such an approach, instead of modeling an acquired blurred image as a convolution of the sharp image with an unknown blur kernel, the captured blurred image is modeled as a particular set of linear measurements of the underlying video. This reformulation provides a significant advantage since complex scenarios such as nonlinear object motion, multiple independent motions, and spatially-invariant blur can be handled easily in a single framework.

Over the last five years, several examples of such a viewpoint have emerged with a host of imaging architectures and corresponding reconstruction algorithms. Existing methods for video CS assume multiple random linear measurements at each time instant, using either a coded aperture (Marcia & Willett 2008) or a single pixel camera (SPC) (Duarte, Davenport, Takhar, Laska, Sun, Kelly & Baraniuk 2008). Given such a sequence of compressive measurements, prior models about the transform domain sparsity of the video are used to reconstruct the videos. Vaswani (2008) shows that videos with slowly changing dynamics need far fewer measurements for subsequent frames once the first frame is recovered using a standard number of measurements. This is achieved by modeling the sparse coefficients in every frame as a dynamical system, allowing the model to predict only the nonzero coefficients in a newly acquired frame, thus requiring fewer measurements. Park and Wakin (Park & Wakin 2009) present an algorithm for compressive video reconstruction by using a motion-compensated wavelet basis to sparsely represent the spatio-temporal volume. Such methods have achieved only moderate success since (a) the temporal redundancy of videos is not explicitly modeled, and (b) the hardware architectures need to be highly engineered and/or are expensive.

Gu *et al.* (2010) proposed a coded rolling shutter architecture for spatio-temporal tradeoff by exploiting the fact that one can proactively control the exposure time and duration of different rows of a rolling shutter sensor. Flutter shutter (Raskar *et al.* 2006), which was originally proposed for better conditioning the spatially-invariant motion blur problem, has since been extended to be applicable to periodic scenes (Veeraraghavan *et al.* 2011) and more general scene motions (Holloway, Sankaranarayanan, Veeraraghavan & Tambe 2012) by posing as a compressive video reconstruction problem. For the class of videos that can be adequately modeled as a linear dynamical system, Sankaranarayanan *et al.* (2010) provide a method for acquiring videos compressively using the SPC architecture. Both approaches can only handle periodic/dynamic texture scenes, while P2C2 (see Section 10.3.1) can capture arbitrary videos. Gupta *et al.* (2010) show how per-pixel temporal modulation allows flexible post-capture spatio-temporal resolution tradeoff. Similarly, Bub *et al.* (2010) propose a spatio-temporal tradeoff of captured videos which have limited light throughput and, unlike Gupta *et al.* (2010), lack flexible resolution tradeoff. Recently, dictionary learning and sparse representations have been shown to be a powerful technique for performing video compressive sensing (Hitomi *et al.* 2011) and such algorithms, while computationally expensive, are making significant performance improvements.

10.3 Imaging architecture

10.3.1 Programmable pixel compressive camera

To handle the challenging scenarios laid out above, in this chapter we present an imaging architecture (Figure 10.1), known as the programmable pixel compressive camera (P2C2). P2C2 consists of a normal 25 fps, low resolution video camera, with a high resolution, high frame-rate modulating device such as a liquid crystal on silicon (LCOS) or a digital micromirror device (DMD) array. The modulating device modulates each pixel independently in a pre-determined random fashion at a higher rate than the acquisition frame rate of the camera. Thus, each observed frame at the camera is a coded linear combination of the voxels of the underlying high-speed video frames. The effect of the spatial light modulator is to make the blur observed at every pixel coded, thereby making the linear mixing better conditioned. The underlying high resolution, high-speed frames are recovered from the captured low resolution frames by exploiting temporal redundancy in the form of brightness constancy, and spatial redundancy through transform domain sparsity in a convex optimization framework. The advantage of recovering a video over recovering a single image is that this allows for easy handling of spatially-varying blur kernels that result due to both complex camera motion and/or subject motion.

10.3.2 Prototype P2C2

We implement P2C2 with an LCOS mirror SXGA-3DM from Forth Dimension Displays as a spatio-temporal modulator. The mirror has 1280×1024 pixels and each pixel can be fluttered (binary) independently at maximum rate of 3.2 kHz. This imposes an upper limit of 3200 fps on the frame rate of the recovered video. LCOS works by flipping the polarization state of incoming light and therefore needs to be used with a polarizing beamsplitter and the necessary relay optics, as shown in Figure 10.2. The scene is focused on the LCOS device which modulates this incoming light. The Point Grey Dragonfly2 sensor (1024×768 pixels at 25 fps) is in turn focused on the LCOS mirror.

10.3.3 P2C2 as an under-determined linear system

Let the intensity of the desired high frame-rate video be $x(s, t)$ where $s = (r, c) \in [1 \;\; N] \times [1 \;\; N]$ are the row and column coordinates respectively, and $t \in [1 \;\; T]$ the temporal coordinates. We call the higher rate frames x_t "sub-frames", since the acquired frames are formed by integrating them. P2C2 captures the modulated intensities $y(s_l, t_l)$ where $s_l = (r_l, c_l) \in [1 \;\; N/L_s] \times [1 \;\; N/L_s]$ and $t_l \in [1 \;\; T/L_t]$ are its spatial and temporal coordinates. L_s and L_t are the spatial and temporal subsampling factors, respectively. The captured frame y_{t_l} is related to sub-frames x_t as

$$y_{t_l} = D\left(\sum_{t=(t_l-1)L_t+1}^{t_l L_t} x_t \phi_t \right) \tag{10.1}$$

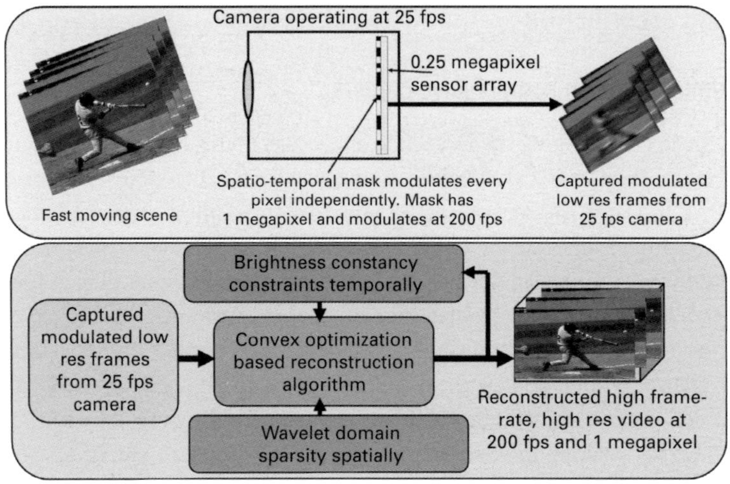

Figure 10.1 Programmable pixel compressive camera (P2C2). Each pixel of a low frame-rate, low resolution camera is modulated independently with a fast, high resolution modulator (LCOS or DMD). The captured modulated low resolution frames are used with accompanying brightness constancy constraints and a wavelet domain sparsity model in a convex optimization framework to recover high resolution, high-speed video. Since deblurring is solved as a video estimation problem, spatially-varying blur can be handled effectively. (Figure reproduced from Reddy *et al.* (2011) with permission from IEEE.)

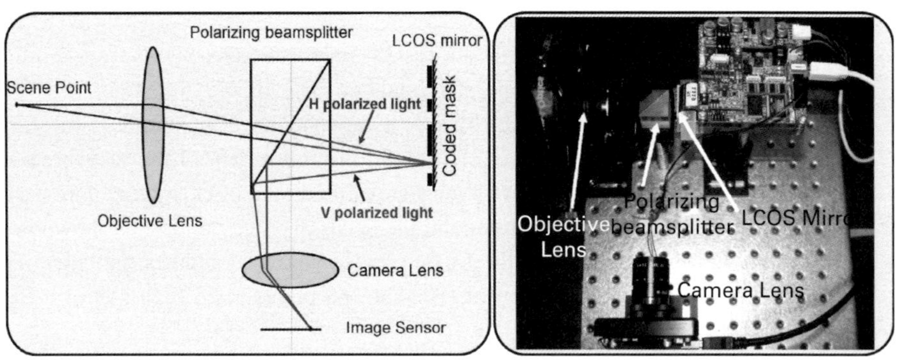

Figure 10.2 Prototype. Illustration of the optical setup. (Figure reproduced from Reddy *et al.* (2011) with permission from IEEE.)

where ϕ is the spatio-temporal modulation function (achieved by LCOS as shown in Figure 10.2) and $x_t\phi_t$ is the modulation of the sub-frame at t with a mask. $D(\cdot)$ denotes a spatial subsampling operation to account for the possibility that the camera could also be spatially modulated at a sub-pixel resolution. Notice that L_t sub-frames are modulated with L_t independent high resolution random masks and then integrated to produce one spatio-temporally subsampled frame of captured video (as shown in Figure 10.3). Moving objects exhibit motion blur in each of the captured frames. The size of this motion blur kernel is proportional to the object velocity, while the kernel shape is controlled by the local per-pixel modulation function applied using the LCOS modulator. As a consequence of the modulation using LCOS, the local blur kernel is no

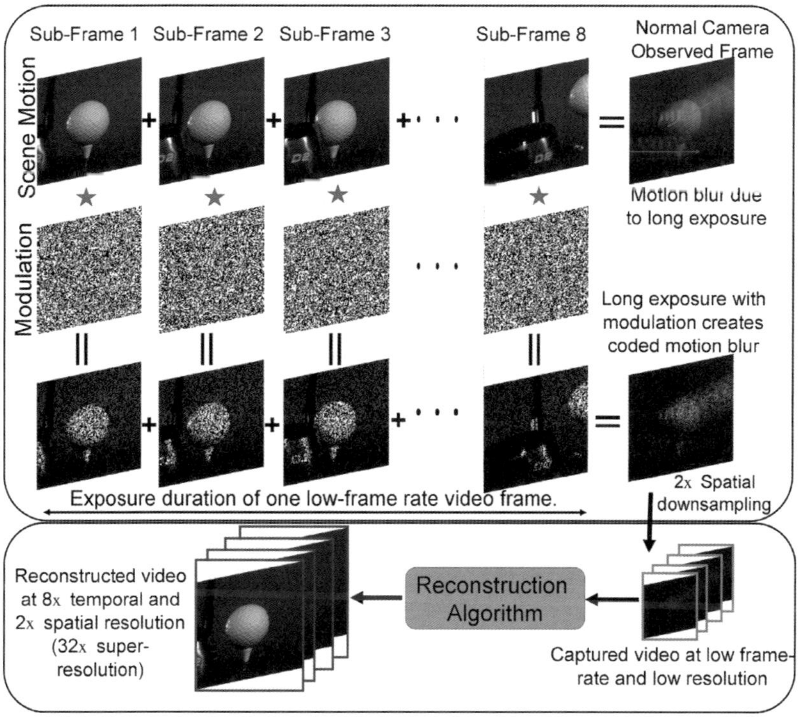

Figure 10.3 Camera architecture. At every pixel the camera independently modulates the incoming light at sub-frame durations and then integrates it. For example, the 3D spatio temporal volume of a golf ball is modulated with a random mask at sub-frame durations and then integrated into a frame. A frame captured by our camera has the code embedded in the blur. (Figure reproduced from Reddy *et al.* (2011) with permission from IEEE.)

longer a box function, but rather becomes better conditioned allowing preservation of high frequency detail.

Since the observed pixel intensities y are linear combinations of the desired voxels x, with the weights given by the modulation function ϕ, Eq. (10.1) can be written in matrix-vector form as

$$\mathbf{y} = \Phi\mathbf{x}, \tag{10.2}$$

where Φ is the matrix representing per-pixel modulation followed by integration in time and spatial subsampling. $\mathbf{x}$ and $\mathbf{y}$ are the vectorized forms of desired high-speed voxels x (e.g. $256 \times 256 \times 32$ voxels) and the captured video y ($128 \times 128 \times 4$ video), respectively, corresponding to a temporal downsampling of 8 and spatial downsampling of 2. The optimization term enforcing fidelity of the recovered sub-frames to the captured frames is given by $E_{\text{data}} = \|\mathbf{y} - \Phi\mathbf{x}\|_2^2$.

10.4 High-speed video recovery

As we described earlier, instead of deblurring the coded blurred images and obtaining a single deblurred image, we attempt to recover a short video. The number of unknown pixel intensities in this short video is much larger than attempted in Eq. (10.2), and therefore the resulting set of linear equations is severely under-determined. To solve for sub-frames $\mathbf{x}$, a prior on spatio-temporal video volume should be incorporated. Most natural video sequences are spatio-temporally redundant. Sparsity of video in some transform domains is an important piece of prior information and is often used in compression, video denoising, and even compressive video acquisition. Temporally, object and/or camera motion preserve the appearance of objects in consecutive frames and this fact is used in video compression schemes such as Moving Picture Experts Group (MPEG). We exploit both forms of redundancy to solve the system of under-determined equations arising from Eq. (10.2). The basic algorithm was originally developed and presented in Reddy *et al.* (2011).

10.4.1 Transform domain sparsity

Each sub-frame is sparse in the appropriate transform domain and we enforce this property in our recovery through ℓ_1 regularization of its transform coefficients. The regularization term enforcing spatial sparsity of sub-frames is $E_{\text{spatial}} = \sum_{t=1}^{T} \beta \|\Psi^{-1}\mathbf{x}_t\|_1$, where $\mathbf{x}_t$ is the vectorized sub-frame x_t and Ψ the transform basis.

10.4.2 Brightness constancy as temporal redundancy

We exploit the brightness constancy (BC) constraint distinctly from, and in addition to, the spatial transform domain sparsity regularization. Consider three consecutive frames of a club hitting a ball in Figure 10.4. The points p_1, p_2, and p_3 correspond to the same point on the golf club in frames x_1, x_2, and x_3, respectively. If relative displacement of these points is estimated, then their pixel intensities in Eq. (10.2) can be constrained to be equal, i.e. the brightness at these pixels is constant $x(p_2, 2) - x(p_1, 1) = 0$ (backward flow) and $x(p_2, 2) - x(p_3, 3) = 0$ (forward flow). This effectively decreases the number of unknowns by 2. The system becomes significantly less under-determined if BC

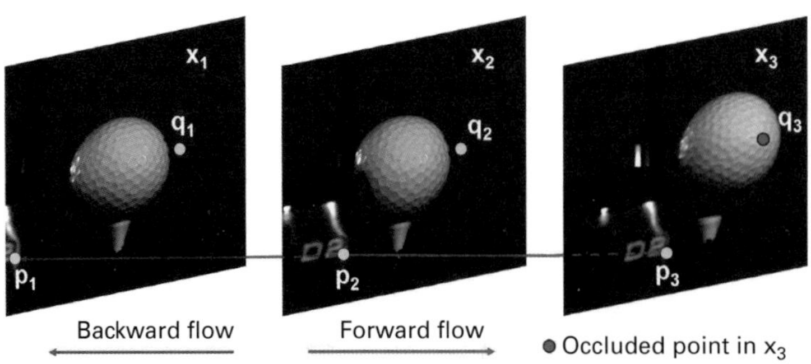

Figure 10.4 Brightness constancy constraints at p_1, p_2, and p_3 and optical flow consistency check at q_2 and q_3. (Figure reproduced from Reddy *et al.* (2011) with permission from IEEE.)

constraints at other points are known as well. The sub-frame BC constraints over the entire video volume are then given by

$$\Omega \mathbf{x} = 0 \tag{10.3}$$

where every row of matrix Ω is the relevant BC equation of a spatio-temporal point (s, t). We incorporate these constraints in the optimization by adding a BC regularization term $E_{BC} = \lambda \|\Omega \mathbf{x}\|_2^2$.

Recovery algorithm

Initialization

Given optical flow, BC constraints are incorporated through E_{BC}. But optical flow can be determined only when the sub-frames are available. Hence, we iteratively determine the sub-frames and the optical flow in an alternating fashion. We begin by estimating the sub-frames without any BC constraints. In the first iteration we tradeoff spatial resolution to recover frames at a desired temporal resolution. We assume that each sub-frame x_t is an upsampled version of a lower spatial resolution frame: $\mathbf{x}_t = U(\mathbf{z}_t)$ where $\mathbf{z}_t$ is a vectorized $\left[\frac{N}{L_s\sqrt{L_t}} \times \frac{N}{L_s\sqrt{L_t}}\right]$ image and $U(\cdot)$ is a linear upsampling operation such as bilinear interpolation. The initial estimate is given by solving

$$\mathbf{z}^0 = \arg\min \sum_{t=1}^{T} \beta \|\Psi^{-1} U(\mathbf{z}_t)\|_1 + \|\mathbf{y} - \Phi U(\mathbf{z})\|_2^2. \tag{10.4}$$

The estimate $\mathbf{x}^0 = U(\mathbf{z}^0)$ does not capture all the spatial detail and is noisy but it preserves the motion information accurately as shown in Figure 10.5(a). We estimate optical flow (Liu 2009) on the initial estimate (Figure 10.5(b)) and perform consistency checks to trim the flow (Figure 10.5(c)) as described in Section 10.4.2. Only the consistent flow is used to build the BC constraint matrix Ω^0 for the next iteration. This flow initialization is noisy but preserves enough detail to improve reconstructions at successive iterations. When the flow at a spatio-temporal pixel (r, c, t) is sub-pixel, we treat the pixel intensity as bilinearly interpolated by the closest pixels in the consequent frame.

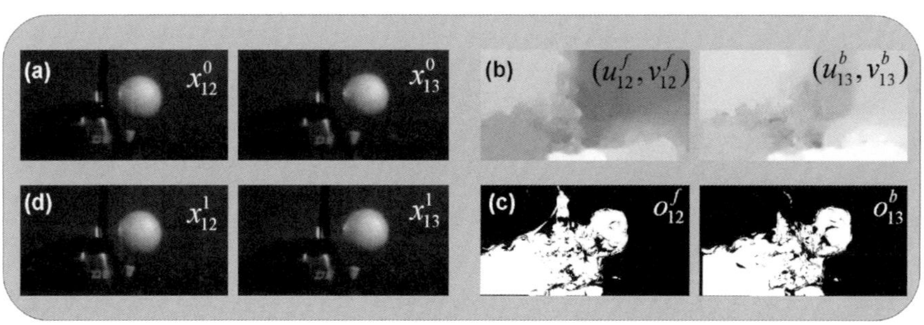

Figure 10.5 Progression of algorithm. Clockwise, from top left: (a) two sub-frames from the initialization; (b) forward and backward OF at respective sub-frames; (c) corresponding forward and backward consistency maps; (d) sub-frames from the next iteration incorporate BC constraints only at white pixels from (c). (Figure reproduced from Reddy *et al.* (2011) with permission from IEEE.)

Optimization

We minimize the total energy function, which also includes the term E_{BC}, built using matrix Ω^{k-1} from previous iterations:

$$\mathbf{x}^k = \arg\min \sum_{t=1}^{T} \beta \|\Psi^{-1}\mathbf{x}_t\|_1 + \|\mathbf{y} - \Phi\mathbf{x}\|_2^2 + \lambda \|\Omega^{k-1}\mathbf{x}\|_2^2. \tag{10.5}$$

The above problem is convex but the problem size is significantly large. Even for a moderate-sized video of 256×256 pixels and 32 frames, we need to solve for 2 million variables. We use a fast algorithm designed for large systems, based on fixed point continuation (Hale, Yin & Zhang 2007), to solve the optimization problem. In all our experiments we fix the parameters at $\beta = 10^{-5}$ and $\lambda = 10^{-1}$. In practice, our algorithm converges in about five iterations.

10.5 Experimental results

10.5.1 Simulation on high-speed videos

Figure 10.6 shows example reconstructions of high-speed sub-frames at $16\times$ temporal super-resolution. Note that while normal camera frames are highly blurred, the reconstruction retains sharpness and high frequency texture detail is maintained. Several of our examples contain complex and nonlinear motion. Most examples also contain several objects moving independently, causing occlusions and disocclusions. To better

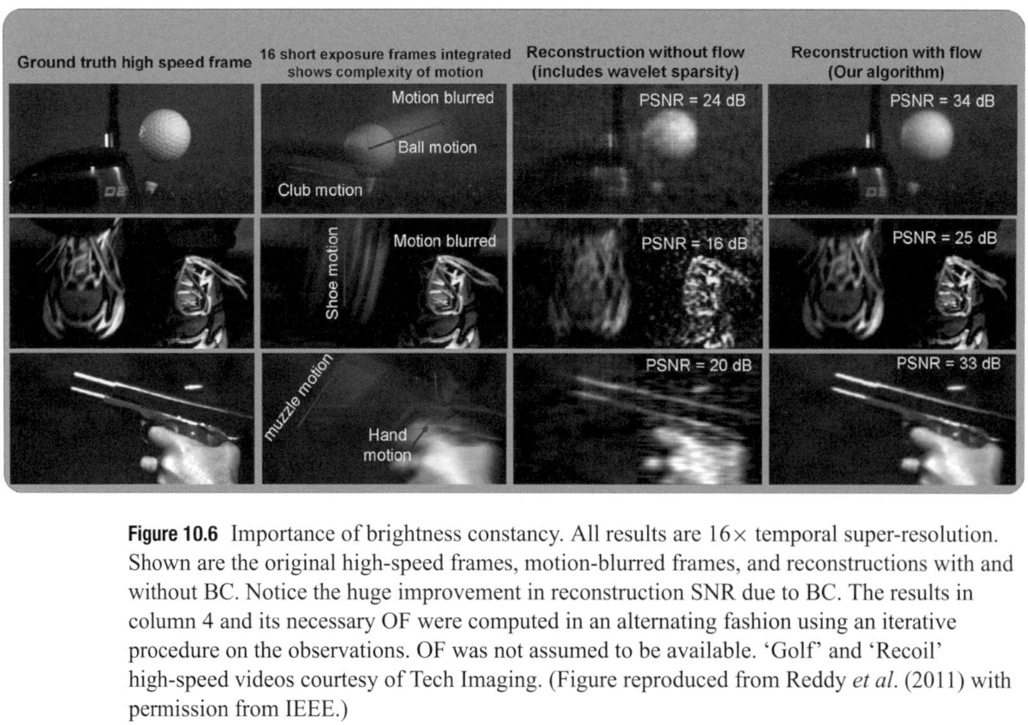

Figure 10.6 Importance of brightness constancy. All results are $16\times$ temporal super-resolution. Shown are the original high-speed frames, motion-blurred frames, and reconstructions with and without BC. Notice the huge improvement in reconstruction SNR due to BC. The results in column 4 and its necessary OF were computed in an alternating fashion using an iterative procedure on the observations. OF was not assumed to be available. 'Golf' and 'Recoil' high-speed videos courtesy of Tech Imaging. (Figure reproduced from Reddy *et al.* (2011) with permission from IEEE.)

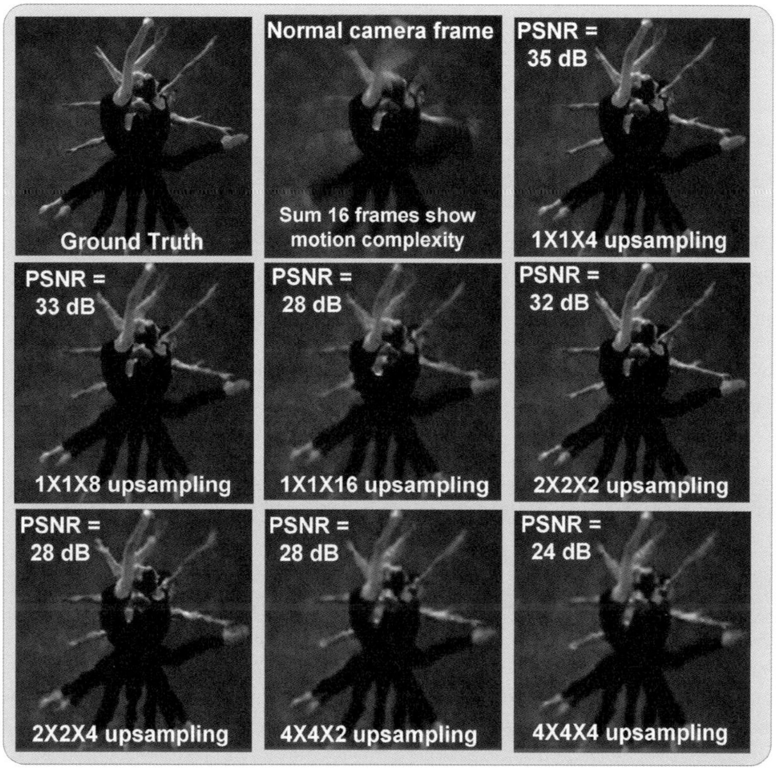

Figure 10.7 Effect of spatio-temporal upsampling factors on Dancers video. Notice that our reconstructions retain visual fidelity even in the presence of complex non-rigid multi-body motions and occlusions. High-speed video courtesy of Tech Imaging. (Figure reproduced from Reddy *et al.* (2011) with permission from IEEE.)

understand the quality of reconstruction with varying spatio-temporal compression factors, examine Figure 10.7. This video contains highly complex motions, where different dancers are performing different movements. There is significant non-rigidity in motion and challenging occlusion effects. Note that our reconstruction retains high fidelity even at high compression factors. Even a compression factor of $4 \times 4 \times 4 = 64$ produces acceptable visual quality and 24 dB PSNR.

10.5.2 Results on P2C2 prototype datasets

We captured several datasets using our prototype device. The camera was operated at 23 fps and the speed of the mask was 8 flips per frame. This allows us to reconstruct the sub-frames at a frame rate of 184 fps (23×8). We note that in our experimental setup we were limited by the field of view since the beamsplitter size forced us to use a lens with a large focal length.

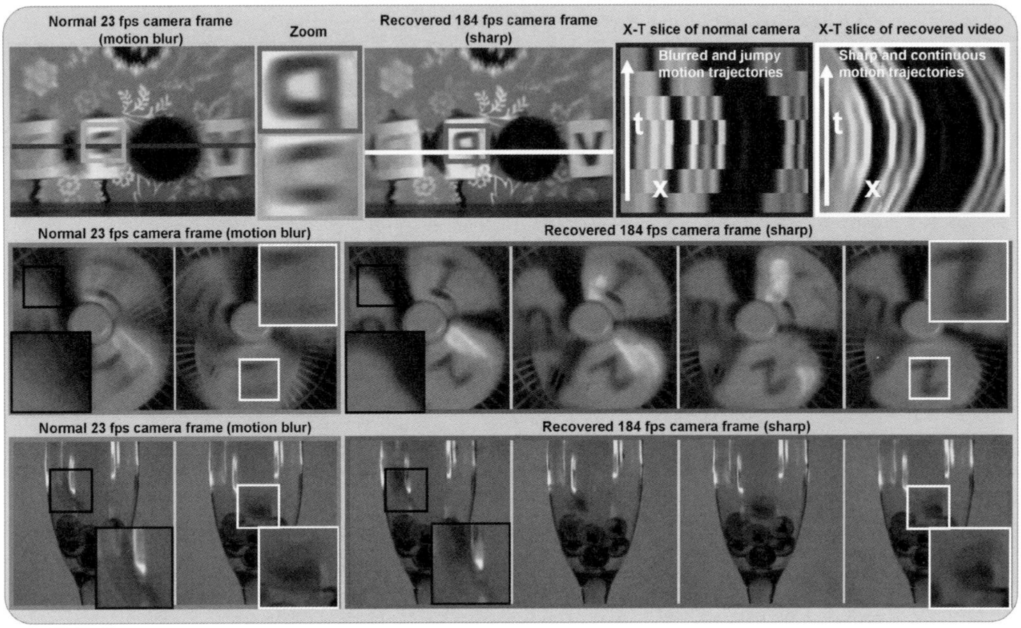

Figure 10.8 Results from LCOS prototype. Top row: using one frame from a normal 23 fps camera, our recovered video and zoom-in insets are shown. The fourth and fifth columns show the x–t slices of the original and recovered videos. Middle and bottom rows: two images from a normal 23 fps camera and four recovered images are shown. (Figure reproduced from Reddy *et al.* (2011) with permission from IEEE.)

In Figure 10.8, we show three different datasets. In the pendulum dataset, the letters C, V, P, R were affixed to four out of five of the balls and the pendulum was swung. As shown in the $x-t$ slices, the balls and the letters had significant acceleration and changes in direction of motion. The recovered frames are much sharper than the original 23 fps frames, as shown in the inset. Note that the characters are much clearer in the reconstructed frame despite a 40 pixel blur in each captured frame. Also, the $x-t$ slice clearly shows reconstruction quality. On the other hand, a fine feature such as the thread is blurred out since the flow corresponding to it is hard to recover.

Next, we rotate a fan and capture it at 23 fps and reconstruct sharper and clearer frames at 184 fps, as shown in the white inset. During recovery, we do not assume that motion is rotational. Note that the normal camera frame has intensities integrated from both the fan blade and the background mesh, as shown in the black inset. We can handle this sub-frame occlusion in our recovery as indicated by the clear background and foreground in the recovered frame.

Finally, we drop a marble in water, capture it at 23 fps, and reconstruct at 184 fps. Again, we do not assume any motion path but still recover the curved path of the marble. Note that despite specularities in the scene, the algorithm is robust.

10.6 Conclusions

In this chapter, we presented an alternative viewpoint for tackling motion blur in images. If we reformulate the motion deblurring problem as one of recovering video from an under-determined set of linear observations, then this naturally allows us to tackle complex scenarios, such as non-uniform motion, multiple independent motions and spatially variant point spread functions, without the need for explicit segmentation. While the contents of this chapter focused exclusively on programmable pixel compressive cameras (Reddy *et al.* 2011), we note that other video compressive sensing techniques (Gupta *et al.* 2010, Bub *et al.* 2010, Gu *et al.* 2010, Sankaranarayanan *et al.* 2010, Veeraraghavan *et al.* 2011, Marcia & Willett 2008, Hitomi *et al.* 2011) inherit similar advantages, resulting in similar performance characteristics.

Acknowledgement

The authors acknowledge that much of the intellectual material included in this chapter was originally developed for a paper presented at the IEEE International Conference on Computer Vision and Pattern Recognition, 2011 (Reddy *et al.* 2011). This work was supported by NSF Grants NSF-IIS:1116718, NSF-CCF:1117939, and by a gift from Samsung Telecommunications America.

References

Agrawal, A., Gupta, M., Veeraraghavan, A. & Narasimhan, S. (2010). Optimal coded sampling for temporal super-resolution. In *IEEE Conference on Computer Vision and Pattern Recognition*, pp. 599–606.

Baker, S. & Kanade, T. (2000). Limits on super-resolution and how to break them. In *IEEE Conference on Computer Vision and Pattern Recognition*, pp. 2372–9.

Ben-Ezra, M. & Nayar, S. K. (2004). Motion-based motion deblurring. *IEEE Transactions on Pattern Analysis and Machine Intelligence*, **26**, 689–98.

Borman, S. & Stevenson, R. (1998). Super-resolution from image sequences – a review. In *IEEE Proceedings of the Midwest Symposium on Circuits and Systems*, pp. 374–8.

Bub, G., Tecza, M., Helmes, M., Lee, P. & Kohl, P. (2010). Temporal pixel multiplexing for simultaneous high-speed, high-resolution imaging. *Nature Methods*, **7**(3), 209–11.

Duarte, M., Davenport, M., Takhar, D., Laska, J., Sun, T., Kelly, K. & Baraniuk, R. (2008). Single-pixel imaging via compressive sampling. *IEEE Signal Processing Magazine*, **25**(2), 83–91.

Fattal, R. (2007). Image upsampling via imposed edge statistics. *ACM Transactions on Graphics*, **26**(3), 95:1–8.

Fergus, R., Singh, B., Hertzmann, A., Roweis, S. & Freeman, W. (2006). Removing camera shake from a single photograph. In *ACM Special Interest Group on Graphics and Interactive Techniques*, **25**(3), 787–94.

Freeman, W., Jones, T. & Pasztor, E. (2002). Example-based super-resolution. *IEEE Computer Graphics and Applications*, **22**(2), 56–65.

Glasner, D., Bagon, S. & Irani, M. (2010). Super-resolution from a single image. In *IEEE 12th International Conference on Computer Vision*, pp. 349–56.

Gu, J., Hitomi, Y., Mitsunaga, T. & Nayar, S. (2010). Coded rolling shutter photography: flexible space-time sampling. In *IEEE International Conference on Computational Photography*, pp. 1–8.

Gupta, M., Agrawal, A., Veeraraghavan, A. & Narasimhan, S. (2010). Flexible Voxels for Motion-Aware Videography. In *European Conference on Computer Vision*, pp. 100–14.

Hale, E., Yin, W. & Zhang, Y. (2007). *A Fixed-Point Continuation Method for ℓ_1-Regularized Minimization with Applications to Compressed Sensing*. CAAM Technical Report TR07-07, Rice University, Houston, Texas, USA.

Hitomi, Y., Gu, J., Gupta, M., Mitsunaga, T. & Nayar, S. K. (2011). Video from a single coded exposure photograph using a learned over-complete dictionary. In *IEEE International Conference on Computer Vision*, pp. 287–94.

Holloway, J., Sankaranarayanan, A., Veeraraghavan, A. & Tambe, S. (2012). Flutter shutter video camera for compressive sensing of videos. In *IEEE International Conference on Computational Photography*, pp. 1–9.

Levin, A., Sand, P., Cho, T., Durand, F. & Freeman, W. (2008). Motion-invariant photography. *ACM Special Interest Group on Graphics and Interactive Techniques*, **27**(3), 71:1–9.

Lin, Z. & Shum, H. (2004). Fundamental limits of reconstruction-based super-resolution algorithms under local translation. In *IEEE Transactions on Pattern Analysis and Machine Intelligence*, pp. 83–97.

Liu, C. (2009). Beyond pixels: exploring new representations and applications for motion analysis, PhD thesis, Massachusetts Institute of Technology, Cambridge, MA, USA.

Marcia, R. F. & Willett, R. M. (2008). Compressive coded aperture video reconstruction. In *Proceedings of the European Signal Processing Conference*, Lausanne, Switzerland, pp. 1–5.

Park, J. & Wakin, M. (2009). A multiscale framework for compressive sensing of video. In *IEEE Picture Coding Symposium*, pp. 1–4.

Park, S., Park, M. & Kang, M. (2003). Super-resolution image reconstruction: a technical overview. *IEEE Signal Processing Magazine*, **20**(3), 21–36.

Raskar, R., Agrawal, A. & Tumblin, J. (2006). Coded exposure photography: motion deblurring using fluttered shutter. In *ACM Special Interest Group on Graphics and Interactive Techniques*, **25**(3), 795–804.

Reddy, D., Veeraraghavan, A. & Chellappa, R. (2011). P2C2: Programmable pixel compressive camera for high speed imaging. In *IEEE Conference on Computer Vision and Pattern Recognition*, pp. 329–36.

Sankaranarayanan, A., Turaga, P., Baraniuk, R. & Chellappa, R. (2010). Compressive acquisition of dynamic scenes. In *European Conference on Computer Vision*, pp. 129–42.

Shan, Q., Jia, J. & Agarwala, A. (2008). High-quality motion deblurring from a single image. *ACM Special Interest Group on Graphics and Interactive Techniques*, **27**(3), 73:1–10.

Shechtman, E., Caspi, Y. & Irani, M. (2005). Space–time super-resolution. *IEEE Transactions on Pattern Analysis and Machine Intelligence*, **27**, 531–45.

Sun, J., Xu, Z. & Shum, H. (2008). Image super-resolution using gradient profile prior. In *IEEE Conference on Computer Vision and Pattern Recognition*, pp. 1–8.

Takeda, H. & Milanfar, P. (2011). Removing motion blur with space–time processing. *IEEE Transactions on Image Processing*, **20**(10), 2990–3000.

Vaswani, N. (2008). Kalman filtered compressed sensing. In *IEEE International Conference on Image Processing*, pp. 893–6.

Veeraraghavan, A., Reddy, D. & Raskar, R. (2011). Coded strobing photography: compressive sensing of high speed periodic videos. *IEEE Transactions on Pattern Analysis and Machine Intelligence*, **33**(4), 671–86.

Wilburn, B., Joshi, N., Vaish, V., Talvala, E., Antunez, E., Barth, A., Adams, A., Horowitz, M. & Levoy, M. (2005). High performance imaging using large camera arrays. *ACM Special Interest Group on Graphics and Interactive Techniques*, **24**(3), 765–76.

www.photron.com.

11 Coded exposure motion deblurring for recognition

Scott McCloskey

This chapter addresses the use of deblurring to improve the performance of image-based recognition, where motion blur may suppress key visual detail. A key difference between recognition and photographic uses of imagery is the extent to which they require a faithful rendering of the real world. For aesthetic photography, where the goal is to produce a pleasant-looking result, image processing does not generally need to produce an accurate rendition. As an example, film and digital camera designers have long preferred to render colors with unrealistically high saturation in order to match the photographer's recollection. Though not intentionally distorting reality, motion deblurring methods using outside information in the form of gradient or edge priors (Jia 2007, Shan, Jia & Agarwala 2008, Fergus, Singh, Hertzmann, Roweis & Freeman 2006, Fergus, Weiss & Torralba 2009, Shan, Xiong & Jia 2007) can be used to produce visually pleasing images.

When an image's ultimate purpose is to facilitate recognition, the use of prior information can be problematic. The methods presented in this chapter are intended to improve the robustness of image-based recognition to subject motion, and we illustrate this using iris-based biometric identification. Because iris biometrics may be used in criminal prosecutions, it requires that the image faithfully renders the real world. Though widely accepted in the computer vision community, even simple concepts such as regularization are problematic when applied to recognition. For motion deblurring, a regularized estimate of the sharp image would satisfy

$$\arg\min_{J} \left(\|f_1(J) - I\| + \lambda f_2(J) \right), \tag{11.1}$$

where I represents the motion-blurred image, f_1 models the blurring process applied to the estimated latent image J, and f_2 evaluates the smoothness of its argument. The problem of using such an approach for biometrics is that, when $\lambda \neq 0$, the estimated sharp image may not be the one that minimizes the data term $\|f_1(J) - I\|$. Put another way, the smoothness term may select one latent image estimate, instead of another, that is more consistent with the observed image. This may be sufficient to create reasonable doubt in the veracity of any identification arising from the deblurred image.

In order to avoid these issues, this chapter develops motion deblurring methods which do not rely on priors for deblurring. In order to avoid the *need* for such priors, we pursue motion deblurring via a computational photographic technique that captures enough iris texture for positive identification. Specifically, we develop an image capture and processing system based on coded exposure. While the basic approach to coded exposure-based image capture had been outlined in previous work (Raskar, Agrawal & Tumblin 2006, Agrawal & Xu 2009, Agrawal & Raskar 2007, Agrawal & Raskar 2009), additional algorithms and hardware implementations were needed in order to automate capture, and to evaluate the method's effectiveness.

Throughout this chapter, we consider shift-invariant blur of objects moving along a linear trajectory. For iris recognition, this arises when a stationary camera captures subjects undergoing motion parallel to the sensor. Though subject motion trajectories are generally 2D, relatively short exposure times allow for good approximations using 1D trajectories. We use the standard shift-invariant convolution model

$$I = J * B + \eta, \tag{11.2}$$

where I represents the captured (motion-blurred) image, J represents the sharp latent image (which we need for identification), B represents the blur point spread function (PSF), and η represents noise. By common convention, we denote image domain quantities (e.g. the blur PSF) as characters such as B and Fourier domain quantities (e.g. the blur optical transfer function) as the corresponding character with a hat, such as $\hat{B}$. The modulation transfer function (MTF), $|\hat{B}|$, is of particular importance in this chapter. One-dimensional functions in the primal domain (e.g. the PSF) are indexed by i, and their Fourier equivalents by k. Two-dimensional functions in the primal domain are indexed by x and y, whereas those in the Fourier domain are indexed by u and v.

While the linear, shift-invariant model does not address shift-variant effects due to camera rotation, it is a good model for a number of iris recognition devices that capture images from people as they stand facing the camera. Even when trying to stand still, people generally sway from side to side. Though this may seem like a trivial amount of motion, the next section describes how this impacts iris recognition performance.

11.1 Motion sensitivity of iris recognition

Iris matching is an excellent application of motion deblurring because the fine texture needed for recognition is easily lost when the subject moves even slightly. The ANSI standard for iris image interchange (ANS ANSI standard INCITS M1/03-0590) requires that the iris image retains at least 60% contrast at spatial frequencies up to two line pairs per millimeter on the iris. Using traditional image capture – where linear, constant velocity motion induces a rectangular PSF – even $\frac{1}{4}$ mm of subject motion during image exposure will reduce contrast to 0% at two line pairs per millimeter. Even if lighting levels are sufficient to use an exposure time of $\frac{1}{400}$ s, the use of a traditional shutter will preclude any subject velocity greater than $\frac{1}{10}$m/s.

In cases where image quality is insufficient for interchange, i.e. saving the iris image and template into an enrollment database, it may still be possible to recognize someone from a traditional, motion-blurred image. For the purpose of future comparison, we begin by evaluating iris recognition performance on deblurred images degraded by linear, constant velocity motion blur observed through a stationary camera with a traditional shutter. We take the iris images used as part of the ICE evaluation (Phillips,

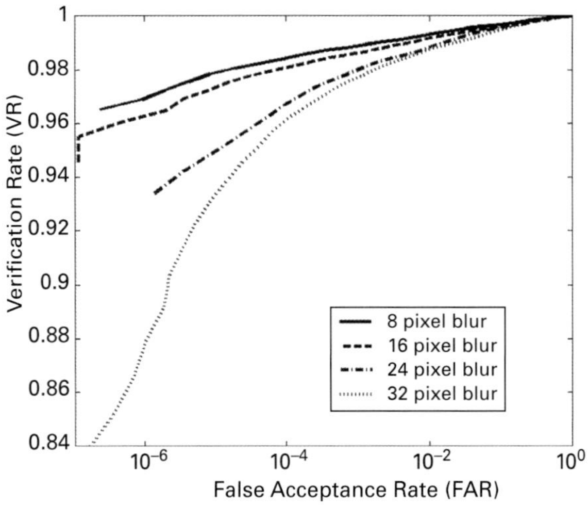

Figure 11.1 ROC curves for synthetically deblurred images from the ICE dataset, using a traditional shutter. Iris matching performance degrades sharply as a function of blur magnitude in traditional shutter images, even when the correct deblurring is applied.

Scruggs, O'Toole, Flynn, Bowyer, Schott & Sharpe 2007), apply convolution with rectangular PSFs of various sizes, add Gaussian noise with $\sigma = 1$ gray level, and deconvolve the PSF. Black-box matching is performed using an algorithm (POSE) similar to the classic Daugman method (Daugman 1993). We measure the verification rate (VR), the false rejection rate (FRR), and the false acceptance rate (FAR) as performance metrics. For a given threshold Hamming distance τ, two iris images are considered a computed match if the Hamming distance computed by POSE is less than or equal to τ. FRR, FAR, and VR are defined as:

$$\mathrm{FRR}(\tau) = \frac{N_r(\tau)}{N_{sr}}, \qquad \mathrm{FAR}(\tau) = \frac{N_a(\tau)}{N_{sa}}, \tag{11.3}$$

$$\mathrm{VR}(\tau) = 1 - \mathrm{FRR}(\tau), \tag{11.4}$$

where

- $N_a(\tau)$ is the number of iris matches that do not belong to the same subject as the test sample and the scores are lower than the threshold τ;
- $N_r(\tau)$ is the number of iris matches that belong to the same subject as the test sample and the scores are greater than the threshold τ;
- N_{sr} is the total number of enrolled samples that belong to the test subject;
- N_{sa} is the total number of enrolled samples that do not belong to the test subject.

Figure 11.1 shows the receiver operator characteristic (ROC) curves. Separate ROC curves are given for different blur extents (8, 16, 24, and 32 pixels) corresponding to a range of subject motions ($\frac{1}{3}$, $\frac{2}{3}$, 1, and $\frac{4}{3}$ mm) over the exposure duration. From this, we can see that traditional motion blur reduces iris matching performance significantly *even when the motion blur is perfectly linear and removed by deconvolution*. This degradation can be understood as a consequence of the spatial frequencies lost due to the use of a traditional shutter. The Fourier transform of its rectangular PSF is a sinc function which equals zero at a number of *lost spatial frequencies*. The original iris texture at these frequencies is not contained in the captured image I; the only content at those frequencies is due to the noise η. The deconvolution process amplifies this noise, but cannot reintroduce the lost iris texture. The number of lost spatial frequencies increases with the number of pixels in the rectangular PSF, which explains the reduced iris matching performance as a function of PSF size.

11.2 Coded exposure

Though not motivated by iris recognition performance, the problem of lost spatial frequencies in motion blur is exactly the problem addressed by the computational photographic technique known as *coded exposure* (Raskar *et al.* 2006). Because losing spatial frequencies is a direct result of the shape of the PSF arising from a traditional shutter, the fundamental idea of coded exposure is to temporally modulate light in order to reshape the PSF and avoid lost frequencies. Doing this does not avoid motion blur in the image, but rather makes the motion blur invertible at all spatial frequencies by a numerically stable deconvolution.

This section develops the steps of a coded exposure imaging system with the intent of capturing sharp images of moving irises (illustrated in Figure 11.2). First, the coded exposure image is captured by manipulating a shutter during image acquisition. Next, the coded exposure image and the shutter's open/close sequence are used to estimate the direction and extent of blur. Finally, the shutter sequence and blur estimate are used to define a stable deconvolution which produces an estimated latent image. Each of these steps is described in the following subsections.

11.2.1 Sequence selection for image capture

Conceptually, the shutter is any mechanism that modulates exposure in the captured image, and may include a mechanical shutter (as used in a single lens reflex camera), an electro-optical mechanism (e.g. the LCD shutter used in Raskar *et al.* (2006)), or image formation from multiple exposures. The shutter's timing sequence is represented as a binary sequence $[S(0)\ S(1)\dots S(N-1)]$ of length N, where $S(i) = 1$ if the shutter is open during the ith *chop*, and $S(i) = 0$ indicates that the shutter is closed. Together with the temporal duration of each chop t_{chop}, the shutter sequence specifies the camera's

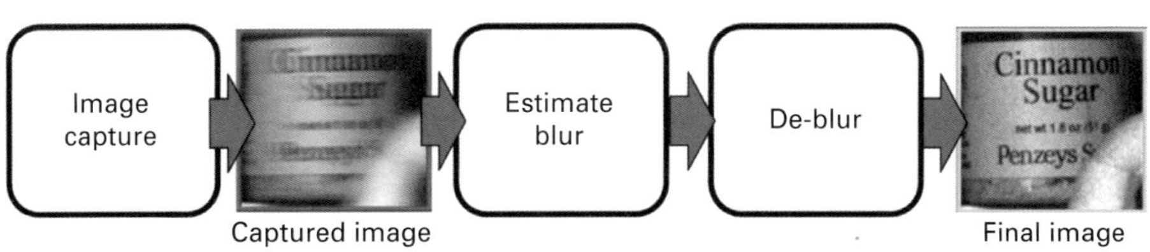

Figure 11.2 The stages of coded exposure image capture and deblurring.

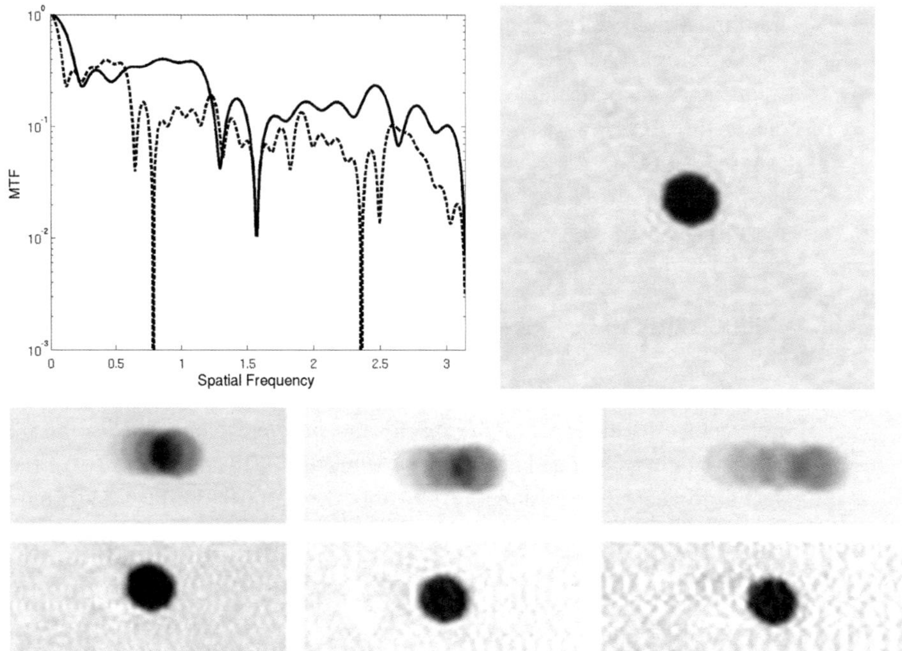

Figure 11.3 The impact of velocity on coded exposure performance. For a particular shutter sequence, the PSF/MTF depends on the subject velocity. Top left: the MTFs of motion through a particular fluttering shutter for object velocities of 3 pixels/ms (solid line) and 6 pixels/ms (dashed line). Top right: reference image of the stationary target. Middle row: coded exposure images of a dot moving left to right at 3, 4, and 6 pixels/ms. Bottom row: deblurred images. Although the reconstructed image quality is good for object speeds of 3 pixels/ms (bottom left), there are lost frequencies at 6 pixels/ms (bottom right) and the reconstruction has noticeable artifacts.

behavior during image capture. Note that the shutter mechanism used may place restrictions on the minimum value of t_{chop}, as discussed in McCloskey (2010).

Since a fluttering shutter is closed for part of the capture time, we distinguish between the *exposure time* (how long the shutter is open during capture, $t_{\text{chop}} \sum_i S(i)$) and the *capture time* (from the first shutter open to the last shutter close, including closed shutter periods, Nt_{chop}). Though the exposure and capture times of a traditional shutter are equal, the flutter shutter's capture time can be as much as twice the exposure time. When comparing traditional and flutter shutter images, one must choose between comparing equal exposure or capture times. In order to keep the magnitude of shot noise fixed, this chapter presents images with equal exposure times. As a result, the coded exposure images will have a greater blur extent than the traditional images.

Though the shutter sequence is one determinant, the effective PSF also depends on the object's velocity[1]. Because of the PSF's dependence on velocity, a particular shutter sequence defines a family of PSFs, as illustrated in Figure 11.3. A notable member of

[1] Though the velocity on the image sensor depends on both the object's real-world velocity and its distance from the camera, it is the image velocity that determines the PSF.

this family, which we refer to as the "nominal" PSF, is effective when the object moves over a range of N pixels during exposure with a fluttering sequence composed of N chops. In this case the effective PSF (we will call it B_N) is equal to a scaled version of the chop sequence S,

$$B_N(i) = \frac{S(i)}{\sum S(i)}, \quad \text{for } i = 0, 1, \ldots, N-1. \tag{11.5}$$

In this case, which has been considered in Raskar *et al.* (2006) and Agrawal & Xu (2009), the Fourier transform $\widehat{B_N}$ of the PSF B_N will be the same as that of the chop sequence S (up to a scale factor). Presuming that S was chosen to preserve all frequencies, this nominal PSF is invertible and the sharp image can be recovered.

In the general case, however, the PSF is a stretched version of the chop sequence and may not be invertible. In fact, no chop sequence can generate a family consisting of invertible PSFs – i.e. those avoiding lost spatial frequencies – for all velocities. In particular, if the object moves over $2N$ pixels during the course of exposure, the effective PSF will be

$$B_{2N} = \frac{1}{2\sum S(t)} \cdot [S(0)\ S(0)\ S(1)\ S(1) \ldots S(N-1)\ S(N-1)], \tag{11.6}$$

where $\cdot$ represents an element-wise multiplication of the sequence. As we will now demonstrate, B_{2N} suffers lost frequencies for any choice of S.

LEMMA 11.1 *Let S be an arbitrary chop sequence of length N. The effective PSF B_{2N} for an object that moves over $2N$ pixels during exposure will have a lost frequency at $k = \frac{N}{2}$.*

Proof Let $A = \frac{1}{2\sum S(t)}$.

$$\begin{aligned}
\widehat{B_{2N}}(k) &= A \sum_{t=0}^{N-1} S(t) \left(e^{-i\frac{2\pi}{N}k(2t+1)} + e^{-i\frac{2\pi}{N}k2t}\right) \\
&= A \sum_{t=0}^{N-1} S(t) e^{-i\frac{2\pi}{N}kt} \left(e^{-i\frac{2\pi}{N}k(t+1)} + e^{-i\frac{2\pi}{N}kt}\right) \\
\widehat{B_{2N}}\left(\frac{N}{2}\right) &= A \sum_{t=0}^{N-1} S(t) e^{-i\pi t} \left(e^{-i\pi(t+1)} + e^{-i\pi t}\right) \\
&= A \sum_{t=0}^{N-1} S(t) e^{-i\pi t} 0 = 0.
\end{aligned} \tag{11.7}$$

□

It can similarly be shown that PSFs of the general form

$$B_{\kappa N} = \frac{1}{\kappa \sum S(t)} * \left[\underbrace{S(0)\ldots S(0)}_{\kappa \text{ times}}, \ldots \underbrace{S(N-1)\ldots S(N-1)}_{\kappa \text{ times}}\right], \tag{11.8}$$

will have at least $\kappa - 1$ lost frequencies. It can also be shown that for any object that moves over more than $2N$ pixels, the MTF will have at least one zero in the range $0 \leqslant k \leqslant \frac{N}{2}$. The implication is that the invertibility of coded exposure blur depends on subject velocity v, and that a coded exposure camera needs to adapt the shutter timing to subject motion. One trivial way to enforce a constant PSF length is to estimate subject velocity and set $t_{\text{chop}} = \frac{N}{v}$, but this has the side effect of reducing the image's exposure time and thus increasing the relative contribution of noise. In order to avoid this, we choose to pre-compute a bank of shutter sequences and use an estimate of subject motion to select between them at capture time.

For any given length, there are many shutter sequences whose nominal PSFs avoid lost frequencies, and we need to choose amongst them. In addition to presenting a 52 chop shutter sequence, Raskar *et al.* (2006) present a method by which other shutter sequences may be selected. Random PSFs are sampled, and the one satisfying $\text{argmax}_B \min_k \|\hat{B}(k)\|$ is chosen as per the code provided by Amit Agrawal. We demonstrate the inadequacy of this max–min criterion using 750 equal exposure PSFs: a traditional shutter PSF with 42 open shutter chops, and 749 random permutations thereof. For each of the 750 PSFs, we synthetically blurred the 100 test images from the Berkeley Segmentation Dataset (BSD) (Martin, Fowlkes, Tal & Malik 2001), added noise (Gaussian with $\sigma = 1$ relative to intensities in [0,255]), deblurred, and computed the RMSE versus the original image. Figure 11.4 shows the histogram of mean RMSE values for these PSFs, and illustrates two important points. First, that *even a random coded exposure PSF will give much lower RMSE than a traditional shutter*. Second, it shows that

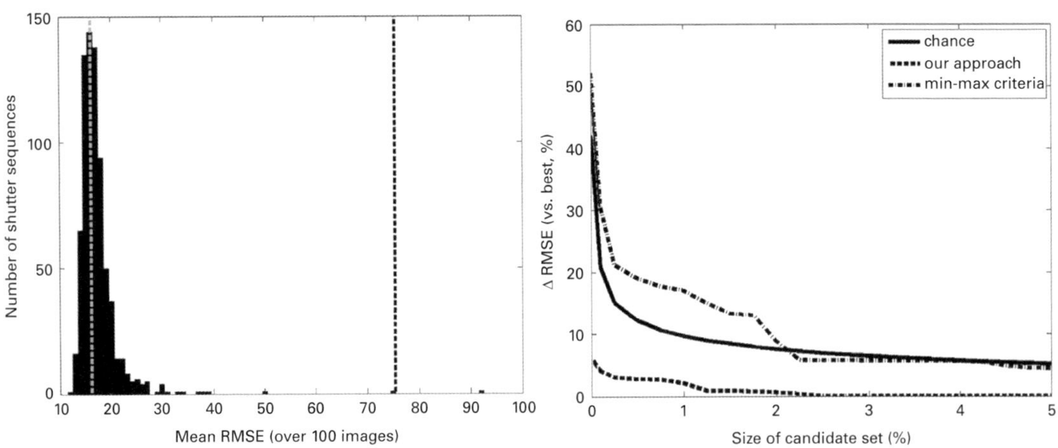

Figure 11.4 Coded exposure PSF design using natural image criteria. Left: the histogram shows the average RMSE over 100 images using the traditional shutter (RMSE = 75; right dashed line) and 749 equal exposure coded PSFs. The PSF satisfying the max–min criterion has RMSE = 16 (left dashed line), about 30% worse than the best. Right: with natural image statistics, we select a PSF providing near-optimal reconstruction while finding the actual RMSE on a small subset of the search space. As compared to chance and max–min sampling, the proposed algorithm produces significantly better results. Note: the leftmost point on the x-axis is at $\frac{1}{50}$, not 0.

Table 11.1 Complete list of metrics collected to find the optimal PSFs.

Number of open shutter periods	m_1
MTF minimum	$m_2 = \min_k \lvert\hat{B}(k)\rvert$
MTF mean	$m_3 = \text{mean}_k(\lvert\hat{B}(k)\rvert)$
MTF variance	$m_4 = \text{var}_k(\lvert\hat{B}(k)\rvert)$
Prior-weighted mean	$m_5 = \text{mean}_k \frac{\lvert\hat{B}(k)\rvert}{k+1}$
Number of lost frequencies	$m_6 = \lVert\{k \text{ where } \lvert\hat{B}(k)\rvert < 10^{-10}\}\rVert$
Weighted lost frequencies	$m_7 = \sum_k \frac{\chi(k)}{k+1}$, where $\chi(k) = 1 \iff \lvert\hat{B}(k)\rvert < 10^{-10}$

the max–min criterion is insufficient, as the PSF satisfying it produces an RMSE 30% worse than the optimal, comparable to random selection.

The failure of the max–min criterion to produce the best PSF can be understood by recalling that natural images have rapidly diminishing power in higher spatial frequencies (our model for natural image statistics is described in the next section). The max–min criterion selects a contrast-preserving PSF *without differentiating, based on which frequency corresponds to the minimum* – despite natural images having highly uneven power distribution. Since natural images have more power in low frequencies, we improve RMSE performance by finding PSFs which preserve higher contrast in those areas.

In addition to the MTF minimum, we computed a number of other metrics on the 750 PSFs. The complete list of metrics collected is shown in Table 11.1. In order to use these metrics to find the optimal PSF, we learn a function mapping a vector of the metrics to average RMSE. With the metrics and RMSE computed for the 750 PSFs, we use linear regression (Matlab's \ operator) to find weights w such that the quantity

$$\sum_{i=0}^{7} w_i m_i \tag{11.9}$$

best fits the mean RMSE, with $m_0 = 1$ included to provide an offset term. We use a combination of natural image-weighted metrics and synthetic blurring/deblurring to choose a PSF that provides low reconstruction RMSE. The high-level steps of our algorithm are outlined in Figure 11.5, as follows.

Training

1. Generate a small set of PSFs.
2. Compute the metrics $m_{1,2,\ldots 7}$ for each PSF.
3. Find the actual RMSE for these PSFs over a representative set of images.
4. Using these actual RMSEs and computed metrics on the PSFs, learn weights $w_{1,2,\ldots,7}$ by regression.

Testing

1. Compute metrics $m_{1,2,\ldots,7}$ for each PSF in the search space.
2. Predict each PSF's RMSE performance using the weighting and Eq. (11.9).

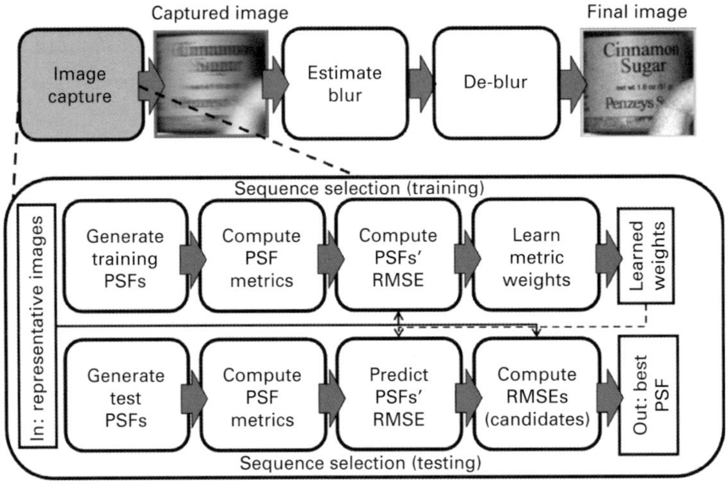

Figure 11.5 The stages of our shutter selection algorithm.

3. Construct a small candidate set containing the best PSFs according to the predicted RMSE.
4. Find the actual RMSEs over the images for the PSFs in the small set of candidates, and choose the one with the lowest actual RMSE.

In order to evaluate this algorithm's effectiveness, we selected an additional 7500 equal-exposure PSFs. For each, we find both the metrics listed in Table 11.1 and the ground truth RMSE over the 100 BSD images.

From these data, we simulate 1000 runs of our algorithm. For each, we randomly select 5000 of the 7500 PSFs as the search space and use our method to find the best one. We compare the actual RMSE of that PSF to the one with the ground truth optimal RMSE in the space. Ideally the difference would be 0, indicating that we have found the best PSF of those in the search space. Figure 11.4 shows a plot of this difference as a function of the size of the candidate set. Over the 1000 trials, the average RMSE of the best PSF was 12.00, and the average RMSE of the PSF with the best weighted combination of metrics was 12.74 (a difference of 0.74, or 6.2%) indicating that, while the weighted metrics out-perform the max–min criterion, the projection does *not* generally produce the optimum PSF. When the 1% of the PSFs with the best projected RMSE are evaluated to find their actual RMSE, the difference narrows to 0.26 (2.2%), and is further reduced as the candidate set is enlarged (the dashed curve). For comparison, we show that if the candidate set is comprised of randomly-selected PSFs instead of using our metric weighting, the comparable algorithm would produce the solid curve. Likewise, if the candidate set were formed using the max–min criterion, the algorithm would produce the dash–dot curve. Our algorithm produces an actual RMSE that is closer to that of the optimal PSF, significantly improving on the criteria from Raskar *et al.* (2006).

In our coded exposure system, we precompute several optimal shutter sequences of varying lengths, and select the required sequence based on an estimate of the subject velocity. This allows us to capture images without losing spatial frequencies, but the velocity estimate is not precise enough to use directly in deblurring. In order to address this, we have developed the image-based blur estimation algorithm detailed in the next section.

11.2.2 Blur estimation

Though the original coded exposure demonstration depended on manual blur estimation by a user, subsequent work has addressed automatic coded blur estimation. Dai and Wu (Dai & Wu 2008) treat motion blur as an alpha matte for estimating the PSF. Like the coded exposure PSF estimation of Agrawal and Xu (Agrawal & Xu 2009), the implicit assumption in alpha-matte-based PSF estimation is the existence of a high-contrast edge in the latent sharp image. We have found that these methods work poorly on iris images, due to a lack of high-contrast edges. Iris images often have high contrast on the pupil/iris boundary but the pupil's diameter can easily be shorter than the blur extent, making alpha matte methods impractical.

In this section, we develop an alternative method to estimate the coded exposure PSF from a single input image. Instead of making a strong assumption about the image's content in the spatial domain, we assume that the image contains only the motion-blurred object, and that the object's texture obeys a natural image prior. Relative to our prior expectation that the object's amplitude spectrum is smooth, we note that the local minima and maxima of a coded exposure MTF (recall Figure 11.3) induce similar minima/maxima in the blurred image's amplitude spectrum, and that these inflection points can be used to estimate the moving object's velocity.

PSF modeling

In order to describe our PSF estimation method, we first formalize the relationship between the PSF and object motion. The normalized PSF $B(x)$ describes what proportion of the exposure of a moving scene point goes to each pixel x. As we have demonstrated, it is a function of both the shutter sequence $S(t)$ and the subject velocity ν. We measure the motion parameters (displacement and velocity) in units of pixels, e.g. velocity as pixels/chop.

A pixel x gets exposed to a moving point when its image passes through x, and exposure duration $w(x)$ is inversely proportional to velocity $\nu(x)$:

$$w(x) = \frac{1}{\nu(x)}. \tag{11.10}$$

Note that while our immediate concern is constant velocity motion, it is natural to describe the velocity and displacement of more general motions in terms of t. Thus, we can rewrite $w(x) = 1/[\nu(t(x))]$, where $t(x)$ is the inverse of the displacement function $x(t)$. For our model, we assume that $x(t)$ is monotonic throughout the shutter sequence, i.e. there is no back and forth motion during exposure.

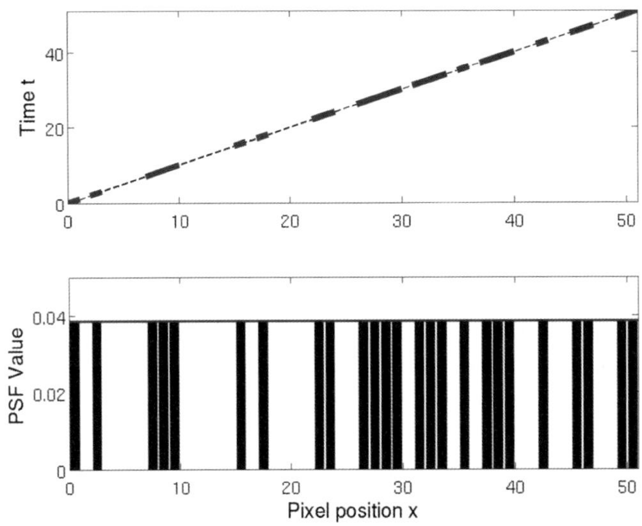

Figure 11.6 PSF modeling for constant velocity motion. Top row: the time–velocity function (thin dashed line) sampled by the shutter (thick solid line representing open shutter segments). Bottom row: the corresponding PSFs.

Finally, we combine the shutter sequence and the exposure $w(x)$ to compute the unnormalized PSF $B_0(x)^2$ as:

$$B_0(x) = S(t(x))w(t(x)) = \frac{S(t(x))}{\nu(t(x))}. \tag{11.11}$$

Equation (11.11) indicates that the PSF can be viewed as an envelope of $w(x)$ sampled by the shutter pattern $S(t)$ as shown in Figure 11.6. To derive the PSF for arbitrary motions, we simply need to derive $t(x)$.

For constant velocity motion at ν_c pixels/chop, we assume the first exposed pixel is the zeroth pixel, and we have $x(t) = \nu_c \cdot t$ and $t(x) = x/\nu_c$. The PSF is thus:

$$B_0(x) = \frac{S(t(x))}{\nu_c} = \frac{S(\frac{x}{\nu_c})}{\nu_c}, \qquad x = 0, \ldots, \nu_c(N-1). \tag{11.12}$$

Since the last exposed pixel coordinate has $x(M_s) = \nu_c M_s$, we can compute the normalized PSF as:

$$B(x) = \frac{B_0(x)}{\sum_{x=0}^{\nu_c(N-1)} B_0(x)} = \frac{S(\frac{x}{\nu_c})}{\nu_c E_s}, \qquad x = 0, \ldots, \nu_c(N-1) \tag{11.13}$$

where $E_s = t_{\text{chop}} \sum_{i=0}^{N-1} S(i)$ is the total exposure time.

Equation (11.13) indicates that varying the velocity ν_c will result in spatial scaling in the PSF $B(x)$, while the envelope of $B(x)$ remains a rectangle. An example is shown in Figure 11.6.

[2] $B_0(x)$ is later normalized to the unit-area PSF $B(x)$.

Our goal is to recover the PSF by analyzing the amplitude spectrum of a blurred image. Recall that the process of motion blur can be modeled as standard convolution:

$$I = J * B + \eta, \tag{11.14}$$

where $*$ is the convolution operator, I is the coded exposure image, J is the latent sharp image, B is the blur kernel, and η is Gaussian noise. By the convolution theorem, the amplitude spectrum of Eq. (11.14) is:

$$|\hat{I}| = |\hat{J}\hat{B}| + |\hat{\eta}| = |\hat{J}||\hat{B}| + |\hat{\eta}|. \tag{11.15}$$

Since $\hat{\eta}$ has constant expected amplitude at all frequencies, noise is an offset in the frequency domain and can be ignored in our analysis of relative spectral power.

Power spectrum statistics

Our PSF estimation algorithm is based on power spectrum statistics in natural images, as modeled by Eq. (11.15). Van der Schaaf and van Hateren (van der Schaaf & van Hateren 1996) have shown that collections of natural images *without* motion blur can be modeled as having circular power spectrum statistics that follow the $1/\omega$-exponent model: if we parameterize $|\hat{J}|$ in polar coordinates (ω, ϕ) where ω is the radius (absolute frequency) and ϕ is the angle, we can average $|\hat{J}|$ over ϕ for every ω and the resulting circular averaged power spectrum $\text{circ}_\omega(|\hat{J}|) \approx C/\omega^m$, where m and C are constants. Statistically, if we assume every frequency is an independent and identically distributed random variable, the expected value of $|\hat{J}(u, v)|$ is:

$$E\left[\left|\hat{J}(u, v)\right|\right] = \frac{C}{(u^2 + v^2)^{m/2}}. \tag{11.16}$$

Figure 11.7 shows example traces of the amplitude spectra of five natural images. From this we can see that the $1/\omega$-exponent model, which is known to hold for *collections* of natural images, also holds for *individual* images.

Our goal is to use power spectrum statistics to recover the coded exposure PSF from the blurred image I. Because we are interested in linear subject motion, we consider natural image statistics after the 2D amplitude spectrum is projected onto a line l that corresponds to the motion direction. We rotate the Fourier plane so that l is aligned with the u axis and apply the projection by integrating over v. This process can be alternatively viewed as applying a Radon transform (Toft 1996) along the v direction. In the discrete case, we can compute the linear averaged amplitude spectrum of an image $|\hat{J}|$ as:

$$R_u[|\hat{J}|] = \frac{1}{V}\sum_{v=0}^{V} |\hat{J}(u, v)|, \tag{11.17}$$

where V is the v-dimension resolution. $R_u[|\hat{J}|]$ represents the horizontal power spectrum statistics and can be approximated using Eq. (11.16) as:

$$R_u[|\hat{J}|] \approx E\left[\frac{1}{V}\sum_{v=0}^{V} |\hat{J}(u, v)|\right] = \frac{1}{V}\sum_{v=0}^{V} \frac{C}{(u^2 + v^2)^{m/2}}. \tag{11.18}$$

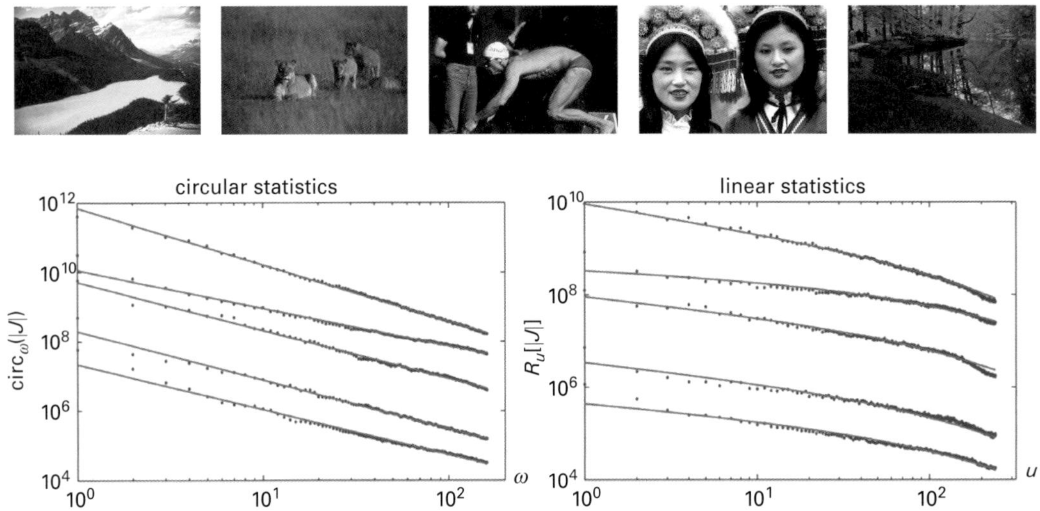

Figure 11.7 Power spectrum statistics on five randomly selected images (top row) from the BSD image set. Bottom left: the circular power spectrum vs the spatial frequency ω in a log–log scale. The lines show the fits of the $1/\omega$-exponent model to points reflecting image data. The scaling of the vertical axis belongs to the top trace. Bottom right: the linear statistics along v vs u in a log–log scale. The curves show our estimated linear statistics from the circular statistics. For clarity, traces in both plots are shifted -1, -2, -3, and -4 log-units.

Figure 11.7 illustrates that our R_u estimation still provides good agreement with an image's spectrum even after projection. We can further apply the R_u operator to the amplitude spectrum model of convolution (Eq. (11.15)):

$$R_u[|\hat{I}|] = \sum_{v=0}^{V} |\hat{J}(u,v)||\hat{B}(u)| = |\hat{B}(u)| \cdot R_u[|\hat{J}|]. \tag{11.19}$$

Equation (11.19) allows us to separate $R_u[|\hat{J}|]$ and $|\hat{B}|$. We can further take the log of Eq. (11.19) as:

$$\log(R_u[|I|]) = \log(|B|) + \log(R_u[|J|]). \tag{11.20}$$

PSF estimation algorithm

Figure 11.8 outlines the steps of our motion estimation algorithm. We first determine the motion direction and rotate the image to align it with the u axis. For every candidate velocity v, we compute its PSF B^v, MTF $|\hat{B}^v|$, and then use them to estimate the latent image amplitude spectrum $|\hat{J}^v| = |\hat{I}|/|\hat{B}^v|$. We then compute the linear statistics $R_u[|\hat{J}^v|]$, and $R_u[|\hat{I}|]$. Finally, we compute the match score μ between $\log(|\hat{B}^v|)$ and $\log(R_u[|\hat{I}|]) - \log(R_u[|\hat{J}^v|])$. The optimal velocity v corresponds to the one that maximizes μ.

Estimating the motion direction

We adopt a similar approach to Oliveira, Figueiredo & Bioucas-Dias (2007) that finds the direction with the most muted high frequencies. This assumes that the latent sharp

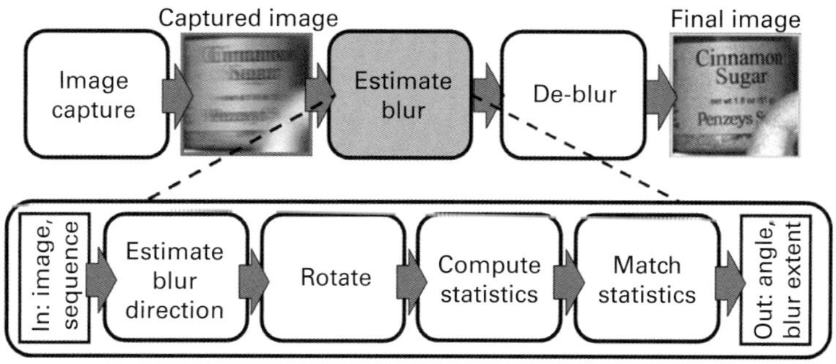

Figure 11.8 Steps of our statistical PSF estimation algorithm.

image is roughly isotropic, i.e. the power spectrum distributions along all directions have similar characteristics (variance, mean values). Because the iris region is circular, the power spectrum is relatively isotropic. Since 1D motion blur attenuates the middle- and high-frequency information in the direction of motion, it detects the direction in which they are most muted. We do this by inspecting the Radon-power spectrum of the blurred image in all directions and choosing the one with the largest variance.

Computing linear statistics of $|\hat{J}^\nu|$

A crucial step in our motion estimation algorithm is to derive the linear statistics of $|\hat{J}^\nu| = |\hat{I}|/|\hat{B}^\nu|$. When $\hat{J}^\nu$ is motion blur free, it should follow $1/\omega$-exponent distribution. To estimate C and m, we compute the discrete circular averaged power spectrum and apply line fitting between $\log(\text{circ}_\omega[|\hat{J}^\nu|])$ and $\log(\omega)$. We then approximate the linear statistics $R_u[|\hat{J}^\nu|]$ using Eq. (11.18).

Matching log-linear statistics

Recall that our ultimate goal is to match $f_1 = \log(|\hat{B}^\nu|)$ and $f_2 = \log(R_u[|\hat{I}|]) - \log(R_u[|\hat{J}^\nu|])$ under some metric μ. A native μ is to measure the squared difference at sampled points on f_1 and f_2. Since the power spectra of images generally have much smaller values in high frequency, directly computing the correlation between the estimate f_1 and f_2 results in unequal contributions from different frequencies.

Instead, we employ a metric based on the signs of the function derivatives to treat all frequencies equally. Specifically, we use a derivative sign function $\Gamma(\cdot)$:

$$\Gamma(f(u)) = \begin{cases} 1, & \dfrac{df}{du} \geqslant 0 \\ -1, & \dfrac{df}{du} < 0 \end{cases} \tag{11.21}$$

where f is a 1D function on u.

Finally, we sample f_1 and f_2 at discrete points $u_1, u_2, \ldots, u_n$, and compute:

$$\mu(f_1, f_2) = \sum_{i=1}^{n} \Gamma(f_1(u_i))\Gamma(f_2(u_i)). \tag{11.22}$$

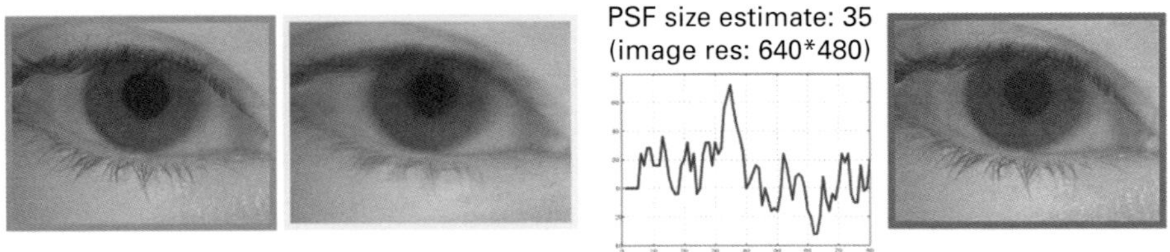

(a) (b) (c) (d)

Figure 11.9 Motion estimation and deblurring results on an iris image. (a) The ground truth blur-free image; (b) the blurred image caused by constant velocity motion under coded exposure; (c) the matching metric μ vs velocity; (d) the deblurred results using our estimated velocity.

PSF estimation results

We have applied our technique to all of the publicly available flutter shutter images (Raskar *et al.* 2006) and have found that our method produces estimates that are within 1 pixel of the ground truth values, giving high quality reconstructions. In order to test additional cases, including irises, we have acquired test images using a Point Grey Flea®2 camera triggered via the serial port of the controlling computer. The camera supports an external shutter mode that accumulates exposure over several chops, after which a single readout produces the flutter shutter image. We captured the images from a fixed camera observing a motion stage to which textured objects were attached. Our motion stage can change velocity via voltage controls. To measure the ground truth velocity, we use a step edge calibration target and measure its blurred width in an image with a known exposure time. We use the 52 chop sequence from Raskar *et al.* (2006) and obtain the ground truth PSF using Eq. (11.13).

Figure 11.9 shows an example acquired using this setup, with an iris moving from right to left with a constant velocity. The motion is axis-aligned horizontally, and our estimated motion direction is within 1° of this ground truth. The plot shows the matching metric μ computed over a range of potential velocities, which has a pronounced peak at exactly the ground truth value (35 pixels). The resulting PSF estimate is used to deblur the captured image, giving a deblurred result that looks very similar to the image captured without motion. The iris template extracted from our deblurred image was successfully matched to a separate image of the same eye. Iris recognition was unsuccessful on a Lucy–Richardson (Lucy 1974) deblurred traditional image captured with the same setup.

11.2.3 Deblurring

The example in the previous section was illustrated using the flutter shutter deblurring code provided in Raskar *et al.* (2006), which performs least squares deconvolution via matrix inversion. For the purposes of iris identification, the simplicity of this deconvolution method is useful. In light of the issues that were discussed in this chapter's introduction, we have avoided using methods that impose prior assumptions about image appearance in the deblurring step. By definition, the solution to the least squares deconvolution avoids this because the deblurred image must agree most with the observed image. When the observed image contains noise, of course, that component will be amplified by deconvolution and the result will not exactly match the desired latent image. Handling this noise component is the reason that prior terms are added to regularized deconvolution so, in exchange for transparency in deblurring, we should expect that a least squares solution will have more noise-related artifacts.

In order to understand how the choice of non-blind deblurring impacts the quality of latent images estimated for other purposes, however, we have performed experiments with two more modern deblurring methods. In particular, we have used the sparse deconvolution technique from Levin *et al.* (Levin, Fergus, Fergus, Durand & Freeman 2007) and the fast deconvolution method of Krishnan and Fergus (Krishnan &

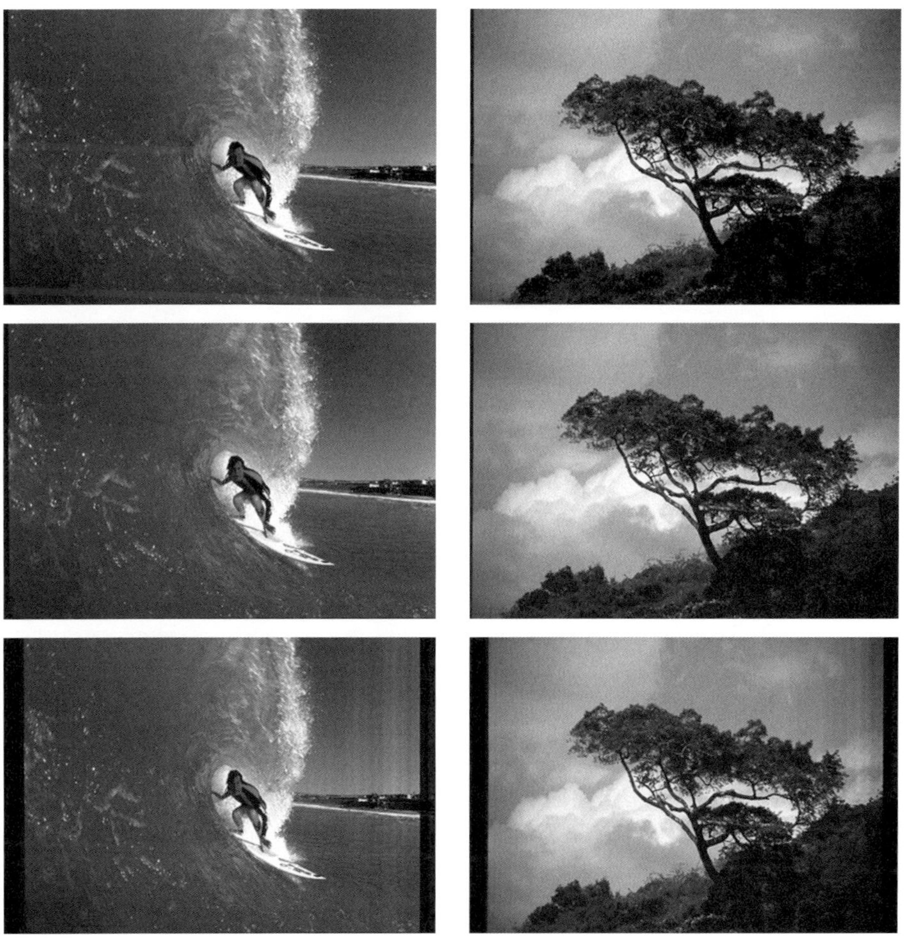

Figure 11.10 Comparison of non-blind deblurring on synthetic coded exposure images. From top to bottom are the outputs from Raskar, Krishnan, and Levin. Execution times are 0.3 s (Raskar), 1.3 s (Krishnan), and 46.3 s (Levin). In raster order, the RMSEs of the first four latent image estimates are 9.3, 9.1, 8.3, and 8.2.

Fergus 2009). These methods are compared to the least squares deconvolution from Raskar *et al.* (2006). In all cases, we use the default parameter values provided with the implementation.

First we tested the three deblurring techniques on synthetically blurred BSD images with several coded exposure PSFs. Figure 11.10 shows the deblurred images from the three methods. The methods used by Raskar and Krishnan produce results that are quite similar to the ground truth, and can be compared quantitatively. On the surfer image, their methods give RMSEs of 9.3 and 8.3, respectively. On the tree image, the RMSEs are 9.1 and 8.2. The output of Levin's method looks enough like the ground truth for us to think that the parameters are correct, but the images are compressed horizontally (preventing a quantitative evaluation) and have banding artifacts near the left and right edges. From a quality point of view, these results (and several others not shown) indicate that Krishnan's method produces RMSEs about 10% lower than Raskar's, and Levin's method has significant artifacts. On these small images (321×481 pixels), the execution times for the methods are 0.3 s (Raskar), 1.3 s (Krishnan), and 46.3 s (Levin).

Figure 11.11 shows deblurring results derived from real coded exposure images taken with our camera. Because these images are larger, we do not evaluate Levin's method

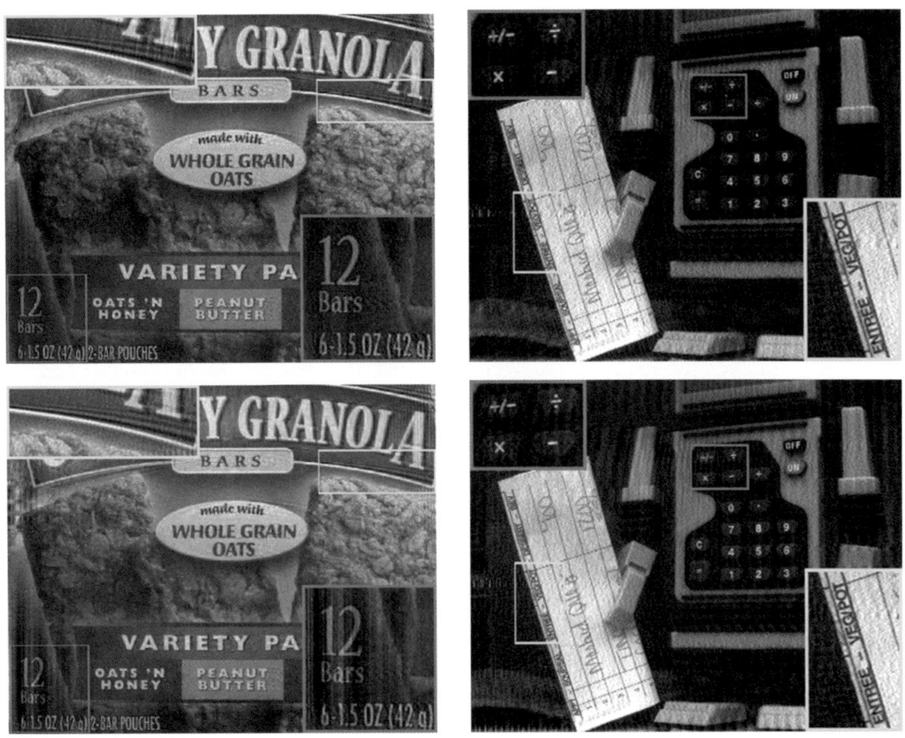

Figure 11.11 Comparison of non-blind deblurring on real coded exposure images. The top row shows the outputs from Raskar, the bottom row shows the results from Krishnan. Using the publicly available code for Levin's method failed to produce a reasonable result on the robot image, so results are omitted.

which takes in excess of 1 minute to deblur. In the granola image (left column), we note that Krishnan's method produces black band artifacts in the bottom right inset. Upon close inspection, we find that Raskar's output has slightly higher noise. Both show a similar amount of ringing in the upper left inset region, likely due to the motion being slightly off-axis. In the robot image (right column), Raskar's result is once again noisier and Krishnan's result still has artifacts near the edges. On these 800×600 pixel images, Raskar and Krishnan's methods run reasonably fast (around 1 s), and Levin's method takes more than 1 minute.

In summary, then, we have not found that these newer deblurring techniques give decisively better latent image estimates than least squares deconvolution. While Krishnan's method produces consistently lower RMSEs on synthetically blurred images (by about 10%), it also produces banding artifacts in real camera images. Levin's method occasionally produces very nice results, but is unacceptably slow.

11.3 Coded exposure performance on iris recognition

11.3.1 Synthetic experiments

In order to illustrate the improved robustness of flutter shutter imagery to motion, we first performed synthetic experiments using the NIST iris challenge evaluation (ICE) dataset (Phillips *et al.* 2007). We simulate motion-blurred capture and deblurring and attempt to match those images to the sharp images of the dataset. We use the same process as in Section 11.1 to synthesize motion-blurred images for coded exposure. For the flutter shutter images, the blur PSF B is a flutter shutter PSF generated using the method of Section 11.2.1, with equivalent exposures to the 8, 16, 24, and 32 pixel rectangle PSFs used in Section 11.1, and images are deblurred using the deconvolution code provided in Raskar *et al.* (2006).

Figure 11.12 shows the ROC curves for the synthetic images. For comparison, we include the ROC curves from traditional imaging presented earlier. The curves in the left plot show the performance of the deblurred traditional shutter images, which degrades sharply as the degree of blur increases. The curves in the right plot show the performance of the deblurred flutter shutter images, and demonstrate consistent performance despite increasing blur. These curves validate our claim that, *when deblurred with the correct blur estimate*, flutter shutter images are more useful than traditional shutter images for iris matching.

11.3.2 Real image experiments

Because the amount of effort involved in capturing an ICE-scale test set for both traditional and coded exposure imaging was too daunting, we carried out only a small scale validation on real-world images. To do this, we made 4×6 in photographic prints of several iris images and taped them to our motion rig. We manually selected the shutter sequence to match the motion velocity (which was set via voltage controls),

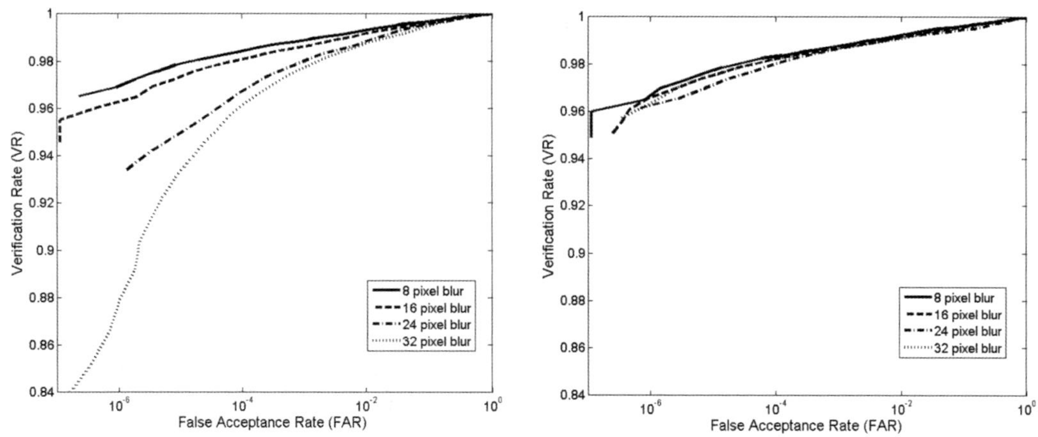

Figure 11.12 ROC curves for synthetically deblurred images from the ICE dataset, using a traditional shutter (left) and a flutter shutter (right).

and captured left-to-right lateral motion. We used the automated blur estimation method from Section 11.2.2, the deblurring code from Raskar *et al.* (2006), and our black-box iris matching method. Figure 11.13 shows a few example results from these tests, where we were able to generate useful iris images from coded exposure in cases where traditional imaging failed to produce a match.

Beyond these anecdotal cases, we have found that coded exposure produces significantly more matches than traditional imaging, but fewer than our simulations would suggest. There are several contributing factors, including non-lateral motion from an imprecise motion rig and image degradation resulting from photographing a print of the eye instead of the eye itself.

11.4 Barcodes

While iris recognition serves as a motivating application of our coded exposure imaging system, there are other uses. We note that many industrial applications of imagery make similar demands with respect to motion tolerance. Barcode scanning, for instance, involves targets with texture on the same fine scale as iris recognition and, though error correction codes provide some robustness, performance degrades significantly with motion blur. Because of the ubiquity of barcode scanning in modern life – Walmart alone was reported to scan 5000 items *per second* during its busiest sales day in 2012 – improved performance on moving objects could save everyone time in the checkout lines.

In Xu & McCloskey (2011), we demonstrated that most of the components of the coded exposure system described in this chapter could be reused for barcode capture. The most interesting difference between the barcode and iris systems arises from the fact that barcode images do not obey our natural image prior. Because 2D barcodes consist

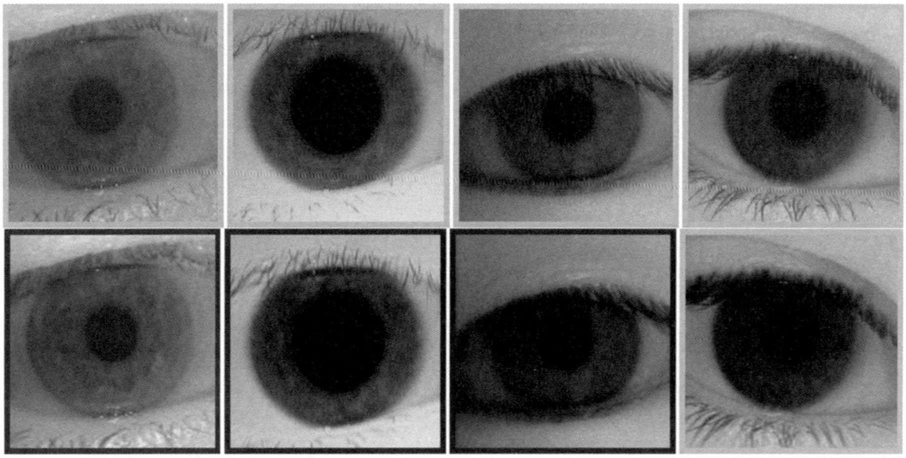

Figure 11.13 Real image results for iris deblurring. Top row: deblurred images from our coded exposure camera, all of which can be matched to another image of the same iris; bottom row: deblurred iris images from a traditional shutter camera, only one of which (furthest right) can be matched to another image of the same iris.

largely of black and white squares on a rectangular lattice, their power spectra are highly anisotropic, and our blur estimation method does not work as reliably as on other image types. In order to address this, we developed a joint blur estimation and deblurring routine that takes advantage of a different aspect of barcode image statistics, namely, that when sharply focused, the intensity histogram of a barcode image is bimodal. We also show that when deblurred with the wrong PSF length, the intensity histogram is smoothed. So, in order to estimate the blur extent without the natural image prior, we deblur the barcode image over a range of PSF lengths and measure an energy function which is low for bimodal intensity distributions. The deblurred image producing the minimum energy is taken as the estimated latent image, which we show can be decoded. Examples for different 2D barcode symbologies are shown in Figure 11.14, illustrating cases where coded exposure produces a decodable image and traditional image deblurring cannot.

11.5 More general subject motion

Having shown that coded exposure deblurring can improve the performance of iris matching and barcode scanning in cases of linear, constant velocity translation, it is natural to consider extensions to more general types of subject motion. Our previous work (Ding, McCloskey & Yu 2010) has addressed blur estimation for other motion types, extending the method from Section 11.2.2 to handle cases of acceleration and harmonic motion. Whereas the constant velocity method searches over different velocities ν for the PSF B^{ν} that best accounts for inflections in the coded exposure image's amplitude spectrum, the more general algorithm searches over motion parameters α

(a) Regular input image (b) Deblurring with LR algorithm (c) Flutter shutter input image (d) Deblurring with flutter shutter

(e) Regular input image (f) Deblurring with LR algorithm (g) Flutter shutter input image (h) Deblurring with flutter shutter

Figure 11.14 Coded exposure deblurring of 2D barcodes. Comparisons between traditional imaging and coded exposure deblurring with Data Matrix (top row) and Aztec symbologies (bottom row).

for the PSF B^{α} that explains the image's inflections. For the case of lateral motion with constant acceleration, values of α form a two-dimensional search space of initial and final speeds. For harmonic motion, α is the tuple of amplitude, angular speed, and initial phase. For both cases, we derive closed form expressions for the PSF similar to the constant velocity case from Eq. (11.13), and show that the matching metric from Eq. (11.22) finds the correct value of α. An example of constant acceleration is shown in Figure 11.15. While this example shows the ability to estimate the correct motion parameters, artifacts are present in the deblurred image. This is because the shutter sequence was designed to optimally preserve image content for constant velocity motion, and the MTF under acceleration may not be invertible. This suggests the need to design shutter sequences for, and detect the presence of, accelerated motion.

11.6 Implications of computational imaging for recognition

Though we have shown that coded exposure motion deblurring improves the performance of black-box iris matching, this work has exposed unforeseen issues with the recognition algorithms themselves. Figure 11.16 shows two images of the same iris captured and deblurred by our coded exposure system. These images are very similar, in that they were acquired with the same camera, show an iris moving with the same velocity, and appear to the human visual system to be nearly identical. Despite this, the two images give very different Hamming distance scores when compared to a third

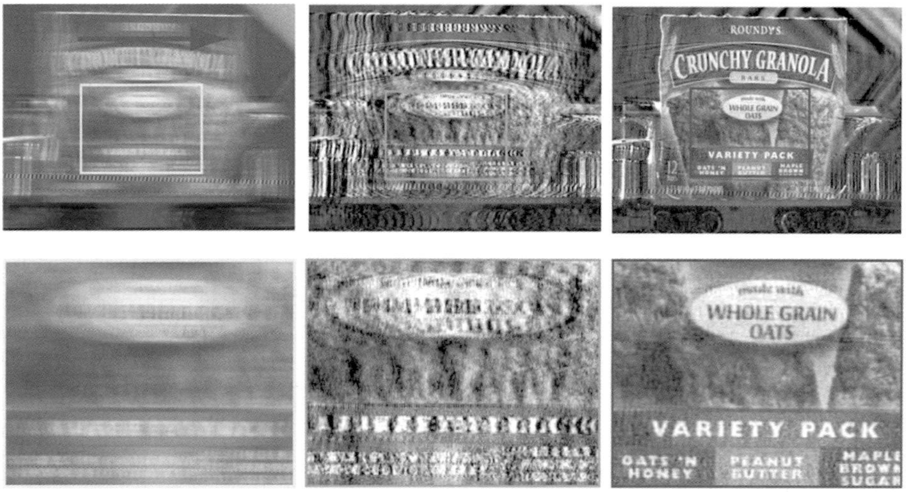

Figure 11.15 PSF estimation and deblurring for constant acceleration. A toy car slides down a slanted track at 55° and the camera is rotated so the motion appears horizontal. For clarity, a piece of textured cardboard was attached to the car. Top left: coded exposure capture of the accelerating train; top center: the best deblurred image using a constant velocity PSF model; top right: the deblurred image using our estimate of the PSF due to acceleration. The bottom row shows the inset regions of the corresponding images.

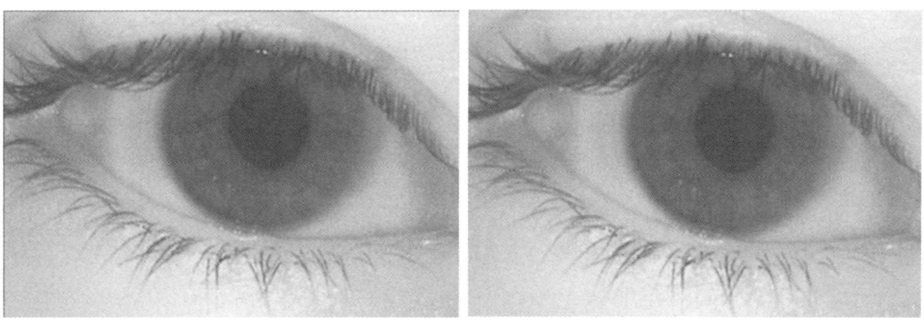

Figure 11.16 Nearly identical deblurred flutter shutter images of the same eye, moving at the same velocity, taken with two different shutter sequences. As a result of differences in their noise spectra, only the left image matches an image of the same iris.

(traditionally acquired) image of the same iris. The left image gives a distance of 0.26 (considered a match), and the right a distance of 0.37 (not a match). The difference between the two images, which results in the divergent match scores, arises from the use of different shutter sequences during capture. Because the two shutter sequences have MTFs with different inflection points, the amplification applied at a given spatial frequency during deblurring will also differ. Presuming that the imaging noise η has a flat power spectrum (which is the case for Gaussian noise) in the coded exposure image, the deblurred images will have different noise power spectra due to the differing amplification. The process of generating an iris template from the deblurred images is sensitive to these differences because it applies a Gabor filter with a certain center frequency and bandwidth. When the two shutters' MTFs differ significantly in this band of spatial frequencies, the iris templates can differ significantly despite high visual similarity.

While this analysis is given for iris recognition, it is possible to show similar issues with other recognition tasks. Person detection and object recognition, to take two examples from modern computer vision, often use a histogram of oriented gradients (HOG) (Kläser, Marszalek & Schmid 2008) feature based on differences in intensity between pixels at a given distance. Because this inter-pixel distance corresponds to a certain spatial frequency, these recognition methods are also sensitive to frequency-specific noise considerations arising from some computational cameras. While most recognition methods do not have an explicit model for the power spectrum of their input images, training and testing them on image sets acquired with a traditional shutter implicitly imposes an assumption that the MTFs smoothly roll off from DC to the Nyquist frequency. As deblurred images from computational cameras become more prevalent, recognition methods that use them should be modified in order to account for these differences.

11.7 Conclusion

In this chapter, we have developed a motion deblurring system based on coded exposure, and have demonstrated that it allows for both iris recognition and barcode scanning in situations where traditional imaging fails to produce a useful image. In order to account for the velocity dependence of coded exposure, we have introduced a method to generate near-optimal shutter sequences by incorporating the statistics of natural images. We have further developed a coded exposure PSF estimation method based on the same model for natural image statistics, which avoids problematic assumptions about the presence of high-contrast edges in the scene. We have also shown how the blur estimation method can be extended to handle more general object motion types, particularly acceleration.

For the problem of designing coded exposure shutter sequences that are optimal for motions other than linear, constant velocity translation remains an open area of research. In addition, our work with iris recognition has presented another open question: how should recognition methods handle imagery from non-traditional cameras? Having shown that imagery from a coded exposure camera produces unexpected results, and anticipating that computational cameras will be increasingly adopted in the future, this is a problem which will eventually need to be addressed.

References

Agrawal, A. & Raskar, R. (2007). Resolving objects at higher resolution from a single motion-blurred image. In *Computer Vision and Pattern Recognition*, pp. 1–8.

Agrawal, A. & Raskar, R. (2009). Optimal single image capture for motion deblurring. In *Computer Vision and Pattern Recognition*, pp. 2560–7.

Agrawal, A. & Xu, Y. (2009). Coded exposure deblurring: optimized codes for PSF estimation and invertibility. In *Computer Vision and Pattern Recognition*, pp. 2066–73.

ANS (ANSI standard INCITS M1/03-0590). Iris image interchange format.

Dai, S. & Wu, Y. (2008). Motion from blur. In *Computer Vision and Pattern Recognition*, pp. 1–8.

Daugman, J. (1993). High confidence visual recognition of persons by a test of statistical independence. *IEEE Transactions on Pattern Analysis and Machine Intelligence*, **15**(11), 1148–61.

Ding, Y., McCloskey, S. & Yu, J. (2010). Analysis of motion blur with a flutter shutter camera for non-linear motion. In *European Conference on Computer Vision*, pp. 15–30.

Fergus, R., Singh, B., Hertzmann, A., Roweis, S. T. & Freeman, W. T. (2006). Removing camera shake from a single photograph. *ACM Transactions on Graphics*, **25**(3), 787–94.

Fergus, R., Weiss, Y. & Torralba, A. (2009). Semi-supervised learning in gigantic image collections. In Y. Bengio, D. Schuurmans, J. Lafferty, C. K. I. Williams & A. Culotta, eds., *Advances in Neural Information Processing Systems 22*, NIPS Foundation, pp. 522–30.

Jia, J. (2007). Single image motion deblurring using transparency. In *Computer Vision and Pattern Recognition*, pp. 1–8.

Kläser, A., Marszalek, M. & Schmid, C. (2008). A spatio-temporal descriptor based on 3D-gradients. In *British Machine Vision Conference*, 275:1–10.

Krishnan, D. & Fergus, R. (2009). Fast image deconvolution using hyper-Laplacian priors. In *Neural Information Processing Systems Conference*, pp. 1033–41.

Levin, A., Fergus, R., Fergus, R., Durand, F. & Freeman, W. T. (2007). Image and depth from a conventional camera with a coded aperture. *ACM Special Interest Group on Graphics and Interactive Techniques*, pp. 841–8.

Lucy, L. B. (1974). An iterative technique for the rectification of observed distributions. *The Astronomical Journal*, **79**(6), 745–54.

Martin, D., Fowlkes, C., Tal, D. & Malik, J. (2001). A database of human segmented natural images and its application to evaluating segmentation algorithms and measuring ecological statistics. In *IEEE International Conference on Computer Vision*, pp. 416–23.

McCloskey, S. (2010). Velocity-dependent shutter sequences for motion deblurring. In *European Conference on Computer Vision*, pp. 309–22.

Oliveira, J. A. P., Figueiredo, M. A. & Bioucas-Dias, J. M. (2007). Blind estimation of motion blur parameters for image deconvolution. In *Proceedings of Iberian Conference on Pattern Recognition and Image Analysis*. Springer, pp. 604–11.

Phillips, P. J., Scruggs, W. T., O'Toole, A. J., Flynn, P. J., Bowyer, K. W., Schott, C. L. & Sharpe, M. (2007). *Large-scale results*, FRVT 2006 and ICE 2006 National Institute of Standards and Technology NISTIR 7408.

Raskar, R., Agrawal, A. & Tumblin, J. (2006). Coded exposure photography: motion deblurring using fluttered shutter. *ACM Transactions on Graphics*, **25**(3), 795–804.

Shan, Q., Jia, J. & Agarwala, A. (2008). High-quality motion deblurring from a single image. *ACM Special Interest Group on Graphics and Interactive Techniques*, **27**(3), 73:1–10.

Shan, Q., Xiong, W. & Jia, J. (2007). Rotational motion deblurring of a rigid object from a single image. In *IEEE International Conference on Computer Vision*, pp. 1–6.

Toft, P. (1996). The radon transform – theory and implementation. PhD thesis, Technical University of Denmark.

van der Schaaf, A. & van Hateren, J. H. (1996). Modelling the power spectra of natural images: statistics and information. *Vision Research*, **36**, 2759–70.

Xu, W. & McCloskey, S. (2011). 2D barcode localization and motion deblurring using a flutter shutter camera. In *Workshop on Applications of Computer Vision*, pp. 159–65.

12 Direct recognition of motion-blurred faces

Kaushik Mitra, Priyanka Vageeswaran, and Rama Chellappa

The need to recognize motion-blurred faces is vital for a wide variety of security applications ranging from maritime surveillance to road traffic policing. While much of the theory in the analysis of motion-blurred images focuses on restoration of the blurred image, we argue that this is an unnecessary and expensive step for face recognition. Instead, we adopt a direct approach based on the set-theoretic characterization of the space of motion-blurred images of a single sharp image. This set lacks the nice property of convexity that was exploited in a recent paper to achieve competitive results in real-world datasets (Vageeswaran, Mitra & Chellappa 2013). Keeping this non-convexity in mind, we propose a bank of classifiers (BoC) approach for directly recognizing motion-blurred face images. We divide the parameter space of motion blur into many different bins in such a way that the set of blurred images within each bin is a convex set. In each such bin, we learn support vector machine (SVM) classifiers that separate the convex sets associated with each person in the gallery database. Our experiments on synthetic and real datasets provide compelling evidence that this approach is a viable solution for recognition of motion-blurred face images.

12.1 Introduction

A system that can recognize motion-blurred faces can be of vital use in a wide variety of security applications, ranging from maritime surveillance to road traffic policing. Figure 12.1 shows two possible maritime surveillance scenarios: shore-to-ship (the camera is mounted on-shore and the subjects are in the ship), and ship-to-shore (the camera is on the ship and the subjects are moving on-shore). A third scenario, not shown, is ship-to-ship (the camera is mounted on a ship and the subjects are on another ship). In all these cases, the face images have significant motion blur due to relative motion between the camera and subjects. In this chapter, we address the problem of recognizing faces from motion-blurred images. An obvious approach of recognizing blurred faces would be to deblur the images first and then recognize them using traditional face recognition techniques (Nishiyama, Hadid, Takeshima, Shotton, Kozakaya & Yamaguch 2010). However, this approach involves solving the challenging problem of blind image deconvolution (Kundur & Hatzinakos 1996, Levin, Weiss, Durand & Freeman 2009). We avoid this step and propose a direct approach for face recognition. Recently, in Vageeswaran *et al.* (2013), we have shown that the set of all images obtained by blurring a given image forms a convex set. Further, this convex set is given

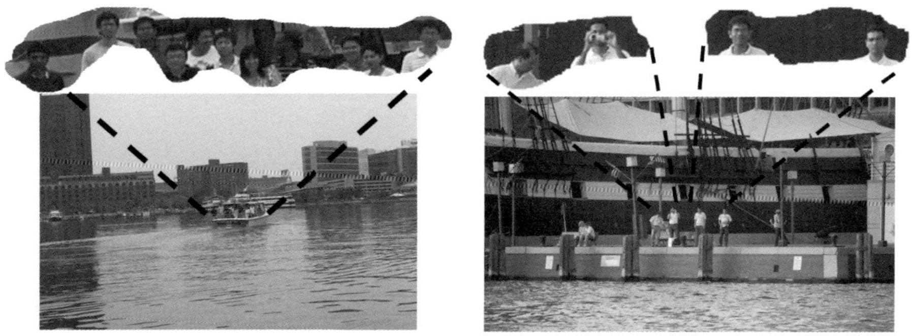

(a) (b)

Figure 12.1 Images captured by a distant camera for maritime surveillance, which could involve one of three scenarios: (a) shore-to-ship (the camera is mounted on-shore and the subjects are in the ship); (b) ship-to-shore (the camera is on the ship and the subjects are moving on-shore); and a third, not shown, ship-to-ship (the camera is mounted on a ship and the subjects are on another ship).

by the convex hull of shifted versions of the original image. Based on this set-theoretic characterization, we proposed a direct approach for recognizing blurred face images. However, if we restrict our attention to only motion-blurred images, then this set is no longer a convex set; hence we cannot use the face recognition algorithm proposed in Vageeswaran *et al.* (2013). In this chapter, we propose a discriminative model-based face recognition algorithm that takes into account the non-convex nature of the motion-blurred image sets.

We specifically consider motion blur along a single direction, which arises when the camera and/or the subject are moving in a linear fashion. This type of motion blur is characterized by the direction and size of the blur kernel. We show that the set of such motion-blurred images (obtained from a sharp image) is a non-convex set. This non-convexity arises mainly because of the directional nature of the motion blur. However, if we fix the direction of blur, then all the blurred images along that direction form a convex set. Keeping this conditional convexity in mind, we propose a bank of classifiers approach for directly recognizing motion-blurred face images. We divide the parameter space of motion blur into many different bins in such a way that the set of blurred images within each bin is a convex set. In each such bin, we learn SVM classifiers that separate the convex sets associated with each person in the gallery database. Given a probe image, we use the SVM classifiers at each bin to find its likely identity at each bin. Finally, we use the majority rule to arrive at the final identity.

To summarize, the main technical contributions of this chapter are:

- We show that the set of motion-blurred images obtained from a single sharp image is a non-convex set. The directional nature of motion blur is responsible for this non-convexity. Given a particular direction, the set of motion-blurred images along that direction is a convex set.
- Based on this set-theoretic characterization, we propose a motion blur robust face recognition algorithm, which avoids solving the problem of blind image deconvolution.

- We use the conditional convexity property of motion blur to propose a bank of classifiers-based face recognition algorithm. This is a discriminative approach and hence it scales well with the number of face classes and training images per class.

12.1.1 Related work

Face recognition from blurred images can be classified into four major approaches. In one approach, the blurred image is first deblurred and then used for recognition. This is the approach taken in Hu & de Haan (2006) and Nishiyama *et al.* (2010). The drawback of this approach is that we first need to solve the challenging problem of blind image deconvolution. Though there have been many attempts at solving this (Richardson 1972, Kundur & Hatzinakos 1996, Levin 2006, Fergus, Singh, Hertzmann, Roweis & Freeman 2006, Levin *et al.* 2009, Shan, Jia & Agarwala 2008, Likas & Galatsanos 2004, Jia 2007, Cho & Lee 2009, Yitzhaky, Mor, Lantzman & Kopeika 1998), it is not an essential step for the face recognition problem. Also, in Nishiyama *et al.* (2010), statistical models are learned for each blur kernel type and amount; this step might become infeasible when we try to capture the complete space of blur kernels.

In another approach, blur-invariant features are extracted from the blurred image and then used for recognition – Ahonen, Hadid & Pietikainen (2004) and Gopalan, Taheri, Turaga & Chellappa (2012) follow this approach. In Ahonen, Rahtu, Ojansivu & Heikkila (2008), the local phase quantization (LPQ) (Ojansivu & Heikkil 2008) method is used to extract blur invariant features. Though this approach works very well for small blurs, it is not very effective for large blurs (Nishiyama *et al.* 2010). In Gopalan *et al.* (2012), a "blur" subspace is associated with each image and face recognition is performed by measuring the distance of the blurred probe image from these subspaces. It has been shown that the "blur" subspace of an image contains all the blurred versions of the image. However, this analysis does not take into account the convexity constraint that the blur kernels satisfy (blur coefficients are non-negative and sum up to 1), and hence the "blur" subspace will include many other images apart from the blurred images.

The third approach is the direct recognition approach. This is the approach taken in Stainvas & Intrator (2000) and by us (Vageeswaran *et al.* 2013). In Stainvas & Intrator (2000), artificially blurred versions of the gallery images are created and the blurred probe image is matched to them. Again, it is not possible to capture the whole space of blur kernels using this method.

Finally, the fourth approach is to jointly deblur and recognize the blurred probe image (Zhang, Yang, Zhang, Nasrabadi & Huang 2011). However, this involves solving for the original sharp image, blur kernel, and identity of the face image, and hence it is a computationally intensive approach.

Set theoretic approaches for signal and image restoration have been considered in Combettes (1996), Trussell & Civanlar (1984), and Combettes & Pesquet (2004). In these approaches the desired signal space is defined as an intersection of closed convex sets in a Hilbert space, with each set representing a signal constraint. Image deblurring has also been considered in this context (Trussell & Civanlar 1984), where the non-negativity constraint of the images has been used to restrict the solution space. We differ from these approaches as our primary interest lies in recognizing blurred faces rather than restoring them.

The rest of the chapter covers the following. In Section 12.2 we provide a set-theoretic characterization of the space of motion-blurred images. In Section 12.3 we present the bank of classifiers approach for recognizing motion-blurred face images, and in Section 12.4 we perform experiments to evaluate our algorithm against other standard algorithms.

12.2 The set of all motion-blurred images

We review the convolution model for blur and, based on this model, we compare and contrast the set of *all* blurred images with the set of *only* motion-blurred images. In Vageeswaran *et al.* (2013), it is shown that the set of all blurred images is a convex set. However, we show here that the set of only motion-blurred images is non-convex.

12.2.1 Convolution model for blur

A pixel in a blurred image is a weighted average of the pixel's neighborhood in the original sharp image. Thus, blur is modeled as a convolution operation between the original image and a blur filter-kernel which represents the weights (Bovik 2009). Let I be the original image and H be the blur kernel of size $(2k+1) \times (2k+1)$, then the blurred image I_b is given by

$$I_b(r,c) = I * H(r,c) = \sum_{i=-k}^{k} \sum_{j=-k}^{k} H(i,j) I(r-i, c-j) \tag{12.1}$$

where $*$ represents the convolution operator and r, c are the row and column indices of the image. Blur kernels also satisfy the following properties: their coefficients are non-negative, $H \geqslant 0$, and sum up to 1 (i.e. $\sum_{i=-k}^{k} \sum_{j=-k}^{k} H(i,j) = 1$). The blur kernel for motion blur has additional structure such as being linear along a certain direction.

12.2.2 The set of all blurred images versus the set of motion-blurred images

The set of all blurred images $\mathcal{B}$ obtained from I is given by:

$$\mathcal{B} \triangleq \{I * H | H \geq 0, \|H\|_1 = 1\}. \tag{12.2}$$

This set is clearly convex since given any two blur kernels H_i and H_j, both satisfying $H \geqslant 0$ and $\|H\|_1 = 1$, their convex combination $(\lambda H_i + (1-\lambda)H_j, 0 \leqslant \lambda \leqslant 1)$ will also satisfy the above mentioned constraints. Moreover, this set is the convex hull of shifted versions of the image I. This is because the set $(H|H \geqslant 0, \|H\|_1 = 1)$ can be expressed as the convex hull of H matrices with a single nonzero entry. These H matrices in turn

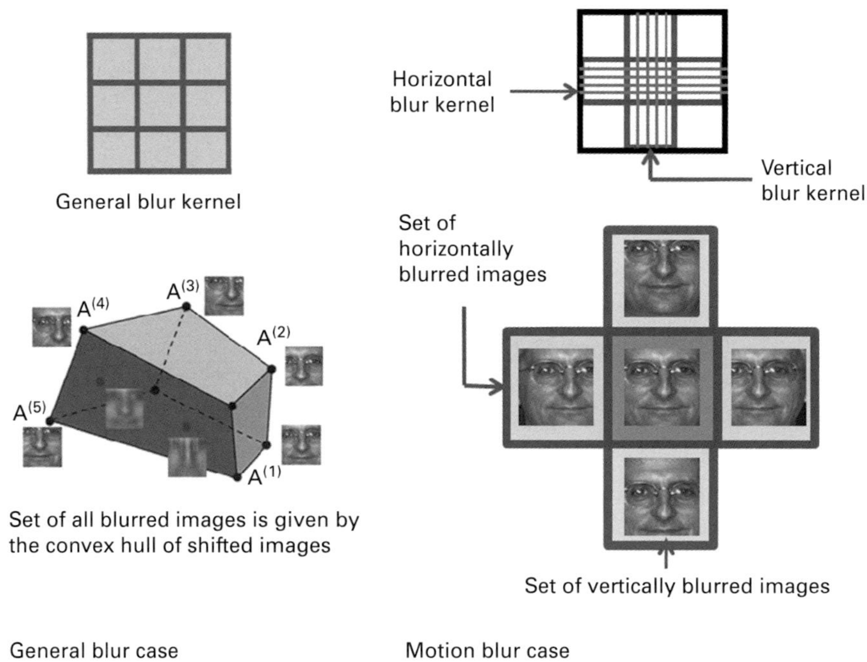

Figure 12.2 The set of all blurred images (which is a convex set) vs the set of only motion-blurred images (which is a non-convex subset of the original convex set).

gives rise to shifted versions of the original image I and thus the set of blurred images is the convex hull of these shifted images; see Vageeswaran *et al.* (2013).

Now, we consider the set of motion-blurred images. Motion blur arises due to the movement of the camera or the subject at the time of exposure. If the relative motion between the camera and the subject is linear, then the blur is along a single direction. If the motion blur is along a particular direction, say θ, then the blur kernel H satisfies the following conditions (Bovik 2009):

$$H(i,j) > 0 \quad \text{if} \arctan(i/j) = \theta, \quad \text{else } H(i,j) = 0. \tag{12.3}$$

With this definition of motion blur, we show that the set of all motion-blurred images is non-convex, see Figure 12.2. We prove this by producing two motion blur kernels, whose convex combination does not satisfy the motion blur kernel constraints, and hence is not a motion blur kernel. Consider the horizontal motion blur kernel

$$H_1 = \begin{pmatrix} 0 & 0 & 0 \\ 1/3 & 1/3 & 1/3 \\ 0 & 0 & 0 \end{pmatrix}$$

and the vertical motion blur kernel

$$H_2 = \begin{pmatrix} 0 & 1/3 & 0 \\ 0 & 1/3 & 0 \\ 0 & 1/3 & 0 \end{pmatrix}.$$

Their convex combination, corresponding to $\lambda = 0.5$, is

$$H_3 = \begin{pmatrix} 0 & 1/6 & 0 \\ 1/6 & 1/3 & 1/6 \\ 0 & 1/6 & 0 \end{pmatrix},$$

which is clearly not a motion blur kernel. Thus the set of motion blur images is not a convex set. However, if we consider a particular direction of motion blur θ, then the set of images blurred along that direction is a convex set. Furthermore, this convex set is given by the convex hull of shifted images along that direction. Again, this is because the blur kernel H is the convex hull of H matrices that are shifted impulse functions along the direction θ, which give rise to shifted images along θ. We use the above set-theoretic analysis to propose a face recognition algorithm that can handle motion-blurred images robustly.

12.3 Bank of classifiers approach for recognizing motion-blurred faces

We design our face recognition algorithm based on the above set-theoretic characterization. We first divide the parameter space of motion blur, direction, and size of blur into a certain number of bins. Within each bin the set of blurred images is a convex set, and is given by the convex hull of appropriately shifted versions of the sharp image. We use these shifted images as training data to learn a *system* of SVM classifiers for each bin, that distinguishes among the number (C) of face classes. Combining these systems of SVM classifiers for each bin, we get a *bank of classifiers*. Given a motion-blurred probe image, we use this entire bank of classifiers to identify the correct class. Figure 12.3 gives an overview of our BoC approach. In the following paragraphs, we describe our algorithm in more detail.

Motion blur is characterized by the direction (θ) and size of blur (S) (Bovik 2009). We divide the size of blur into a certain number of bins, say N_s. For each blur size, we divide the angle space (between 0 and 180°) into a certain number of equally spaced bins, say N_{θ_s}. We choose the number of angle bins as a function of the blur size. This is because we are working with discrete kernels. For instance, for a kernel size of 3 we choose the angle bin centers at $0, 45, 90, 135, 180°$; for size 5 we choose bin centers at $0, 22.5, 45, \ldots, 180°$. The total number of bins in the parameter space is $N = \sum N_{\theta_s}$, for $s = 1, \ldots, N_S$.

From our analysis in Section 12.2.2, the set of motion-blurred images obtained from a sharp image within each bin is a convex set. This set is a convex hull of shifted images which depends on the direction and size of motion blur kernel determined by the particular bin. To find these shifted images we find the nonzero entries of the blur kernel H. If the blur angle of the bin is between $\theta_1 \leqslant \theta < \theta_2$ and the blur size is $K \times K$, then the nonzero entries of the blur kernel are given by:

$$H(i,j) > 0 \quad \text{if } \theta_1 \leqslant \arctan(j/i) < \theta_2. \tag{12.4}$$

There are as many shifted images as the number of nonzero entries in H and each shifted image is obtained by convolving the sharp image with a kernel that is a shifted delta

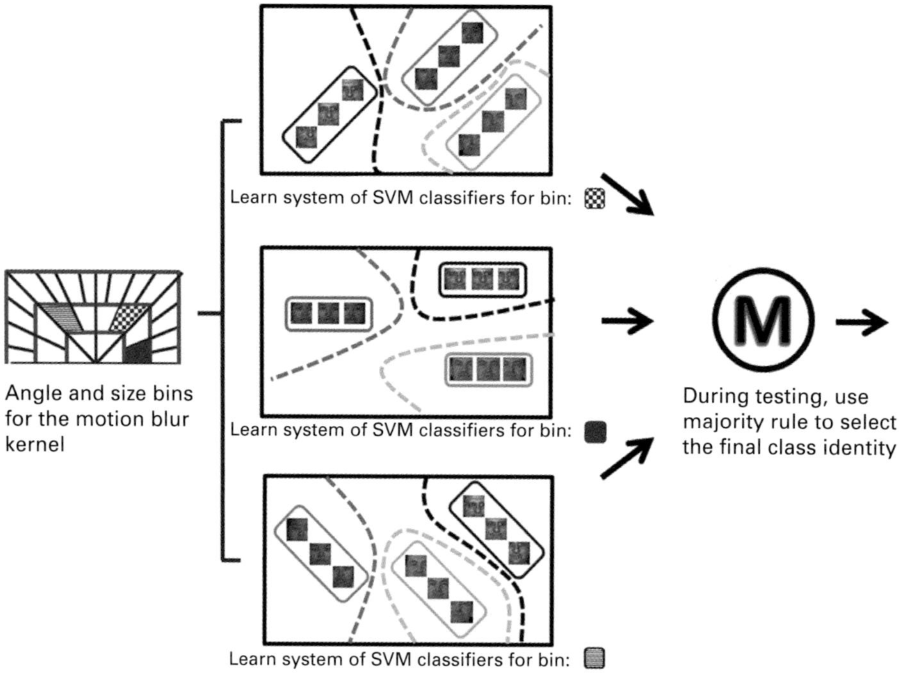

Figure 12.3 The bank of classifiers (BoC) approach for recognizing motion-blurred face images. Here we show three such systems corresponding to the appropriately shaded bin.

function. These shifted images then define the convex hull for the bin. We use these shifted images as the training data for learning SVM classifiers as described below.

For each bin of the blur parameter space we learn a system of SVM classifiers to distinguish between the *C*-face classes: one-versus-all SVM classifiers (Hsu & Lin 2002). Thus for each face class, we have a bank of *N* classifiers. Given a test image, we query it against this bank of classifiers and find the likely class for each bin. We finally choose the class that occurs the most number of times.

12.4 Experimental evaluation

In this section, we first analyze the performance sensitivity of our proposed algorithm BoC with the blur angle and size bins. We then study the recognition performance of BoC on synthetically generated motion-blurred images from the PIE dataset (Sim, Baker & Bsat 2002) and on a real dataset REMOTE (Ni & Chellappa 2010). We compare it with competing algorithms for blur robust recognition such as direct recognition of blurred faces (DRBF) (Vageeswaran *et al.* 2013) and LPQ (Ojansivu & Heikkil 2008).

12.4.1 Sensitivity analysis of the BoC approach

In our BoC approach we divide the parameter space of motion blur, and the direction and size of blur, into many bins and learn the SVM classifier for each bin. Here

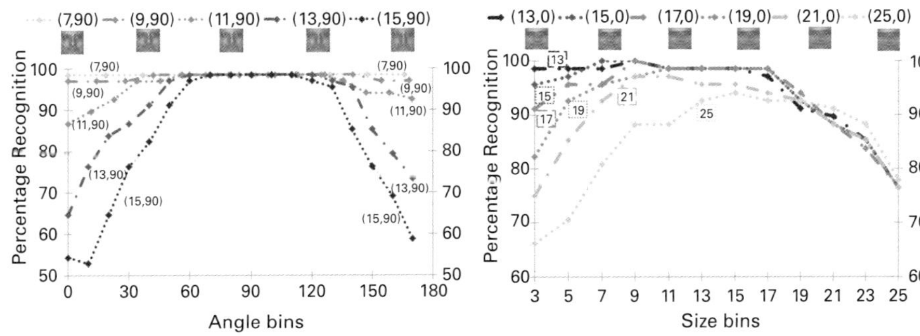

(a) (b)

Figure 12.4 Sensitivity of the recognition rate of BoC with blur angle and size bins, (a) angle sensitivity plot; (b) size sensitivity plot.

we study the sensitivity of the recognition rate with respect to the blur angle and size bins. For this we use the PIE dataset (Sim *et al.* 2002) which has 68 subjects and the face images are of size 64×64. The training and testing sub-folders are chosen such that their pose and illumination conditions are roughly constant. We choose sub-folders ($c27$–$f21$, $c27$–$f20$) for training and ($c27$–$f9$, $c27$–$f11$) for testing. We bin the kernel size into 7 bins corresponding to sizes $(3, 5, \ldots, 15)$. For each kernel size, we bin the angle space appropriately to get a total of $N = 89$ bins. At each bin and for each face class, we generate training images, which are shifted versions of the original sharp image and learn SVM classifiers based on them. To study the sensitivity of recognition rate with respect to blur angle, we blur the test images with a blur kernel of angle 90° and a sequence of different blur sizes: $7, 9, 11, 13, 15$. We then use the learned SVM classifiers at different angle bins (with the correct kernel size) to compute the recognition rate, see Figure 12.4(a). The recognition rate is quite good and is almost 100% for the correct angle bin of 90°. As we go further away from the correct angle bin, the performance falls off. The decrease in performance also depends on the size of blur kernel. For small blur sizes $\leqslant 9$, the performance is largely insensitive to the bin angles, but for larger blur sizes ($\geqslant 11$) we observe a bell-shaped curve. In Figure 12.4(b), we study the sensitivity of recognition rate with blur size. We blur the test images with a blur kernel of angle 0° and a sequence of blur sizes: $13, 15, 17, 19, 21, 23, 25$. We then use the learned SVM classifiers at different blur size bins (of angle 0°) to compute the recognition rate. For small blur sizes $\leqslant 17$, the recognition rate is almost 100% for the correct blur size bins, but for large blur sizes $\geqslant 19$, performance falls off even at the correct bin. This is because at large blur sizes, convex hulls corresponding to different face classes either intersect or are very close to one another. Hence, it becomes difficult for the SVM classifier to learn an appropriate decision surface.

12.4.2 Performance evaluation on synthetically generated motion-blurred images

We synthetically blur the PIE dataset with many motion blur kernels and compare our BoC approach with other algorithms such as DRBF (Vageeswaran *et al.* 2013), FADEIN+LPQ (Nishiyama *et al.* 2010), and local binary patterns (LBP) (Ahonen *et al.*

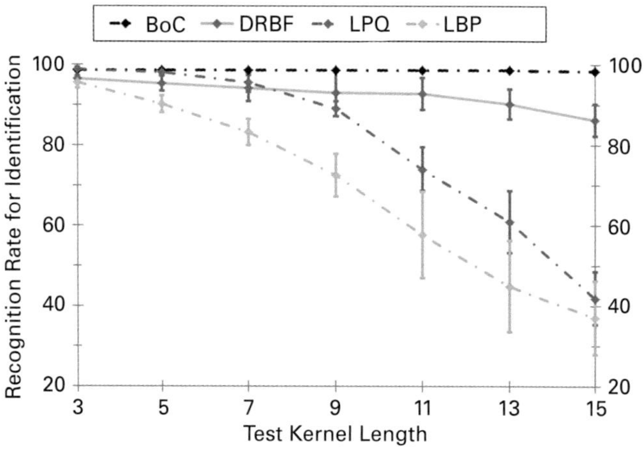

Figure 12.5 Recognition results for synthetically generated motion-blurred images.

2004). We use face images with a frontal pose (c_{27}) and good illumination (f_{21},f_{20}) as our gallery and c_{27}–(f_9,f_{11},f_{12}) as the probe. We vary the blur size from 3 to 15, and at each blur size we randomly generate 10 different blur angles. Figure 12.5 shows the mean recognition rate versus blur size. BoC gives the best performance followed by DRBF, LPQ, and LBP. At each blur size, we also show the standard deviation as error bars. Since we model the blur direction explicitly in BoC, the standard deviation for BoC is almost zero, whereas for the other algorithms it is much larger.

12.4.3 Performance evaluation on the real dataset REMOTE

We test our algorithm on the REMOTE dataset where the images have been captured in an unconstrained manner (Ni & Chellappa 2010, Chellappa, Ni & Patel 2011). The images were captured in two settings: ship-to-shore and shore-to-ship. The distance between the camera and the subjects ranges from 5 m to 250 m. Since there are moving objects in the scene and images were captured from a moving camera, the images have significant motion blur. This dataset has 17 subjects. The gallery folder of the dataset has 5 frontal, sharp and well-illuminated images for each subject. The probe folder is divided into three sub-folders: (i) blur only, (ii) illumination only, and (iii) blur and illumination. We use the blur sub-folder for our experiment. This sub-folder has 75 probe images, see Figure 12.6. We register the images as a preprocessing step and normalize the size of the images to 120×120 pixels. We then evaluate the recognition performance of our algorithm and compare it with other state of the art algorithms such as DRBF (Vageeswaran *et al.* 2013), the partial least squares (PLS)-based face recognition algorithm (Chellappa *et al.* 2011), the sparse representation-based face recognition algorithm (SRC) (Wright, Yang, Ganesh, Sastry & Ma 2009), LPQ, and PCA+LDA+SVM (Ni & Chellappa 2010). We plot the recognition rate versus the number of gallery images in Figure 12.7. Our algorithm BoC outperforms the other algorithms significantly when the number of gallery of images is three or less. It is noteworthy that the

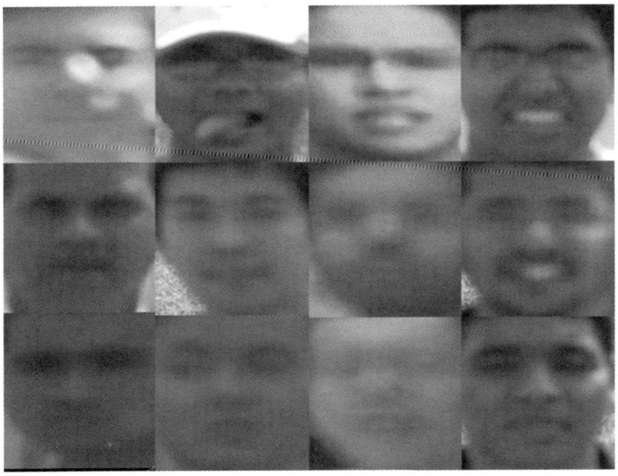

Figure 12.6 Sample probe images from REMOTE dataset (Ni & Chellappa 2010).

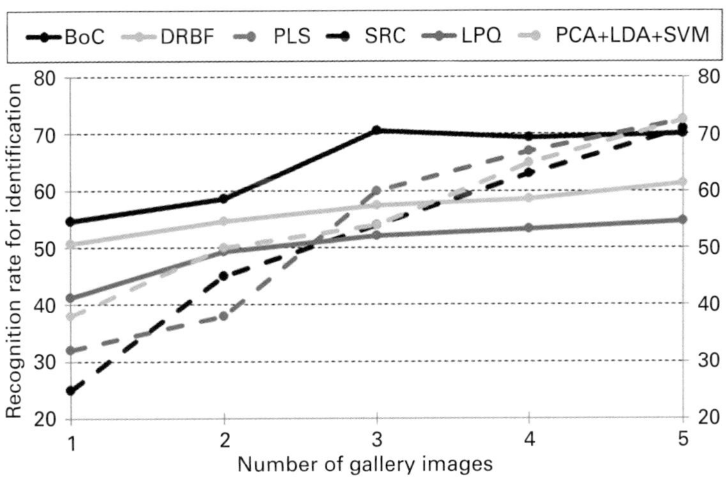

Figure 12.7 Recognition results on the unconstrained dataset REMOTE. We plot the recognition rate vs number of gallery images for BoC, DRBF (Vageeswaran *et al.* 2013), PLS-based face recognition algorithm (Chellappa *et al.* 2011), a sparse representation-based face recognition algorithm SRC (Wright *et al.* 2009), LPQ, and PCA+LDA+SVM (Ni & Chellappa 2010).

other algorithms use more informative features (e.g. PLS uses Gabor-Jets, LBP, and histograms of oriented gradients (HOG)), which likely makes them robust to illumination and registration effects. Hence, it would be interesting to see the effects of modeling illumination and pose in our BoC framework.

12.5 Discussion

We address the problem of recognizing faces from motion-blurred images. We show that the set of motion-blurred images obtained from a single sharp image is a non-convex set. The main reason for this non-convexity is the directional nature of motion blur. However, if we fix the direction of blur, then all the blurred images along that direction form a convex set. Keeping this conditional convexity in mind, we propose a bank of classifiers approach for directly recognizing motion-blurred faces. This is a discriminative approach and hence it scales well with the number of face classes and training images per class. Our experiments on both synthetic and real datasets show that our algorithm gives state of the art performance. In future, we would like to model the effect of illumination and pose in our framework.

Acknowledgements

This work was partially supported by a MURI from the Office of Naval Research under the Grant N00014-08-1-0638. It was also supported by NSF Grants NSF-IIS:1116718 and NSF-CCF:1117939, and by a gift from Samsung Telecommunications America.

References

Ahonen, T., Hadid, A. & Pietikainen, M. (2004). Face recognition with local binary patterns. In *European Conference on Computer Vision*, pp. 469–481.

Ahonen, T., Rahtu, E., Ojansivu, V. & Heikkila, J. (2008). Recognition of blurred faces using local phase quantization. In *International Conference on Pattern Recognition*, pp. 1–4.

Bovik, A. C. (2009). *The Essential Guide to Image Processing*. Elsevier.

Chellappa, R., Ni, J. & Patel, V. M. (2011). Remote identification of faces: problems, prospects, and progress. *Pattern Recognition Letters*, **33**(14), 1849–59.

Cho, S. & Lee, S. (2009). Fast motion deblurring. *ACM Transactions on Graphics*, **28**(5), 145: 1–8.

Combettes, P. L. (1996). The convex feasibility problem in image recovery. *Advances in Imaging and Electron Physics*, **95**, 155–270.

Combettes, P. & Pesquet, J.-C. (2004). Image restoration subject to a total variation constraint. *IEEE Transactions on Image Processing*, **13**(9), 1213–22.

Fergus, R., Singh, B., Hertzmann, A., Roweis, S. T. & Freeman, W. T. (2006). Removing camera shake from a single photograph. *ACM Transactions on Graphics*, **25**(3), 787–94.

Gopalan, R., Taheri, S., Turaga, P. & Chellappa, R. (2012). A blur-robust descriptor with applications to face recognition. *IEEE Transactions on Pattern Analysis and Machine Intelligence*, **34**(6), 1220–6.

Hsu, C.-W. & Lin, C.-J. (2002). A comparison of methods for multiclass support vector machines. *IEEE Transactions on Neural Networks*, **13**(2), 415–25.

Hu, H. & de Haan, G. (2006). Low cost robust blur estimator. In *IEEE International Conference on Image Processing*, pp. 617–20.

Jia, J. (2007). Single image motion deblurring using transparency. In *IEEE Conference on Computer Vision and Pattern Recognition*, pp. 1–8.

Kundur, D. & Hatzinakos, D. (1996). Blind image deconvolution. *IEEE Signal Processing Magazine*, **13**(3), 43–64.

Levin, A. (2006). Blind motion deblurring using image statistics. *Advances in Neural Information Processing Systems*, pp. 841–8.

Levin, A., Weiss, Y., Durand, F. & Freeman, W. T. (2009). Understanding and evaluating blind deconvolution algorithms. In *IEEE Conference on Computer Vision and Pattern Recognition*, pp. 1964–71.

Likas, C. & Galatsanos, N. (2004). A variational approach for Bayesian blind image deconvolution. *IEEE Transactions on Signal Processing*, **52**(8), 2222–33.

Ni, J. & Chellappa, R. (2010). Evaluation of state-of-the-art algorithms for remote face recognition. In *IEEE International Conference on Image Processing*, pp. 1581–4.

Nishiyama, M., Hadid, A., Takeshima, H., Shotton, J., Kozakaya, T. & Yamaguch, O. (2010). Facial deblur inference using subspace analysis for recognition of blurred faces. *IEEE Transactions on Pattern Analysis and Machine Intelligence*, **33**(4), 838–45.

Ojansivu, V. & Heikkil, J. (2008). Blur insensitive texture classification using local phase quantization. In *Image and Signal Processing*. Berlin/Heidelberg: Springer.

Richardson, W. H. (1972). Bayesian-based iterative method of image restoration. *Journal of the Optical Society of America*, **62**(1), 55–9.

Shan, Q., Jia, J. & Agarwala, A. (2008). High-quality motion deblurring from a single image. *ACM Transactions on Graphics*, **27**(3), 73:1–10.

Sim, T., Baker, S. & Bsat, M. (2002). The CMU pose, illumination, and expression (PIE) database.

Stainvas, I. & Intrator, N. (2000). Blurred face recognition via a hybrid network architecture. In *International Conference on Pattern Recognition*, pp. 2805–8.

Trussell, H. & Civanlar, M. (1984). The feasible solution in signal restoration. *IEEE Transactions on Acoustics, Speech and Signal Processing*, **32**(2), 201–12.

Vageeswaran, P., Mitra, K. & Chellappa, R. (2013). Blur and illumination robust face recognition via set-theoretic characterization. *IEEE Transactions on Image Processing*, **22**(4), 1362–72.

Wright, J., Yang, A., Ganesh, A., Sastry, S. & Ma, Y. (2009). Robust face recognition via sparse representation. *IEEE Transactions on Pattern Analysis and Machine Intelligence*, **31**(2), 210–27.

Yitzhaky, Y., Mor, I., Lantzman, A. & Kopeika, N. S. (1998). A direct method for restoration of motion blurred images. *Journal of the Optical Society of America A*, **15**(6), 1512–19.

Zhang, H., Yang, J., Zhang, Y., Nasrabadi, N. M. & Huang, T. S. (2011). Close the loop: joint blind image restoration and recognition with sparse representation prior. In *IEEE Proceedings of the International Conference on Computer Vision*, pp. 770–7.

13 Performance limits for motion deblurring cameras

Oliver Cossairt and Mohit Gupta

13.1 Introduction

A number of computational imaging (CI) based motion deblurring techniques have been introduced to improve image quality. These techniques use optical coding to measure a stronger signal level instead of a noisy short exposure image. However, the performance of these techniques is limited by the decoding step, which amplifies noise. While it is well understood that optical coding can increase performance at low light levels, little is known about the quantitative performance advantage of computational imaging in general settings.

In this chapter, we derive the performance bounds for various computational imaging-based motion deblurring techniques. We then discuss the implications of these bounds for several real-world scenarios. The scenarios are defined in terms of real-world lighting (e.g. moonlit night or cloudy day, indoor or outdoor), scene properties (albedo, object velocities), and sensor characteristics. The results show that computational imaging techniques do not provide a significant performance advantage when imaging with illumination brighter than typical indoor lighting. This is illustrated in Figure 13.1. These results can be readily used by practitioners to decide whether to use CI and, if so, to design the imaging system. We also study the role of image priors on the decoding steps. Our empirical results show that the use of priors reduces the performance advantage of CI techniques even further.

Scope

The analysis in this chapter focuses on techniques that use optical coding to preserve high frequencies in the blur kernels so that deblurring becomes a well-conditioned problem. These techniques assume that the blur kernel is known a priori. The analysis is limited to techniques that acquire a single image and follow a linear imaging model. In particular, we consider two classes of techniques: (a) coded exposure based techniques (e.g. Raskar, Agrawal & Tumblin (2006)), and (b) camera motion based techniques (Levin, Sand, Cho, Durand & Freeman 2008, Cho, Levin, Durand & Freeman 2010). The coded exposure based techniques selectively block light during camera exposure, whereas the camera motion based techniques move the camera along a particular trajectory during image capture.

Our analysis assumes the motion blur point spread function (PSF) is known. Methods for determining the PSF include estimation using blind deconvolution

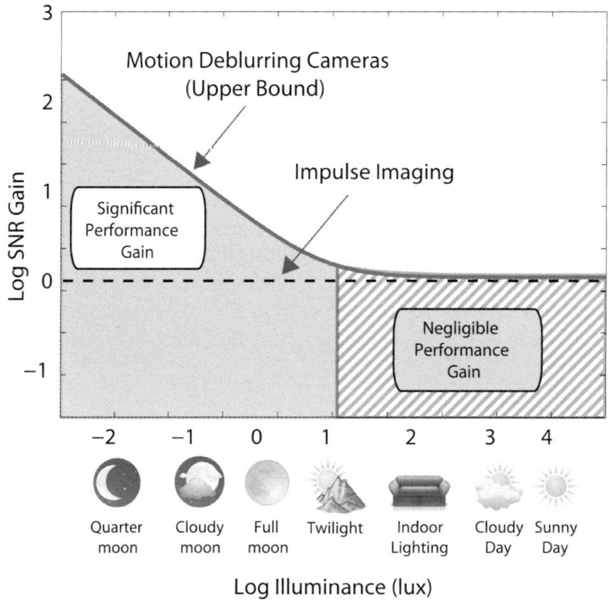

Figure 13.1 Performance of motion deblurring cameras for naturally occurring lighting conditions. We show that motion deblurring cameras (solid curve) give a negligible performance gain over conventional (impulse) imaging (dotted line) if the illumination level is higher than that of a typical living room. The plot assumes the following scene and sensor characteristics: average scene reflectivity is 0.5, object motion is 50 pixels/s, aperture setting is $F/2.1$, pixel size is $1\,\mu$m, quantum efficiency is 0.5, and read noise standard deviation is $4e^-$.

(Fergus, Singh, Hertzmann, Roweis & Freeman 2006, Shan, Jia & Agarwala 2008), and measurement using inertial sensors (Joshi, Kang, Zitnick & Szeliski 2010). We do not consider techniques that acquire multiple images, such as a blurred/non-blurred pair (Yuan, Sun, Quan & Shum 2007), a flash/no-flash pair (Krishnan & Fergus 2009), or an image stack (Cai, Ji, Liu & Shen 2009, Agrawal, Xu & Raskar 2009). Finally, our analysis does not consider techniques that require nonlinear computations such as motion estimation between frames (Zhang, Deshpande & Chen 2010).

Evaluation methodology

We evaluate the performance of different techniques compared to the corresponding *impulse imaging* technique that acquires a single short exposure blur-free image. In the rest of the chapter, we refer to a conventional camera that captures short exposure, motion blur-free images as an *impulse camera*. Impulse imaging measures the desired signal directly without the need for any computational decoding, although the image may be noisy due to low signal strength. Figure 13.2 illustrates the comparison between computational (coding-based) and impulse imaging techniques for motion deblurring.

We use image quality as a metric for performance. The theoretical performance bounds are derived in terms of the SNR metric. In addition, we provide empirical

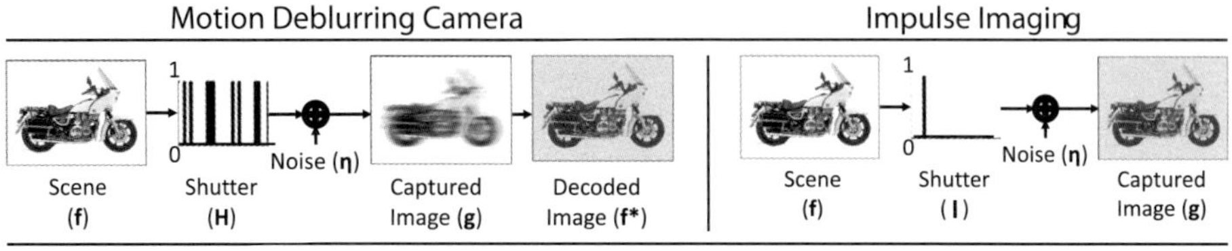

Figure 13.2 Motion deblurring versus impulse imaging. Left: the motion deblurring cameras discussed here can be modeled using the linear image formation model given by Eq. (13.1). In order to recover the desired image, these techniques require an additional decoding step, which amplifies noise. Right: impulse imaging (capturing a short exposure, motion blur free image) measures the signal directly without requiring any decoding.

results for several other perceptually motivated metrics (Wang, Bovik, Sheikh & Simoncelli 2004, Wang & Bovik 2002, Sheikh & Bovik 2006).

Assumptions

The analysis in this chapter assumes that the motion blur kernel (resulting from camera or object motion) is spatially invariant across the entire image. The coded exposure based techniques require the motion to be the same for all scene objects in order to achieve a spatially-invariant motion blur kernel. On the other hand, the camera motion based techniques are designed to achieve a spatially-invariant motion blur kernel even in the case of scene objects having different velocities. We assume an affine model for image noise; both the signal-independent and the signal-dependent components are modeled as additive Gaussians.

Related work

Harwit and Sloane analyzed optically coded image acquisition in the context of spectrometers (Harwit & Sloane 1979). They showed that in the absence of photon noise, the optimal measurement scheme corresponds to Hadamard matrices, and provides a significant performance advantage over impulse imaging (impulse imaging in the context of spectrometers corresponds to taking measurements over a narrow spectral band, so that no decoding is required post-capture). Ratner and Schechner (Ratner & Schechner 2007, Ratner, Schechner & Goldberg 2007) extended these results to derive optimal measurement matrices in the presence of photon noise. Ihrke *et al.* (Ihrke, Wetzstein & Heidrich 2010) analyzed the noise performance of different light field cameras and color filter array imagers. The performance gain of multiplexing, as derived in these papers, depends on the measurement matrices. Recently, Tendero (Tendero 2012) studied the performance gain of coded exposure cameras (Raskar *et al.* 2006) with respect to conventional impulse imaging, for motion deblurring. The general conclusion of the above mentioned works was that CI techniques do not give a significant performance advantage at high light levels. Agrawal and Raskar also analyzed the performance of CI motion deblurring cameras (Agrawal & Raskar 2009), but they analyzed how performance varies with increasing exposure time. In contrast, we analyze how performance varies with increasing light levels, which allows us to determine practical settings when CI gives a notable advantage. In this chapter, we focus on deriving theoretical performance bounds for CI based motion deblurring cameras. For similar bounds on CI techniques applicable in different domains (e.g. defocus deblurring, light-field imaging), the reader is referred to Cossairt, Gupta & Nayar (2012), and Cossairt (2011).

13.2 Performance bounds for flutter shutter cameras

To begin this section, we consider scenes that consist of objects moving with constant linear motion, resulting in images with a spatially-invariant motion blur. The motion may be the result of camera motion, object motion, or a combination of the two. However, for the time being, we assume that the motion is identical for each point in the scene, and furthermore that the direction and magnitude of motion are known. Note that this implicitly assumes that we have measured the motion blur PSF at the time of capture. This restriction will be somewhat relaxed in the next section when we consider motion-invariant cameras, which are designed to work over a range of object motions.

Since the motion blur is identical for each pixel, and the blur is inherently 1D, we may use a linear image formation model (see Figure 13.2) (Cossairt *et al.* 2012, Cossairt 2011):

$$\mathbf{g} = H\mathbf{f} + \eta, \tag{13.1}$$

where $\mathbf{g}$ is the vector of measurements of size N representing a scanline of pixels in the image along the direction of motion. The same equation can be used to express image formation for each scanline in the image. In this equation, $\mathbf{f}$ is the vector of pixel values corresponding to an all-in-focus scanline and η is a vector of noise values measured at each pixel. H is the measurement matrix which is a Toeplitz (convolution) matrix. We adopt the convention where the entries of H are between 0 and 1. For the impulse camera the exposure time is set so that motion blur is less than one pixel. Then the measurement matrix is the identity ($H = I$), and the camera measures the signal $\mathbf{f}$ directly. If we increase the exposure time by a factor of C, we introduce a box-shaped motion blur function, corresponding to a banded-diagonal measurement matrix with entries of 1 all contained within a bandwidth of C. This measurement matrix is poorly conditioned with several empty singular values. The flutter shutter camera was introduced to solve this exact problem (Raskar *et al.* 2006). This camera temporally modulates the exposure to effectively zero out some of the diagonals in the banded diagonal matrix. Exactly which diagonals are zeroed depends on the flutter sequence that is used.

There are several important factors to consider when designing a flutter shutter camera. The first is the choice of the temporal sequence of on/off patterns that is used over the duration of the exposure (see Figure 13.2). The flutter sequence determines how the motion blur is encoded in captured images, and will have a dramatic effect on motion deblurring performance. The length L of the flutter sequence results in an exposure time that is L times greater than an impulse camera. The sum of "on" values in the flutter sequence C determines the amount of light captured by the flutter camera. Greater values for C imply both more captured light (which increases SNR), and a larger shutter sequence (which amplifies more noise after deblurring). As we will see, this quantity will play an important role in determining the optimal flutter shutter performance.

Flutter sequences have been identified that significantly improve the conditioning of the measurement matrix H (Raskar *et al.* 2006), but a globally optimal flutter sequence has yet to be discovered. Initially, this may seem like a major limitation that prevents comprehensive analysis of flutter shutter cameras. However, as we show in this chapter, it is possible to derive an upper bound on the best possible performance that can be achieved with a flutter shutter camera *without* explicit knowledge of optimal flutter sequences.

In Eq. (13.1), each element of the noise vector η is assumed to be independently sampled from a zero mean Gaussian distribution $\mathcal{N}(0,\sigma^2)$. We consider an affine noise model where there are two sources of noise, signal-independent *read noise*, and signal-dependent *photon noise*[1]. The photon noise can be approximated by a Gaussian with variance equal to the measured signal level J (in photons). Let the variance of the read noise be σ_r^2. The total noise variance is:

$$\sigma^2 = J + \sigma_r^2. \tag{13.2}$$

An estimate of the signal can be found as:

$$\mathbf{f}^* = H^{-1}\mathbf{g}. \tag{13.3}$$

The mean-squared error (MSE) for the estimate $\mathbf{f}^*$ is given by Harwit & Sloane (1979):

$$\mathrm{MSE} = \frac{\sigma^2}{N}\mathrm{Tr}(H^{-t}H^{-1}), \tag{13.4}$$

where Tr() is the matrix trace operator.

Performance gain

In order to compute the performance gain of a motion deblurring camera, we compare the signal-to-noise ratio (SNR) of the recovered signal with the signal captured using impulse imaging (i.e. using a shorter exposure). The SNR is defined as $\mathrm{SNR} = J/\sqrt{\mathrm{MSE}}$.

Denoting σ_i^2 as the noise variance for the impulse camera, the MSE is just equal to the variance $\mathrm{MSE}_i = \sigma_i^2$. Let σ_c^2 be the noise variance for the measurement made with the motion deblurring camera. The performance gain G is the ratio of the SNR for the motion deblurring camera to the SNR of the impulse camera:

$$G = \sqrt{\frac{\mathrm{MSE}_i}{\mathrm{MSE}_c}} \tag{13.5}$$

$$= \sqrt{\frac{N}{\mathrm{Tr}(H^{-t}H^{-1})}}\,\frac{\sigma_i}{\sigma_c}. \tag{13.6}$$

[1] We ignore the effect of dark current noise, which is typically negligible when exposure times remain less than around one second.

When the noise is signal independent ($\sigma_i = \sigma_c$), the matrix that maximizes the gain for masking-based techniques is the S-matrix (Harwit & Sloane 1979) (note that the optimal matrix may not always be realizable). However, when the noise is signal dependent, the optimal measurement matrix, and hence the performance gain, depend on the *matrix light throughput* $C(H)$, which is the sum of elements in each row of the measurement matrix[2] H, and is a measure of the amount of light captured if H is used as the measurement matrix. For example, if H is the identity matrix, $C(H) = 1$. On the other hand, if H is the S-matrix, $C(H) \approx N/2$. Consequently, the S-matrix captures significantly more light. Similarly, if H is a flutter coding matrix, $C(H)$ represents the total "on" time of the flutter sequence, indicating that the flutter camera captures $C(H)$ more light than an impulse camera. In the remainder of the chapter, we drop the argument H from $C(H)$ for brevity.

Optimal measurement matrices

The problem of identifying optimal measurement matrices (that result in maximum gain) for masking-based CI techniques was explored by Ratner *et al.* (Ratner & Schechner 2007, Ratner *et al.* 2007). They found an analytic expression for the lower bound of the trace term:

$$\mathrm{Tr}(H^{-t}H^{-1}) \geqslant \frac{N(CN - 2C + 1)}{(N - C)C^2}. \tag{13.7}$$

Suppose the average signal level for the impulse camera is J, so that the total noise $\sigma_i^2 = J + \sigma_r^2$. The CI technique captures C times more light. Hence, the total noise is $\sigma_c^2 = CJ + \sigma_r^2$. Substituting these and Eq. (13.7) in Eq. (13.6), we get an expression for the maximum SNR gain G for matrices with a given light throughput C:

$$G(C) \leqslant \sqrt{\underbrace{\frac{(N - C)C^2}{CN - 2C + 1}}_{\text{decoding term}} \; \underbrace{\frac{J + \sigma_r^2}{CJ + \sigma_r^2}}_{\text{noise term}}}. \tag{13.8}$$

The righthand side of Eq. (13.8) consists of two competing terms. As light throughput is increased, the noise-dependent term decreases while the decoding-dependent term increases. There is an optimal light throughput $C_{\max}$ for which the SNR gain achieves the maximum value of $G_{\max}$.

13.2.1 Optimal flutter shutter performance

In this section, we derive a performance bound for flutter shutter cameras that is independent of the signal size N and the measurement matrix H (defined by the flutter sequence). The bound depends only on the signal strength, and the camera read noise levels. In comparison, the previous result (Eq. (13.8)) gives the maximum SNR gain G

[2] We consider matrices for which all the rows have the same sum, which is always the case for flutter shutter coding matrices.

for coding matrices with a light throughput C and a signal size N. We first derive an upper bound on the decoding term in Eq. (13.8):

$$\begin{aligned}\frac{C^2(N-C)}{NC-2C+1} &= \frac{(N-C)C^2}{NC-C^2+C^2-2C+1}\\ &= \frac{(N-C)C^2}{C(N-C)+(C-1)^2}\\ &\leqslant \frac{(N-C)C^2}{C(N-C)}\\ &\leqslant C. \end{aligned} \tag{13.9}$$

Next, we derive an upper bound on the noise term:

$$\begin{aligned}\frac{J+\sigma_r^2}{CJ+\sigma_r^2} &= \frac{1}{C}\frac{CJ+C\sigma_r^2}{CJ+\sigma_r^2}\\ &\leqslant \frac{1}{C}\frac{CJ+\sigma_r^2+C\sigma_r^2}{CJ+\sigma_r^2}\\ &\leqslant \frac{1}{C}\left(1+\frac{C\sigma_r^2}{CJ+\sigma_r^2}\right)\\ &\leqslant \frac{1}{C}\left(1+\frac{\sigma_r^2}{J}\right). \end{aligned} \tag{13.10}$$

By substituting the bounds in Eqs. (13.9) and (13.10) in Eq. (13.8), we get the upper bound on the performance gain of any flutter shutter camera:

$$G < \sqrt{1+\frac{\sigma_r^2}{J}}. \tag{13.11}$$

Equation (13.11) expresses the best possible SNR gain that can be achieved for any flutter shutter camera, measured relative to an impulse camera with a sufficiently shorter exposure to remove motion blur.

As a final note, we point out that the same analysis applies if we consider constant speed spatially-invariant motion along an arbitrary 2D trajectory. Then the PSF becomes 2D instead of 1D. The coding matrix H in Eq. (13.1) becomes block Toeplitz, and the quantities $\mathbf{f}$ and $\mathbf{g}$ both represent a lexicographical reordering of the pixels in a 2D image. As a result, the same upper bound on performance given by Eq. (13.11) holds.

It is also possible to relax the assumption of objects moving at a constant speed. Let v_{imp} be the speed at which a scene object was moving during the impulse imaging exposure. The speed is sufficiently low to prevent any motion blur. Then, if the object moves with a non-constant speed (but always higher than v_{imp}) during the exposure of the motion deblurring camera, the entries of the measurement matrix H will be in the continuous range [0–1], instead of being limited to only 0 or 1. It can be shown (Harwit & Sloane 1979) that even in this case, the bound in Eq. (13.11) holds. Finally, we point out that while this chapter assumed that the flutter sequence is the same for all pixels on the sensor, it is also feasible to consider the possibility that a unique flutter sequence is applied to each pixel. This would correspond to a nondiagonal coding matrix H. However, the derivation in this chapter holds for arbitrary coding matrices, and therefore the bound in Eq. (13.11) will continue to hold in this case as well.

13.3 Performance bound for motion-invariant cameras

In the previous section, we derived an upper bound on the performance for masking-based multiplexing systems, which allowed us to determine the best possible performance that can be achieved for a flutter shutter camera. In this section, we extend this analysis to motion-based cameras, where the camera is moved along a specified trajectory during image capture. The performance bound for this class of cameras is given as:

$$G < \sqrt{2\left(1 + \frac{\sigma_r^2}{J}\right)}. \tag{13.12}$$

Note that the performance bound is higher than exposure-coding-based cameras (Eq. (13.11)) since motion-based cameras do not block light. In the following, we derive this bound. The analysis closely follows the treatment given in Levin, Fergus, Durand & Freeman (2007), Cho *et al.* (2010), Cossairt *et al.* (2012), Cossairt (2011). Readers not interested in the detailed derivation may skip directly to Section 13.4.

13.3.1 Space–time analysis

Motion-invariant cameras do not attenuate any of the light reaching the sensor. Instead, they utilize camera motion to encode motion blur in a manner that can easily be removed via deconvolution. The camera motion is chosen so that blur becomes independent of object velocity, resulting in an image formation model that is entirely shift invariant. In practice, this means that either the direction of motion must be known a priori (Levin *et al.* 2007), or multiple images must be captured with complimentary camera motion (Cho *et al.* 2010). For simplicity, we begin by considering the case of 1D motion blur where objects in the scene move with varying speed, but in the same (known) direction. We model our scene as a 2D space–time volume $l(x, t)$ (where x is measured in pixels) consisting of objects moving at different speeds s (measured in pixels/s). We assume the speed of objects is bounded so that $|s| < S_{\max}$. Ignoring the effects at occlusion boundaries, the scene can be expressed as a summation of point sources moving at different speeds. The motion blur induced by object motion will depend on the speed of the point source. We characterize the motion blur by considering the space–time volume produced by a point source l_p with unit flux moving at a constant velocity s:

$$l_p(x, t) = \delta(x - s \cdot t). \tag{13.13}$$

A stationary camera captures a photograph by integrating the space–time volume along the temporal direction with an exposure duration of T. The image produced by a moving point source is the motion-dependent 1D PSF $h_s(x)$, where the subscript s denotes dependence on object speed:

$$h_s(x) = \int_{-T/2}^{T/2} l_p(x,t)\,dt \tag{13.14}$$

$$= \int_{-T/2}^{T/2} \delta(x - s \cdot t)\,dt \tag{13.15}$$

$$= \frac{1}{|s|} \sqcap\left(\frac{x}{|s|T}\right), \tag{13.16}$$

where the rect function is defined as

$$\sqcap(x) = \begin{cases} 1 & \text{if } |x| < 1/2 \\ 0 & \text{otherwise.} \end{cases} \tag{13.17}$$

Thus we are left with the familiar line-shaped motion blur PSF with length $|s|T$. In particular, we note that if we set the exposure time such that $T = 1/S_{\text{max}}$, this will ensure that the motion blur will always be less than a pixel. The PSF will be a (Dirac) delta function with a magnitude T for each object in the scene, regardless of the exact speed at which it is moving.

In contrast, a motion-invariant camera uses a longer exposure time in conjunction with camera motion in an attempt to produce a blur that is independent of object motion. In this case, we use an exposure time that is C times greater, and therefore collects C times as much light. The quantity C is analogous to the matrix light throughput in Section 13.2. Thus, we expect to be able to follow a similar procedure where we derive an expression for the SNR gain that is decomposed into decoding and noise-dependent terms (see Eq. (13.8)). This expression will allow us to derive the best possible motion-invariant performance that occurs with the optimal choice of C.

A motion-invariant camera moves along the motion path $f(t)$ during exposure. The PSF of this camera is then given by

$$h_s(x) = \int_{-C \cdot T/2}^{C \cdot T/2} l_p(x - f(t), t)\,dt. \tag{13.18}$$

$$= \int_{-\infty}^{\infty} k(x - s \cdot t, t)\,dt, \tag{13.19}$$

where we have introduced the 2D integration kernel for the motion-invariant camera

$$k(x,t) = \delta(x - f(t)) \sqcap\left(\frac{t}{C \cdot T}\right). \tag{13.20}$$

Referring to Eq. (13.19), the motion-invariant camera PSF is a 1D projection of the 2D integration kernel. According to the Fourier slice theorem, the optical transfer function (OTF) is then given by a slice of the Fourier transform of the integration kernel $K(\omega_x, \omega_t)$ (see (Wikipedia 2013*b*) for an illustration):

$$H_s(\omega_x) = K(\omega_x, s\omega_x), \tag{13.21}$$

where ω_x are spatial frequencies measured per-pixel. Equation (13.21) tells us how to relate the motion-dependent OTF to the Fourier transform of the integration kernel, which is itself determined by the camera motion path. If we consider the set of objects moving at velocities in the range $|s| < S_{\max}$, we can see that the OTF will be maximized when energy in $K(\omega_x, \omega_t)$ is concentrated within the double wedge in frequency space given by $|\omega_t| < S_{\max}|\omega_x|$ (see Figure 5 in Levin *et al.* (2007) for an illustration). Furthermore, to ensure that the OTF (and consequently also the PSF) is motion invariant, we want to enforce that $K(\omega_x, \omega_t)$ remains constant while speed s is allowed to vary, but spatial frequency ω_x remains fixed. However, it can be shown that (Levin *et al.* 2007)

$$\int |K(\omega_x, \omega_t)|^2 \, d\omega_t \leqslant C \cdot T. \tag{13.22}$$

The importance of Eq. (13.22) can be understood as follows. For a fixed ω_x, the optimal kernel K will only contain energy along a line in frequency space of length $2S_{\max}|\omega_x|$. However, if we integrate $|K|^2$ along this line, the result can be no greater than $C \cdot T$. Therefore, the best possible motion-invariant camera will be one that produces an integration kernel given by

$$|K(\omega_x, \omega_t)|^2 \leqslant \frac{C \cdot T}{2S_{\max}|\omega_x|}. \tag{13.23}$$

Since the optimal kernel is independent of ω_t, the slicing operation given by Eq. (13.21) does not alter the result. The OTF then becomes invariant to object motion and is simply given by

$$|H(\omega_x)|^2 \leqslant \frac{C \cdot T}{2S_{\max}|\omega_x|}. \tag{13.24}$$

Note that while the righthand side of Eq. (13.24) represents the upper bound on the OTF for any motion-invariant camera, it may not be possible to achieve this upper bound exactly. However, Levin *et al.* show that the upper bound can be approximately met when a motion path of constant acceleration is used $f(t) = a_0 \cdot t^2$ (Levin *et al.* 2007).

For the general case of 2D motion blur, we consider 3D integration kernels given by $k(x, y, t)$, arbitrary object motion given by the velocity vector $\mathbf{s} = (s_x, s_y)$, and a 2D camera motion path given by $\mathbf{f}(t) = (f_x(t), f_y(t))$. Again we assume the maximum object speed is bounded so that $|\mathbf{s}| < S_{\max}$. In the Fourier domain, the energy of the optimal 3D integration kernel $K(\omega_x, \omega_y, \omega_t)$ is concentrated on the double inverted cone where $\omega_t \leqslant S_{\max}\sqrt{\omega_x^2 + \omega_y^2}$ (Cho *et al.* 2010). The upper bound on the OTF for a 2D motion-invariant camera is then given by

$$|H(\omega_x)|^2 \leqslant \frac{C \cdot T}{2S_{\max}\sqrt{\omega_x^2 + \omega_y^2}}. \tag{13.25}$$

13.3.2 Optimal motion-invariant performance

Because the motion-invariant camera produces a blur that is independent of object motion, the image formation of Eq. (13.1) can be described as a convolution

$$g(x, y) = h(x, y) * f(x, y) + \psi(x, y), \tag{13.26}$$

where g is the captured blurred image, f is the focused image, and ψ is the sensor noise. In the Fourier domain, this equation becomes:

$$G(\omega_x, \omega_y) = H(\omega_x, \omega_y)F(\omega_x, \omega_y) + \Psi(\omega_x, \omega_y), \tag{13.27}$$

where F is the focused image spectrum, H is the optical transfer function (OTF) of the camera, Ψ is the noise spectrum, and G is the captured image spectrum. An estimate of the focused image can be found as

$$F^*(\omega_x, \omega_y) = \frac{G(\omega_x, \omega_y)}{H(\omega_x, \omega_y)}, \tag{13.28}$$

and the expected MSE can be written as

$$\text{MSE} = E\left[\|F^*(\omega_x, \omega_y) - F(\omega_x, \omega_y)\|^2\right] \tag{13.29}$$

$$= \sigma^2 \int_{-1/2}^{1/2} \int_{-1/2}^{1/2} \frac{1}{\|H(\omega_x, \omega_y)\|^2} \, d\omega_x \, d\omega_y, \tag{13.30}$$

where E denotes expectation with respect to the noise Ψ, and σ^2 is the noise variance. Equation (13.30) is analogous to Eq. (13.4), except here MSE is calculated in the continuous domain. The bounds in the integral are set to $(\omega_x, \omega_y) \in \{-1/2, 1/2\}$ because we have implicitly assumed the function F to be bandlimited. From this expression, we can derive the best possible performance that can be achieved for any motion-invariant camera.

As mentioned previously, an impulse camera will remove motion blur by setting the exposure time T so that the maximum blur size is equal to one pixel and $T = 1/S_{\text{max}}$. To make our notation consistent with Section 13.2, we normalize our temporal coordinates so that $T = 1$. For the impulse camera, the OTF is then given by $\text{OTF}(\omega_x, \omega_y) = 1$ so that the MSE is simply

$$\text{MSE}_i = \sigma_i^2. \tag{13.31}$$

Substituting the upper bound on the 2D motion-invariant OTF, given by Eq. (13.25), into Eq. (13.30) gives a lower bound on the MSE for any motion-invariant camera:

$$\text{MSE}_m \geqslant \frac{\sigma_m^2\left(\sqrt{2} + \text{asinh}(1)\right)}{3C} \geqslant \frac{\sigma_m^2}{2C}. \tag{13.32}$$

Substituting Eqs. (13.32) and (13.31) into Eq. (13.5) and using the bound expressed by Eq. (13.10), results in the bound given by Eq. (13.12). This bound expresses the best possible SNR gain that can be achieved for any motion-invariant camera, measured relative to a static camera with a sufficiently shorter exposure to remove motion blur.

As a final note, we mention that while the theoretical analysis in this section assumed constant motion, in practice, the conclusions are applicable for small deviations from these assumptions.

13.4 Simulations to verify performance bounds

Equations (13.11) and (13.12) are noteworthy because they provide the maximum possible SNR gain for any motion deblurring camera. In Figure 13.3, we show the simulated performance for motion deblurring cameras. The SNR gain of each technique is calculated using Eq. (13.6), and the result is plotted against the ratio of photon to read noise variance (J/σ_r^2). Figure 13.4 shows motion deblurring performance for the flutter shutter (Raskar *et al.* 2006) and motion-invariant (Levin *et al.* 2008) cameras. For the motion-invariant camera, the optimal value of C was chosen for each signal level. For the flutter shutter camera, the optimal 52-length code for each signal level was found using brute force search (note that for very small signal levels, the length of the optimal flutter code may be larger than 52). As expected, both the techniques performed at or below the performance bound given by Eq. (13.12).

Implication of the bounds

The bounds in Eqs. (13.11) and (13.12) imply that the *performance gain for computational imaging is significant only when the average signal level J is considerably smaller than the read noise variance σ_r^2*. The read noise variance for currently available sensors ranges from less than one gray level (on a scale of [0–255]) for high quality DSLR cameras to approximately five gray levels for low quality machine vision cameras (Schechner, Nayar & Belhumeur 2007). Typical daylight illumination results in signal strengths that are considerably greater than these read noise variance values.

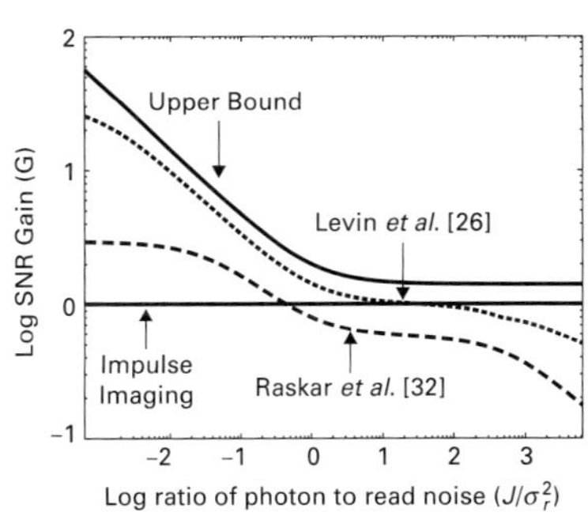

Figure 13.3 Verification of the performance bound using simulations: motion deblurring performance for the flutter shutter (Raskar *et al.* 2006) and motion-invariant cameras (Levin *et al.* 2008). Both techniques perform at or below the performance bound given by Eq. (13.12).

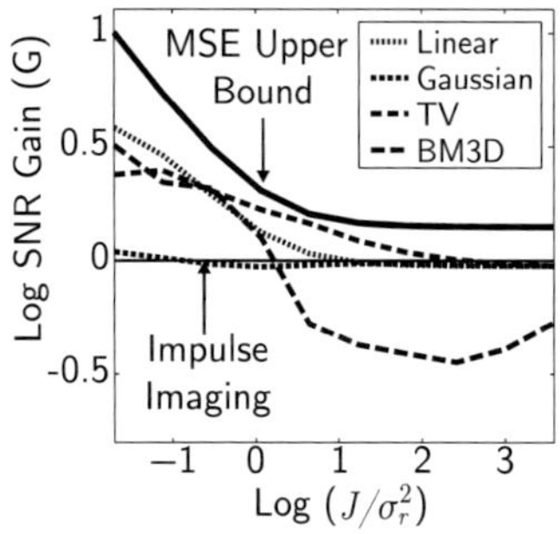

(a) Performance using MSE metric

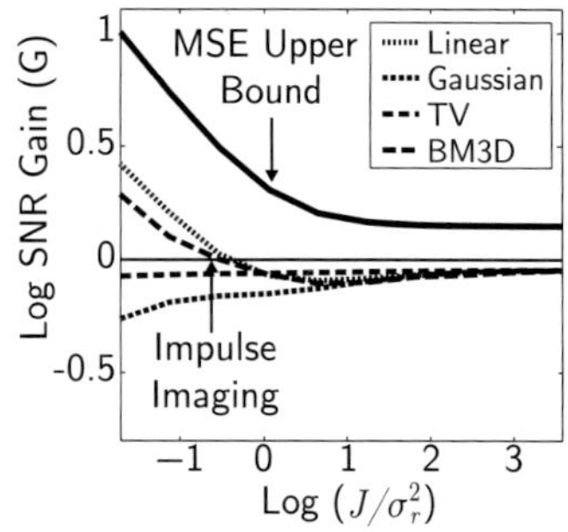

(b) Performance using SSIM metric

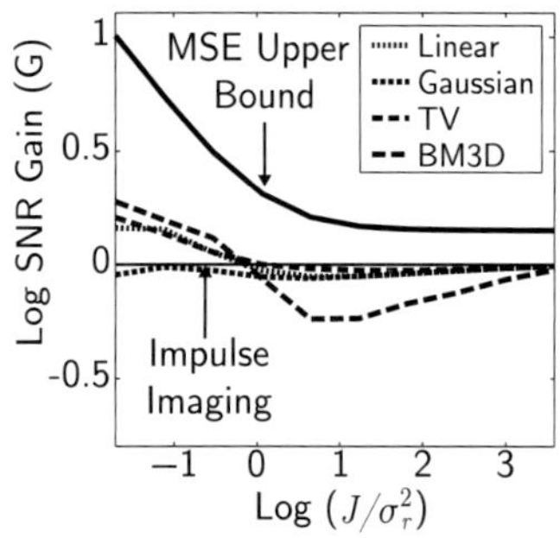

(c) Performance using VIF metric

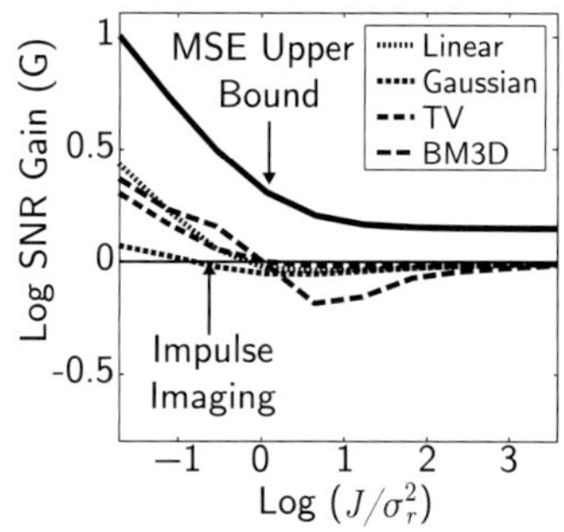

(d) Performance using UQI metric

Figure 13.4 Simulated performances for the flutter shutter camera using: (a) the mean-squared error metric; (b) the structural similarity metric; (c) the visual information fidelity metric; (d) the universal quality index metric. Several reconstruction techniques are applied, including linear inversion, Gaussian prior on derivatives, total variation prior, and BM3D prior. The SNR gain is always less than the bound given by Eq. (13.12), regardless of the prior or metric used.

13.5 Role of image priors

Thus far, we have not considered the role of image priors. Priors can be used to improve image quality, for both computational and impulse imaging (Levin *et al.* 2007, Zhou & Nayar 2009). The improvement depends on the type of priors and image coding used. In addition, our analysis thus far used MSE as the quality metric for images because it makes the derivation of performance bounds tractable. However, a number of metrics have been introduced which measure the perceived image quality (Wang *et al.* 2004, Sheikh & Bovik 2006, Wang & Bovik 2002, Gaubatz 2013). In this section we analyze the effect of various priors and metrics on performance.

Image priors

We can think of the estimate $\mathbf{f}^*$ given in Section 13.2 as the maximum likelihood (ML) estimate of the following optimization problem:

$$\mathbf{f}^*_{\mathrm{ML}} = \arg\max_{\mathbf{f}} P(\mathbf{f}|\mathbf{g}) \tag{13.33}$$

$$= \arg\min_{\mathbf{f}} \|\mathbf{f} - H\mathbf{g}\|^2, \tag{13.34}$$

where $P(\mathbf{f}|\mathbf{g})$ is the probability of the unknown image $\mathbf{f}$ given the measurement $\mathbf{g}$, which is Gaussian due to the properties of the measurement noise vector η. If we have knowledge of the probability distribution of our unknown image $P(\mathbf{f})$, then we can improve performance by finding the maximum a posteriori (MAP) estimate:

$$\mathbf{f}^*_{\mathrm{MAP}} = \arg\max_{\mathbf{f}} P(\mathbf{g}|\mathbf{f})P(\mathbf{f}) \tag{13.35}$$

$$= \arg\min_{\mathbf{f}} \|\mathbf{f} - H\mathbf{g}\|^2 + \lambda \log(P(\mathbf{f})), \tag{13.36}$$

where the constant λ determines how much significance to attach to the prior. The image prior can essentially be thought of as a way to coerce the optimization problem to produce more probable estimates of the unknown image. In this section, we consider three image priors. Firstly, we consider a Gaussian prior on the distribution of gradients in the image (Levin *et al.* 2007, Hasinoff, Kutulakos, Durand & Freeman 2009) of the form:

$$\log(P_{\mathrm{Gauss}}(\mathbf{f})) = -\|\nabla\mathbf{f}\|_2^2, \tag{13.37}$$

where ∇ is the first order finite difference matrix, which calculates a discrete approximation to the gradient operator. In this case, because the prior is Gaussian, the MAP estimate can be calculated directly as:

$$\mathbf{f}^*_{\mathrm{MAP}} = (\nabla^t\nabla + H^tH)^{-1}H^t\mathbf{g}. \tag{13.38}$$

We also consider the total variation (TV) prior (Bioucas-Dias & Figueiredo 2007, Chambolle 2004) which has the form:

$$\log(P_{\mathrm{TV}}(\mathbf{f})) = -\|\nabla\mathbf{f}\|_1. \tag{13.39}$$

In this case, there is no direct way to calculate the MAP estimate, and an iterative algorithm must be used. We use the TwIST algorithm (Bioucas-Dias & Figueiredo 2007) to solve Eq. (13.36) and find a MAP estimate.

Lastly, we consider the current state-of-the-art BM3D algorithm (Dabov, Foi, Katkovnik & Egiazarian 2007). Although it is hard to write a single linear expression for this prior (because of nonlinear shrinkage in the 3D spectrum domain), it is possible to calculate a MAP estimate numerically.

Image quality metrics

In Section 13.2, we established MSE as the metric for evaluating images. We now extend this formalization to more general metrics. We can define a general form for any similarity metric $S(\mathbf{f}, \mathbf{f}^*)$ that measures how close our estimate $\mathbf{f}^*$ is to the actual image $\mathbf{f}$. The MSE metric can then be defined as

$$S_{\mathrm{MSE}}(\mathbf{f}, \mathbf{f}^*) = \frac{1}{\|\mathbf{f} - \mathbf{f}^*\|_2^2}. \tag{13.40}$$

The interpretation here is that the smaller the MSE, the more similar the estimate $\mathbf{f}^*$ relative to the true image $\mathbf{f}$. The performance gain for any metric can then be written as

$$G = \sqrt{\frac{S_c(\mathbf{f}, \mathbf{f}^*)}{S_i(\mathbf{f}, \mathbf{f}^*)}}, \tag{13.41}$$

where S_c is the metric applied to the computational camera, and S_i is the metric applied to the corresponding impulse camera. For the MSE metric, the definition remains the same as the expression given in Eq. (13.6). However, with this general definition in place, we are now free to use other metrics to evaluate performance. Note that because we have defined SNR gain in terms of a similarity metric instead of an error metric, the term for the computational camera appears in the numerator instead of the denominator. In addition to MSE, we use the following image quality metrics to measure performance: structural similarity (SSIM) (Wang *et al.* 2004), visual information fidelity (VIF) (Sheikh & Bovik 2006), and universal quality index (UQI) (Wang & Bovik 2002). We use the MeTriX MuX visual quality assessment package to calculate performance using these metrics (Gaubatz 2013).

Simulations

Figure 13.4 shows the simulated performance of a flutter shutter camera. Since the performance of reconstruction algorithms can be image dependent, we report the performance averaged over a large dataset of images. For this simulation, we use the Caltech 101 image database of 9140 different images (L. Fei-Fei & Perona 2004). For each image, we simulate the performance under ten different photon to read noise ratios (J/σ_r^2). Performance is shown for the MSE, SSIM, VIF, and UQI metrics, and for each plot, the performance gain G is plotted on a log scale. Thus, a value of zero corresponds to a performance gain of $G = 1$, meaning that both computational and impulse imaging have the same performance. The solid black lines correspond to the performance bound expressed by Eq. (13.12). The dotted lines correspond to performance gain using direct linear inversion (i.e. estimating the image using Eq. (13.3)), as well as Gaussian, TV, and BM3D priors.

There are two interesting observations to be made from these plots. First, in most cases, image priors boost the performance of impulse imaging more than motion deblurring cameras. As a result, the performance advantage of motion deblurring over impulse imaging is reduced even further, especially at low light levels. Thus, the performance bound expressed by Eq. (13.12) is the tightest when no prior is used.

The second observation is that the bound derived using linear inversion and the MSE metric (solid black curve) appears to be an upper bound for performance across all metrics and priors. This is surprising because it is well known that MSE does not accurately measure perceived image quality. Nonetheless, the upper bound expressed by Eq. (13.12) does appear to provide a consistent upper bound on performance regardless of the image quality metric used.

	Starry night	Quarter moon	Full moon	Twilight	Indoor Lighting	Cloudy Day	Sunny Day
I_{src} (lux)	2×10^{-3}	10^{-2}	1	10	10^2	10^3	10^4
J (e^-)	7×10^{-4}	4×10^{-3}	.39	3.85	38.49	384.9	3,849

Figure 13.5 Relating typical lighting conditions for photography to average photon counts. The top row shows typical illuminance values in lux (Wikipedia 2013*a*). The bottom row shows the photon counts calculated using Eq. (13.42) assuming an average reflectivity of $R = 0.5$, quantum efficiency of $q = 0.5$, and exposure time of $t = 1/50$ s, aperture setting of $F/2.1$, and pixel size of $\Delta = 1\,\mu$m.

13.6 When to use computational imaging

Equations (13.11) and (13.12) provide performance bounds for motion deblurring cameras in terms of the sensor read noise σ_r and the average signal level J of the impulse image. In order to determine when motion deblurring is advantageous, we have derived an expression for the signal level J in terms of the scene and sensor-dependent parameters (see appendix for a derivation):

$$J \approx 10^{15} \underbrace{(F/\#)^{-2}\, t\, I_{\text{src}}\, R}_{\substack{\text{scene}\\ \text{dependent}}} \quad \underbrace{q\, \Delta^2}_{\substack{\text{sensor}\\ \text{dependent}}}, \tag{13.42}$$

where $F/\#$ is the ratio of focal length to aperture size of the lens, t is the exposure time, I_{src} is the incident illuminance given in lux, R is the average reflectivity of the scene, q is the quantum efficiency of the sensor, and Δ is the pixel size in meters. In Figure 13.5, we give values of J corresponding to several commonly encountered lighting conditions.

Scene-dependent parameters

The $F/\#$ of the camera will typically depend on the depth range of the scene. Assuming we want the entire depth range to be in focus, a larger depth range will require stopping down to smaller apertures to reduce defocus blur to within one pixel. Similarly, the exposure time t depends on the range of scene velocities. Higher velocities will require the impulse camera to have a short exposure to reduce motion blur to within one pixel. Finally, the signal level is directly proportional to the illumination brightness I_{src} and the object albedo R.

Sensor-dependent parameters

Different sensors have different quantum efficiencies and pixel sizes. Quantum efficiency q for commercially available sensors is quite high, usually greater than 0.5. For today's sensors, the size of pixels Δ has a wide range, from 1 μm for small cell phone sensors to nearly 10 μm for large format sensors.

13.6.1 Rule of thumb

When using one of today's commercial grade image sensors (DSLRs, cell-phone cameras, machine vision sensors), motion deblurring cameras will only yield significant performance benefits when the illuminance is less than 100 lux (typical living room lighting).

This implies that when the illuminance is above 100 lux, it is better to capture the impulse image without any blur (using a small exposure time).

We support this rule of thumb with several simulations. We consider both the lighting conditions (e.g. moonlit night or cloudy day, indoor or outdoor) and scene properties (albedo, speed, range of object velocities). In all our examples, we assume an average reflectivity of $R = 0.5$, quantum efficiency of $q = 0.5$, and read noise of $\sigma_r = 4e^-$, which is typical for today's CMOS sensors (Clark 2013).

Motion deblurring simulations

Here we show a simulation of both flutter shutter and motion-invariant cameras. For these simulations, we used a pixel size of $\Delta = 5\,\mu\text{m}$, aperture setting of $F/20$ and the impulse camera exposure time $t = \frac{1}{50}$ s. For the flutter shutter camera, we used the 52 digit long sequence given in Raskar *et al.* (2006) to simulate a fluttered image captured with an exposure time of $52 \cdot t = 1.02$ s. For the motion-invariant camera, we synthesized the PSF produced from a camera moving with constant acceleration and exposure time of 1.02 s. We simulated the effect of photon and read noise, and decoded the captured image using Eq. (13.3). The simulated images are shown in Figures 13.6 and 13.7. Both the flutter shutter and motion-invariant cameras perform worse than impulse imaging when the illuminance is greater than 100 lux.

In Figure 13.8, we show a contour plot of the SNR gain bound[3] given by Eq. (13.12) versus the illuminance (I_{src}) and the maximum speed of objects (given in pixels/s). Note that this bound holds for flutter shutter and motion-invariant cameras alike, and it is independent of the particular flutter sequence, motion trajectory, and exposure time of the camera. As the maximum object speed increases, the exposure time of the impulse camera must be reduced to avoid motion blur. We can observe that motion deblurring cameras never give an SNR gain greater than 2 when the illuminance is greater than 100 lux.

13.7 Relationship to other computational imaging systems

In addition to motion deblurring cameras, a number of other computational cameras have been introduced to improve performance in terms of image quality. Like motion deblurring cameras, they use optical coding to increase light throughput to obtain a stronger signal. These techniques follow the linear imaging model of Eq. (13.1).

[3] Although the SNR bound is derived assuming no priors and MSE metric, we have observed empirically that it bounds the performance of CI techniques irrespective of the prior and the image quality metric (see Section 13.5).

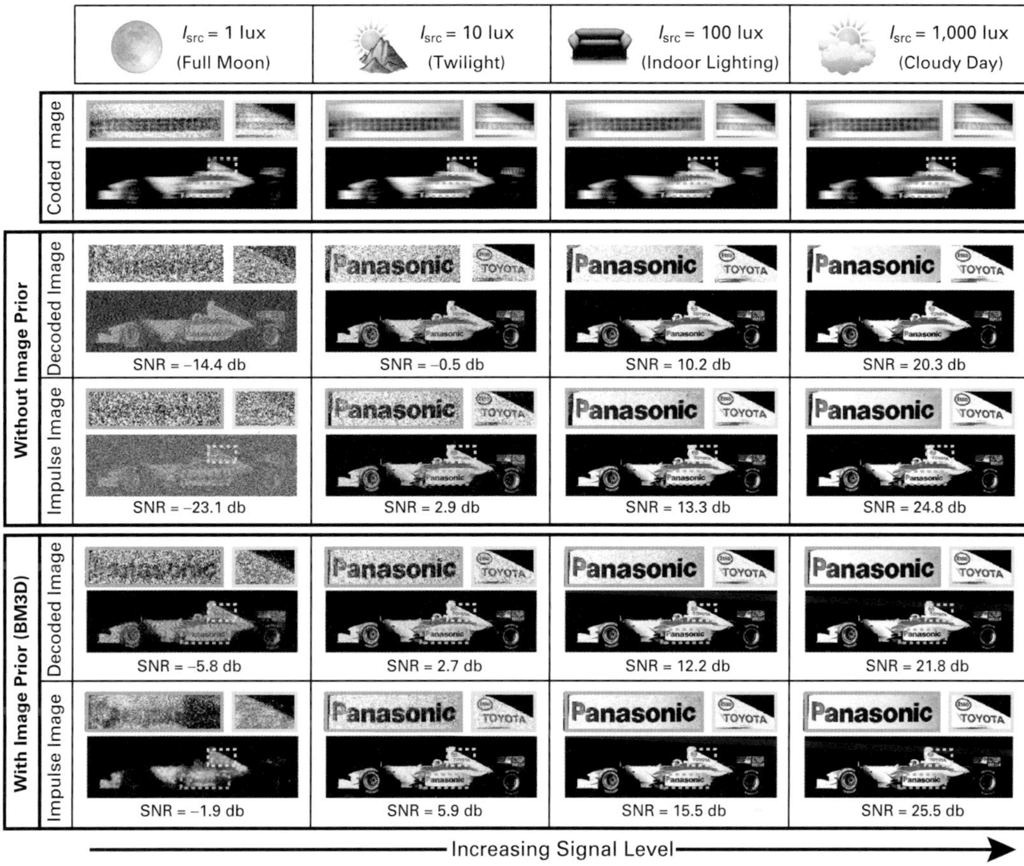

Figure 13.6 Simulated performance for the flutter shutter camera. The parameters for the simulation are given in Section 13.6.1. Top row: image blurred by the flutter sequence given in Raskar *et al.* (2006); second row: results after linear deblurring; third row: results from the impulse camera (shorter exposure); fourth row: results after deblurring the images in the first row with the BM3D algorithm (Dabov *et al.* 2007); last row: results for denoising the images in the third row with the BM3D algorithm. Gaussian distributed read noise and Poisson distributed photon noise is added to each image. The illumination I_{src} increases from left to right. Without the prior, the flutter shutter camera has higher SNR when $I_{src} < 100$ lux.

In addition, for each of these cameras, there is a corresponding conventional imaging technique that can measure the signal directly without the need for any decoding (impulse imaging). Indeed, it has been shown that many computational cameras obey the same performance bound as motion deblurring cameras (Cossairt *et al.* 2012, Cossairt 2011). The following are a few examples.

13.7.1 Example computational cameras

Defocus blur

Coded aperture masks have been used to deblur defocused images (Levin *et al.* 2007, Veeraraghavan, Raskar, Agrawal, Mohan & Tumblin 2007, Zhou & Nayar 2009, Zhou,

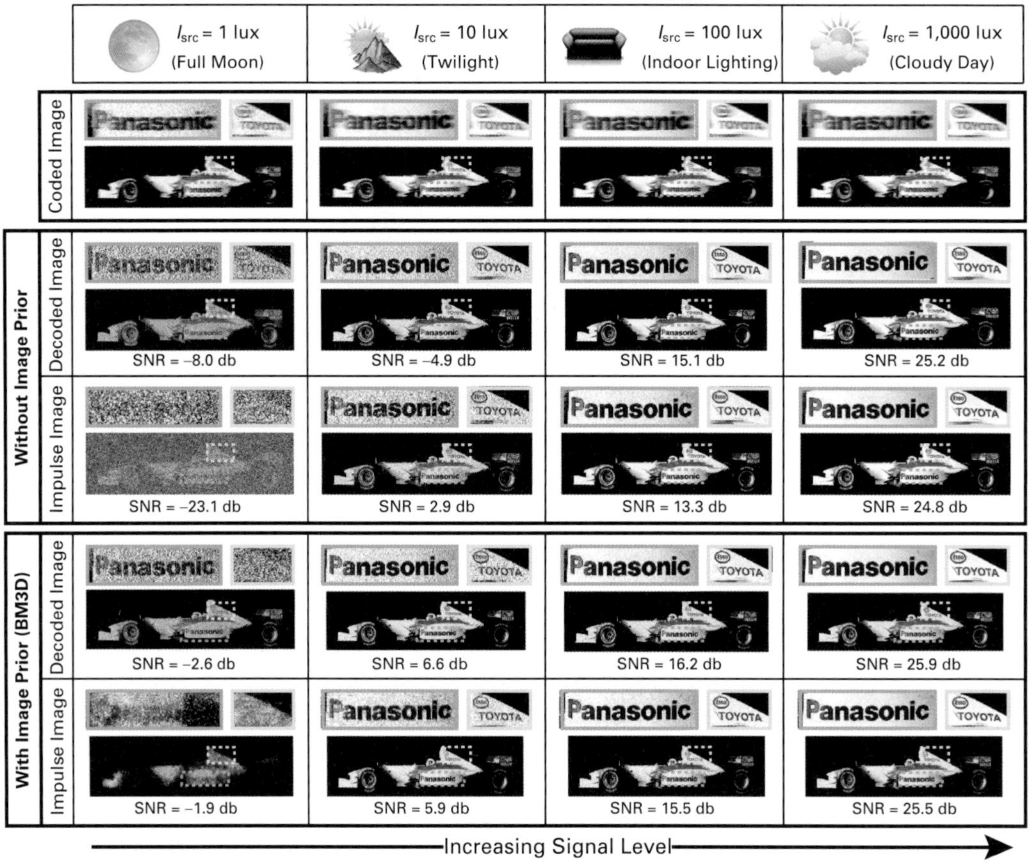

Figure 13.7 Simulated performance for a motion-invariant camera. The parameters for the simulation are given in Section 13.6.1. Top row: image blurred by the 1D motion-invariant camera from Levin *et al.* (2007); second row: results after linear deblurring; third row: results from the impulse camera (shorter exposure); fourth row: results after deblurring the images in the first row with the BM3D algorithm (Dabov *et al.* 2007); last row: results for denoising the images in the third row with the BM3D algorithm. Gaussian distributed read noise and Poisson distributed photon noise is added to each image. The illumination I_{src} increases from left to right. Without the prior, the motion-invariant camera has higher SNR when $I_{src} < 100$ lux.

Lin & Nayar 2011). There are also several techniques that extend the depth of field (DOF) by producing a depth-independent blur that can be inverted without the need for depth estimation (Chi & George 2001, Dowski & Cathey 1995, Ojeda-Castaneda, Landgrave & Escamilla 2005, Häusler 1972, Kuthirummal, Nagahara, Zhou & Nayar 2010, Cossairt, Zhou & Nayar 2010, Cossairt & Nayar 2010, Guichard, Nguyen, Tessières, Pyanet, Tarchouna & Cao 2009). Assuming the blur kernel to be shift invariant, the measurement matrix H is a circulant matrix, where each row of the matrix encodes the defocus blur kernel. $\mathbf{g}$ is the captured blurred image and $\mathbf{f}$ is the extended depth of field (EDOF) image. The corresponding impulse imaging technique is to capture images with a stopped down aperture (H is equal to the identity matrix I).

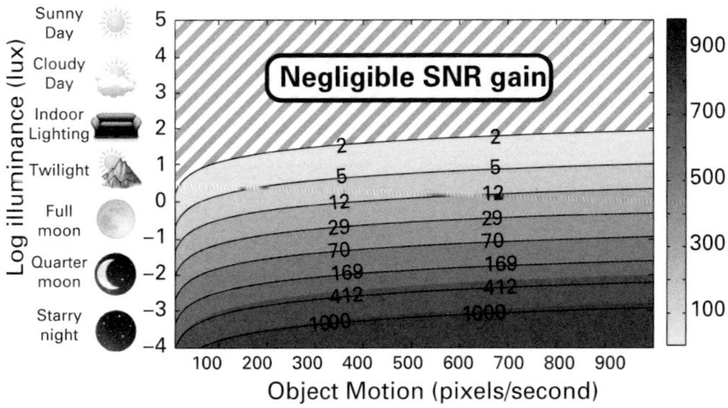

Figure 13.8 When to use motion deblurring. The average signal level is calculated using Eq. (13.42) with the parameters outlined in Section 13.6.1. Shown is a contour plot of the SNR gain bound vs the illuminance and the exposure time of the impulse camera. The SNR gain is always negligible when the illuminance is greater than 100 lux (typical indoor lighting).

Multiplexed light fields

CI techniques for capturing light fields include placing a transmissive mask either at the lens aperture (Liang, Lin, Wong, Liu & Chen 2008), or near the sensor (Veeraraghavan *et al.* 2007, Lanman, Raskar, Agrawal & Taubin 2008). The measurement matrix H is again block circulant. Each pixel measures a linear combination of ray intensities (**g**), and the light field (**f**) must be recovered by demultiplexing the captured data. In contrast, a light field camera can also be built by placing a mask consisting of an array of pinholes near the sensor (Horstmeyer, Euliss, Athale & Levoy 2009), or by capturing a sequence of images with a shifting pinhole in the aperture (Liang *et al.* 2008). These techniques are the impulse imaging counterparts of multiplexed light field capture.

Multiplexing color and spectrum

Mask-based Hadamard multiplexing is used for point (Harwit & Sloane 1979) and imaging (Hanley, Verveer & Jovin 1999) spectrometers. Here, H is the spectral mixing matrix, **g** is the vector of multiplexed spectral samples and **f** is the vector of narrowband spectral samples (desired). The impulse imaging counterpart is capturing narrowband spectral samples (Chakrabarti & Zickler 2011). Color filter arrays (CFAs) that multiplex color have been proposed to capture three color images with more light throughput than RGB Bayer filters (Baer, Holland, Holm & Vora 1999, Ihrke *et al.* 2010) (impulse imaging).

Multiplexed illumination

Measuring the appearance of a scene under varying illumination is useful for scene relighting and estimating depth. These measurements can be multiplexed by measuring the appearance of a scene when illuminated by linear combinations of light sources (Schechner *et al.* 2007, Ratner & Schechner 2007, Ratner *et al.* 2007). Here, H is the measurement ensemble, **g** is the vector of acquired (multiplexed) image intensities and **f** is the vector of intensities corresponding to only single sources. Here, the impulse imaging counterpart is capturing images by turning on only one light source at a time.

13.8 Summary and discussion

In this chapter, we analyzed the performance of a variety of computational imaging-based motion deblurring techniques (coded exposure and camera motion-based techniques), and derived a bound on their performance in terms of SNR. Implementing a motion deblurring camera can add a further, often significant, cost to a conventional imaging system[4]. In order to justify the extra cost, a practitioner may ask the question: What is the performance advantage of a motion deblurring camera with respect to a corresponding impulse camera? Since motion deblurring cameras capture more light than impulse imaging, it may appear that they should result in a higher signal-to-noise ratio (SNR). However, motion deblurring cameras require a computational decoding step (see Figure 13.2) which amplifies noise, thereby lowering the SNR. We have shown that motion deblurring cameras provide a significant performance advantage only if the average signal level is significantly lower than the sensor read noise variance. These results can be readily used by practitioners. For instance, it was shown that motion deblurring cameras do not provide a significant performance advantage when imaging with illumination brighter than typical indoor lighting.

The role of sensor quality

High quality sensors are carefully engineered to have low read noise, albeit with an added cost. In applications where low quality (high read noise) sensors are used, motion deblurring cameras can enhance performance even when illuminance is greater than 100 lux. In these situations, however, the additional cost required to implement coding should be weighed against the cost of improving performance by simply switching to a high quality sensor.

Task-specific imaging

Our aim in this chapter is to analyze the performance of motion deblurring cameras insofar as the final goal is to capture high quality images. If a further task is to be performed on the captured images (e.g. tracking, face recognition, intrusion detection), reconstruction algorithms can benefit from task-specific priors (as opposed to priors based on natural image statistics). Moreover, in this case, the performance should be evaluated in terms of task-specific metrics. While such task-specific priors and image quality metrics are beyond the scope of this chapter, we believe they form an interesting direction for future research.

[4] While this is generally true, flutter shutter cameras have been implemented without significant hardware modifications using specific industrial camera models (Agrawal & Raskar 2009).

Appendix

Lighting conditions

For a Lambertian scene with average reflectance R that is lit by a source with illuminance I_{src} (given in units of lux), the average illuminance falling on the detector (also in units of lux) is (Horn 1986)

$$I_{det} = \frac{1}{4}\frac{1}{F/\#^2}E_{src}R\,.$$

Given a quantum efficiency q, an exposure time of t s, and a pixel size of Δ m, the average energy in joules collected by a pixel is (DeCusatis 1997)

$$E = K\frac{1}{4}\frac{1}{F/\#^2}I_{src}Rq\Delta^2 t,$$

where $K = \frac{1}{680}$ W/lm is the conversion factor between photometric and radiometric units when the detector spectral response is matched to the photopic spectral response of the standard human observer. The energy (in joules) of a single photon is given by $\hbar c/\lambda$, where $\hbar$ is Planck's constant, and c is the speed of light. The average number of photons collected by a pixel is then

$$J = K\frac{\lambda}{4\hbar c}\frac{1}{F/\#^2}I_{src}R\,q\,\Delta^2\,t.$$

Assuming a mean wavelength of $\lambda = 0.55$ µm, the average number of photons becomes

$$J \approx 10^{15}\frac{1}{F/\#^2}I_{src}R\,q\,\Delta^2\,t.$$

References

Agrawal, A. & Raskar, R. (2009). Optimal single image capture for motion deblurring. In *IEEE Conference on Computer Vision and Pattern Recognition*, pp. 2560–7.

Agrawal, A., Xu, Y. & Raskar, R. (2009). Invertible motion blur in video. In *ACM Special Interest Group on Graphics and Interactive Techniques*, **28**(3), 95:1–8.

Baer, R., Holland, W., Holm, J. & Vora, P. (1999). A comparison of primary and complementary color filters for CCD-based digital photography. In *SPIE Electronic Imaging Conference*. Citeseer, pp. 16–25.

Bioucas-Dias, J. & Figueiredo, M. (2007). A new twist: two-step iterative shrinkage/thresholding algorithms for image restoration. *IEEE Transactions on Image Processing*, **16**(12), 2992–3004.

Cai, J., Ji, H., Liu, C. & Shen, Z. (2009). Blind motion deblurring using multiple images. *Journal of Computational Physics*, **228**(14), 5057–71.

Chakrabarti, A. & Zickler, T. (2011). Statistics of real-world hyperspectral images. In *IEEE Conference on Computer Vision and Pattern Recognition*, pp. 193–200.

Chambolle, A. (2004). An algorithm for total variation minimization and applications. *Journal of Mathematical Imaging and Vision*, **20**(1), 89–97.

Chi, W. & George, N. (2001). Electronic imaging using a logarithmic asphere. *Optics Letters*, **26**(12), pp. 875–7.

Cho, T., Levin, A., Durand, F. & Freeman, W. (2010). Motion blur removal with orthogonal parabolic exposures. In *IEEE International Conference on Computational Photography*, pp. 1–8.

Clark, R. N. (2013). Digital camera sensor performance summary. "http://www.clarkvision.com/articles/digital.sensor.performance.summary/#model".

Cossairt, O. (2011). *Tradeoffs and Limits in Computational Imaging*. Ph.D. Thesis. Technical report, Department of Computer Science, Columbia University.

Cossairt, O., Gupta, M. & Nayar, S. (2012). When does computational imaging improve performance? *IEEE Transactions on Image Processing*, **22**(2), 447–58.

Cossairt, O. & Nayar, S. K. (2010). Spectral focal sweep: extended depth of field from chromatic aberrations. In *IEEE International Conference on Computational Photography*, pp. 1–8.

Cossairt, O., Zhou, C. & Nayar, S. K. (2010). Diffusion coding photography for extended depth of field. In *ACM Special Interest Group on Graphics and Interactive Techniques*, **29**(4), 31:1–10.

Dabov, K., Foi, A., Katkovnik, V. & Egiazarian, K. (2007). Image denoising by sparse 3-D transform-domain collaborative filtering. *IEEE Transactions on Image Processing*, **16**(8), 2080–95.

DeCusatis, C. (1997). *Handbook of Applied Photometry*. New York: AIP-Press.

Dowski, E. R., Jr. & Cathey, W. T. (1995). Extended depth of field through wave-front coding. *Applied Optics*, **34**(11), 1859–66.

Fei-Fei, L., Fergus, R. & Perona, P. (2004). Learning generative visual models from few training examples: an incremental Bayesian approach tested on 101 object categories. In *IEEE Conference on Computer Vision and Pattern Recognition*, pp. 1–9.

Fergus, R., Singh, B., Hertzmann, A., Roweis, S. T. & Freeman, W. T. (2006). Removing camera shake from a single image. In *ACM Special Interest Group on Graphics and Interactive Techniques*, **25**(3), 787–94.

Gaubatz, M. (2013). MeTriX MuX visual quality assessment package. "http://foulard.ece.cornell.edu/gaubatz/metrix_mux/".

Guichard, F., Nguyen, H., Tessières, R., Pyanet, M., Tarchouna, I. & Cao, F. (2009). Extended depth-of-field using sharpness transport across color channels. In *Digital Photography V*, vol. 7250, SPIE.

Hanley, Q., Verveer, P. & Jovin, T. (1999). Spectral imaging in a programmable array microscope by Hadamard transform fluorescence spectroscopy. *Applied Spectroscopy*, **53**(1), 1–10.

Harwit, M. & Sloane, N. (1979). *Hadamard Transform Optics*. New York: Academic Press.

Hasinoff, S., Kutulakos, K., Durand, F. & Freeman, W. (2009). Time-constrained photography. In *IEEE International Conference on Computer Vision*, pp. 1–8.

Häusler, G. (1972). A method to increase the depth of focus by two step image processing. *Optics Communications*, **6**(1), 38–42.

Horn, B. (1986). *Robot Vision*. MIT Press.

Horstmeyer, R., Euliss, G. W., Athale, R. A. & Levoy, M. (2009). Flexible multimodal camera using a light field architecture. In *IEEE International Conference on Computational Photography*, pp. 1–8.

Ihrke, I., Wetzstein, G. & Heidrich, W. (2010). A theory of plenoptic multiplexing. In *IEEE Conference on Computer Vision and Pattern Recognition*, pp. 483–90.

Joshi, N., Kang, S. B., Zitnick, C. L. & Szeliski, R. (2010). Image deblurring using inertial measurement sensors. In *ACM Special Interest Group on Graphics and Interactive Techniques*, **29**(4), 30:1–9.

Krishnan, D. & Fergus, R. (2009). Dark flash photography. In *ACM Special Interest Group on Graphics and Interactive Techniques*, **28**(3), 96:1–11.

Kuthirummal, S., Nagahara, H., Zhou, C. & Nayar, S. K. (2010). Flexible depth of field photography. In *IEEE Transactions on Pattern Analysis and Machine Intelligence*, **33**(1), 58–71.

Lanman, D., Raskar, R., Agrawal, A. & Taubin, G. (2008). Shield fields: modeling and capturing 3D occluders. In *ACM Special Interest Group on Graphics and Interactive Techniques*, **27**(5), 131:1–10.

Levin, A., Fergus, R., Durand, F. & Freeman, W. T. (2007). Image and depth from a conventional camera with a coded aperture. In *ACM Special Interest Group on Graphics and Interactive Techniques*, **26**(3), 70:1–9.

Levin, A., Sand, P., Cho, T., Durand, F. & Freeman, W. (2008). Motion-invariant photography. In *ACM Special Interest Group on Graphics and Interactive Techniques*, **27**(3), 71:1–9.

Liang, C., Lin, T., Wong, B., Liu, C. & Chen, H. (2008). Programmable aperture photography: multiplexed light field acquisition. In *ACM Special Interest Group on Graphics and Interactive Techniques*, **27**(3), 55:1–10.

Ojeda-Castaneda, J., Landgrave, J. E. A. & Escamilla, H. M. (2005). Annular phase-only mask for high focal depth. *Optics Letters*, **30**(13), 1647–9.

Raskar, R., Agrawal, A. & Tumblin, J. (2006). Coded exposure photography: motion deblurring using fluttered shutter. In *ACM Special Interest Group on Graphics and Interactive Techniques*, **25**(3), 795–804.

Ratner, N. & Schechner, Y. (2007). Illumination multiplexing within fundamental limits. In *IEEE Conference on Computer Vision and Pattern Recognition*, pp. 1–8.

Ratner, N., Schechner, Y. & Goldberg, F. (2007). Optimal multiplexed sensing: bounds, conditions and a graph theory link. *Optics Express*, **15**, 17072–92.

Schechner, Y., Nayar, S. & Belhumeur, P. (2007). Multiplexing for optimal lighting. *IEEE Transactions on Pattern Analysis and Machine Intelligence*, **29**(8), 1339–54.

Shan, Q., Jia, J. & Agarwala, A. (2008). High-quality motion deblurring from a single image, In *ACM Special Interest Group on Graphics and Interactive Techniques*, **27**(3), 73:1–10.

Sheikh, H. & Bovik, A. (2006). Image information and visual quality. *IEEE Transactions on Image Processing*, **15**(2), 430–44.

Tendero, Y. (2012). Mathematical theory of the flutter shutter. Ph.D. thesis, École normale supérieure de Cachan.

Veeraraghavan, A., Raskar, R., Agrawal, A., Mohan, A. & Tumblin, J. (2007). Dappled photography: mask enhanced cameras for heterodyned light fields and coded aperture refocusing. In *ACM Special Interest Group on Graphics and Interactive Techniques*, **26**(3), 69:1–11.

Wang, Z. & Bovik, A. (2002). A universal image quality index. *Signal Processing Letters*, **9**(3), 81–4.

Wang, Z., Bovik, A., Sheikh, H. & Simoncelli, E. (2004). Image quality assessment: from error visibility to structural similarity. *IEEE Transactions on Image Processing*, **13**(4), 600–12.

Wikipedia (2013*a*). Lux "http://en.wikipedia.org/wiki/Lux".

Wikipedia (2013*b*). Projection slice theorem. "http://en.wikipedia.org/wiki/Projection-slice_theorem".

Yuan, L., Sun, J., Quan, L. & Shum, H.-Y. (2007). Image deblurring with blurred/noisy image pairs. In *ACM Special Interest Group on Graphics and Interactive Techniques*, **26**(3), 1:1–10.

Zhang, L., Deshpande, A. & Chen, X. (2010). Denoising versus deblurring: HDR techniques using moving cameras. In *IEEE Conference on Computer Vision and Pattern Recognition*, pp. 522–9.

Zhou, C., Lin, S. & Nayar, S. (2011). Coded aperture pairs for depth from defocus and defocus deblurring. *International Journal of Computer Vision*, **93**(1), 53–72.

Zhou, C. & Nayar, S. (2009). What are good apertures for defocus deblurring? In *IEEE International Conference on Computational Photography*, pp. 1–8.

Index

L_0-norm, 20
L_1-norm, 8, 14, 163
L_2-norm, 3
1D, 261, 264–266, 276
2D, 264, 265, 267
2D rigid, 61, 67
3D, 139

a priori, 265
aberration, 161
absolute, 157, 188
acceleration, 16, 37–41, 62, 111, 218, 241–244
accelerometer, 37–43, 106, 111, 170
accessories, 184
accuracy, 73
acquisition, 100 101, 112, 184, 207–209, 211, 214, 225, 260
adaptive, 11, 24, 141
additive, 101, 117, 118, 169, 171, 260
affine, 32, 148, 260, 262
albedo, 258, 273, 274
algorithm, 61, 63, 70
aliased, 209
align, 60, 61, 66, 79, 80, 106, 124, 127, 233, 234, 236
alpha, 32, 71, 151–155, 163, 231
alpha matte, 71
alternate, 12, 16, 26, 36, 47, 48, 84, 103, 104, 150, 191, 215, 216
alternating, 66
alternating minimization, 66
alternation, 66, 72
ambiguity, 13, 14, 21, 85
amplify, 111, 146, 147, 156, 157, 174, 225
amplitude, 174, 231, 233, 234, 241
angle, 79, 155
angular speed, 111, 117, 242
angular velocity, 37–40, 43, 111, 112, 118
anisotropic, 106, 241
aperture, 101, 113, 123, 134, 136–138, 154, 184, 259, 273–277
apparent motion, 46
approximate, 50, 51, 53, 75, 76, 87–91, 96, 97, 175–177
arbitrary, 67, 80, 82, 87, 89, 108, 109, 227, 232
architecture, 208, 211
area, 63
array, 61, 142, 145–147
artifact, 1, 2, 7, 12, 18, 35, 46, 52, 67, 68, 108, 109, 113, 115, 119, 129, 132, 136, 161, 163, 171, 176, 179, 207, 226, 237–239, 242
astronomy, 1, 142
asymmetric, 60
atmosphere, 101, 105
attenuate, 145, 147, 154, 207, 265
auto-flutter, 146, 159
autocorrelation, 5
autofocus, 146, 159
autoregressive moving average, 105
auxiliary, 10, 57, 106, 141, 142, 159, 164, 170, 171
axis, 59–62, 67, 107, 111, 115, 185, 193, 195

back-projection, 31, 65, 73
background, 32, 46, 70, 71, 144, 147, 152, 218
backward flow, 214
banded-diagonal, 261
bandlimited, 2, 268
bandwidth, 123, 141, 209, 261
bank of classifiers, 246–249, 251, 252, 256
barcode scanning, 240, 241, 244
Bartlett–Hann window, 88
baseline, 124, 129, 135
basis, 34, 35, 209, 210, 214
Bayer, 134, 137, 159, 277
Bayes' rule, 34
Bayesian, 16, 103, 119, 163
Bayesian matting, 72
beamsplitter, 59–61, 124, 211, 217
bell-shaped, 253
Bezout's identity, 105
bias, 4, 155
bicubic, 150, 168, 181
bilateral, 7, 84, 125, 129–132
bilinear, 34, 82–84, 96, 97, 108, 129, 130, 194, 215
bimodal, 241
bin, 246, 247, 251–253
binary, 17, 71, 105, 142–145, 152, 156, 159, 225
binning, 60

biometric, 222
blend, 17, 70, 108, 112, 113, 147, 189
blind, 75, 77, 83, 84, 86–88, 91, 94, 96, 97
blind deblurring, 83, 84, 86, 88, 93–95
blind deconvolution, 1, 12–16, 18–20, 25–28, 31, 32, 36, 40, 100, 103–109, 113, 117, 119, 120, 185, 259
block, 102, 108, 135, 137, 277
blur, 31–38, 40, 41, 43–52, 54, 55, 57, 61–66, 69–72, 75–79, 81–92, 94–97, 100–110, 112, 113, 115–117, 119, 120, 131, 141–155, 157–159, 161–165, 167–180, 184–187, 189, 191–204, 207–213, 216, 218, 222–226, 228, 229, 231, 233, 235, 239–241, 244, 246–254, 256, 259, 268, 275, 276
blur angle, 252–254
blur coefficient, 248
blur function, 57, 61, 62, 70, 72
blur invariant, 248
blur kernel, 31–37, 43–47, 49–52, 83–85, 89, 92–94, 96, 105, 116, 118, 119, 127, 132, 161, 163, 164, 178, 179, 181, 185, 192, 195, 196, 233, 247–251, 253, 258, 260, 276
blur matrix, 33, 34, 38, 41
blur parameter, 252
blurred face image, 246, 247, 249
blurring, 62, 72
BM3D, 135–137, 270–272, 275, 276
book, 70
borders, 113, 115
bound, 91, 258, 259, 261, 264, 265, 267–270, 272–274
boundary, 16, 17, 90, 129, 131, 173
box, 142, 143, 146, 152, 153, 213
brightness, 211, 212, 214, 273
broadband, 142, 145, 146
Bussgang algorithm, 105

calibrate, 38, 43, 79, 81, 127, 132, 133, 236
calibration, 59, 60
camcorder, 60
camera, 57–62, 65, 66, 71, 72, 75–83, 85, 86, 88, 91, 93–95, 97, 123–127, 129–138, 141–143, 146–148, 150–159, 161–165, 170, 171, 173–180
camera array grid, 133
camera motion, 2, 31–33, 35, 37, 38, 40, 41, 44–48, 52, 53, 75, 76, 78, 79, 82, 100, 101, 106, 109, 111, 115, 117, 119, 185, 193, 195, 207, 211, 214, 258, 260, 261, 265–267, 278
camera response, 2, 54, 187–192, 196–199
camera rig, 60
camera settings, 66, 108
camera shake, 31, 32, 37, 40, 43–45, 54, 72, 75–77, 81–83, 88, 91, 100, 108, 117, 119, 161, 185, 193, 195, 199, 201, 204
camera shutter, 114
capture, 143, 151, 152, 155, 158, 159, 226, 228
CCD, 60
center, 127, 131, 148, 150
centripetal acceleration, 38
centroid, 61, 62
chip, 59, 60
chirp, 142, 145
chop, 144–146, 157, 158, 227, 228, 236
chromatic aberration, 32
circulant, 142–145, 233–235, 276, 277
classical, 101, 103, 104, 106
clear, 70, 71
closed-form, 3, 10, 16, 23, 72, 104
closeup, 72
cluster, 154, 155
cluttered background, 136
CMOS, 60, 115, 141
co-axial, 137
coarse, 20, 23, 66, 84, 88, 96, 104, 105
coarsest, 66
code, 16, 26, 141–159, 258, 260, 278
coded aperture, 210, 275
coded blur, 143, 144, 148, 155
coded exposure, 141–145, 147, 149–153, 155–159, 170, 180, 222, 223, 225, 226, 228, 230, 231
coherence, 141
colour, 2, 57, 71, 72, 131, 133–135, 138, 184, 260, 277
commercial, 32, 46
commute, 107
compensate, 100, 141, 142
complement, 70
complex, 12, 59, 64, 67, 68, 76, 102, 120, 126, 152, 207–209, 211, 217
complexity, 64, 65, 72
component, 70, 71
compose, 71
composite, 126, 127
compressed, 7, 114, 184, 191, 208–210, 214, 217
computation, 1, 6, 10–12, 15, 61, 63, 64, 75, 76, 87, 91, 92, 95–97, 123, 141, 153, 159, 208, 258, 260, 261, 263, 269, 273, 274, 277, 278
computational cameras, 244
concatenate, 8–10, 12, 102, 126, 127, 198
concave, 167
concentric, 150
condition, 7, 35, 85, 103–105, 144–146, 156, 166, 167, 211, 213
configuration, 42
conjugate, 11, 23, 25, 87, 90–92, 197, 199
connected, 124, 125, 133
conservation, 62, 63
consistent, 115, 222, 239
constant velocity, 51, 223, 231, 232, 236, 241–244

constraint, 3, 9, 11, 13, 14, 23, 31, 32, 40, 62, 63, 76, 85, 96, 105, 106, 109, 145, 146, 151–153, 155, 212, 214, 215, 248–250
consumer, 81, 156, 159
continuous, 61–63, 142, 151, 153, 154, 156, 264, 268
contour, 132, 135–137
contrast, 123, 135–138, 223, 229, 231
conventional, 142, 145, 159, 161, 163–166, 168–171, 176, 179–181, 259, 260, 275, 278
converge, 6, 10, 11, 18, 20, 51, 52, 84, 162, 167, 169, 171, 173, 179, 216
convex, 8, 211, 212, 216, 246–253, 256
convolution, 2–4, 6, 10, 11, 17, 18, 25, 33, 62, 64, 68, 70, 75, 76, 81–85, 87–91, 100–102, 106–109, 142, 143, 145, 150, 161, 163, 166–169, 173, 185, 191, 196, 202, 203, 208, 210, 223, 224, 233, 234, 249, 261, 268
coordinate, 79, 80, 82, 83, 102, 165, 168, 211, 233
coprime polynomials, 105
correction, 164, 179
correlation, 6, 90, 167–169
correspondence, 66, 125–127, 129, 171, 178
couple, 57, 186
covariance, 143, 146, 157
crop, 149
cross-correlation, 90
cuboidal, 82
cumulative, 158
curve, 62
cyclical, 144

data, 85, 90–92, 134, 170, 175, 217, 218, 223, 224, 228, 239, 240, 246, 247, 252–256, 272
deblur, 1, 12, 14–16, 21, 24, 26, 31, 32, 34–38, 40, 41, 43–46, 49, 51–54, 57, 63–67, 70–73, 75–77, 83, 84, 87, 88, 90–97, 100, 102, 106, 107, 112, 114, 115, 118–120, 123–125, 127, 128, 130–132, 134–139, 141–143, 145–159, 161–164, 168, 170, 171, 173–181, 207–209, 212, 213, 222–225, 228, 229, 231, 236–244, 246, 248, 258–262, 264, 269, 272–278
deblurring, 57–61, 63–67, 70–73
deblurring stage, 61
decimation, 150
decode, 142, 143, 145, 146, 148, 158, 241, 258–260, 263, 264, 266, 275, 278
decompose, 68
decomposition, 68, 69
deconvolution, 63, 65, 70
deconvolve, 1–7, 9, 12–16, 18–20, 23, 25–28, 31, 32, 34, 36, 40, 41, 44, 45, 47–49, 51, 52, 75, 84, 90, 92, 100, 104, 106, 113, 114, 118, 119, 127, 132, 135, 142, 144, 146, 150, 151, 153, 154, 159, 162, 163, 166, 171, 178, 179, 209, 224, 225, 236–239, 246–248, 258, 265
defocus, 32, 46, 161, 163, 164, 179, 261, 273, 275, 276
deformable, 180
degradation, 75, 80, 100–102, 105, 106, 115, 223, 224, 240
degrees of freedom, 106
delta, 14, 18, 19, 68, 69, 95, 101–104
denoise, 84, 134–138
density, 33, 35
depth, 36, 41, 45, 57, 73, 95, 106, 107, 110, 123–126, 128–132, 134, 137–139, 180, 184, 185, 273, 276, 277
depth-dependent, 73
derivative, 47, 49–51, 84, 87, 90, 103, 104, 107, 111, 163, 167, 235
descriptor, 76, 77, 82
design, 59–61, 72
detector, 57–61, 63–65, 70–72
determinant, 104, 226
DFT, 145–147, 153, 154
diagonal, 25, 48, 146, 147
dictionary learning, 139, 210
difference, 2, 4, 5, 16, 24, 40
differences, 25
differentiable, 62
diffraction, 113
digitization, 60
dimensionality, 105, 106
Dirac delta, 266
direct, 64, 67, 87, 91, 170, 246–248
direction, 150, 154, 155, 218, 247, 249–252, 254, 256, 261, 265, 266, 278
discontinuity, 68, 152, 153, 170
discrete, 15, 33, 36, 61–63, 82, 83, 88, 90, 95, 102, 107, 113, 127, 143, 145–147, 153, 154, 164, 165, 168, 180, 194, 233, 235, 251
discrete time, 61
discriminative, 247, 248, 256
disocclusion, 180, 216
disparity map, 125, 129
displacement, 148, 214, 231
display, 189, 199
distance, 106, 107, 110, 111, 141, 226, 243, 244, 247, 248, 254
distort, 180, 222
distribution, 62, 64, 65, 166, 169, 229, 235, 241
domain, 103, 143, 146, 161, 163, 179, 180, 223
downsample, 41, 66, 106, 113, 125, 129, 132
DRBF, 252–255
drift, 40–42, 45
DSLR, 269, 274
dual, 18
duration, 81, 142–145, 156, 158, 159
duty cycle, 32, 47–54
dynamic, 41, 123, 127, 128, 131, 134, 138, 184, 185, 187, 188, 191, 203, 204, 210

E-step, 16
edge, 7, 84, 94, 115, 130, 154, 155, 163, 222, 231, 238, 239, 244
effectiveness, 173, 178, 179
efficient, 9, 10, 12, 13, 16, 26
egomotion, 161–164, 179
eigenvalue, 144
element-wise multiplication, 6
embedded, 72
empirically, 6, 171
energy, 3, 7–9, 11, 15–17, 20, 21, 26, 35, 37, 40, 41, 47–51, 57, 59, 62–65, 126, 216, 267
energy level, 62
enhance, 72, 131, 132, 135–138, 150, 151, 184
ensemble, 277
error, 61, 66, 67
estimate, 77, 84–86, 88, 91, 94–96
estimation, 62–67, 70, 73, 141, 143, 147, 151–157, 159, 163, 167, 169–171, 178–180
evaluation, 253, 254
executable, 12, 95
executables, 26
EXIF, 81
expectation–maximization, 16, 36
exponent, 233–235
exposure, 31–33, 37, 40, 42, 43, 45–47, 57, 58, 62, 64, 75–77, 79, 80, 100, 101, 109–112, 116, 117, 123, 130, 132, 134, 141–145, 147–159, 161, 163–165, 170, 171, 178, 179, 184, 186–189, 192–195, 198–204, 207, 209, 210, 223–228, 230–232, 236, 239, 258–262, 264–266, 268, 269, 273–278
extrinsic, 38, 41, 110, 129

face, 246–249, 251–256, 278
factorization, 15
failures, 94, 95
false acceptance rate, 224
false rejection rate, 224
fast, 88, 90, 123, 133, 141, 142, 149, 154
fast Fourier transform, 23, 108, 113
feature, 32, 58, 70, 126, 134, 188, 248, 255
field of view, 94, 217
filter, 83, 88, 89, 130, 164, 249
fine, 94, 129, 132, 135, 240
finer, 66
first-order, 8, 101
fixed point continuation, 216
flash, 61, 130, 141
flat, 21, 143–146, 148, 157, 159
flip, 6, 167
flow, 214, 215, 218
flow field, 57
flutter, 141–144, 159, 209–211, 227, 261–265, 269, 270, 272, 274, 275, 278
flux, 58, 59
focal length, 35, 40, 73, 78, 80–82, 94, 97, 101, 109–111, 124, 135, 217
focus, 101, 123, 129, 134–136, 261, 268, 273
foreground, 32, 70–72, 125–127, 129, 131, 136, 147, 152, 153, 218
formulation, 164, 165, 179, 180
forward model, 88–92, 96
Fourier, 2, 6, 10, 11, 75, 87, 143, 145, 223, 267, 268
Fourier slice theorem, 266
fractional, 71
frame, 32, 47, 51, 57, 58, 61–66, 79, 80, 124–127, 134, 138, 139, 141, 142, 171, 178, 184, 188–190, 195–197, 201, 203, 208–218
frame rate, 57, 58, 64, 127
framework, 63–65, 69
frequency, 2, 4, 5, 11, 16, 32, 87–90, 113, 142–146, 149, 154, 158, 159, 161, 163, 172, 173, 179, 180, 207, 213, 216, 227, 229, 233–235, 243, 244, 267
frontal pose, 254
fronto-parallel, 193
fuse, 108, 123, 124, 134, 137, 138, 188
fusing, 73

Gabor, 244, 255
gain, 259, 260, 262–264, 266, 269, 270, 272, 274, 277
gallery, 246–248, 254, 255
gamma, 54, 81, 142, 184, 187
Gauss–Newton, 126
Gaussian, 3, 7–9, 12, 15, 16, 20, 21, 31, 33, 48, 65, 96, 103, 104, 118, 131, 132, 135, 143, 157, 169, 173, 176, 224, 233, 243, 260, 262
geometric, 59, 76, 97
global, 10, 32, 43, 61, 67, 72, 75–77, 82, 94, 161, 163, 168, 171, 178, 179
gradient, 3, 8, 14, 16, 18–23, 25, 34, 36, 47, 48, 50, 51, 85, 89, 103, 163, 169, 188, 271
graph-cut, 125, 129
gravity, 38, 39, 43
gray, 134, 198
greatest common divisor, 105
grid, 95
ground truth, 4, 5, 14, 19, 21, 42–45, 51, 52, 64, 65, 70, 72, 151, 155, 156, 171, 175, 178, 179, 230, 236, 238
gyroscope, 37–43, 106, 111, 112, 114, 116–120

Hadamard, 88, 260, 277
half-quadratic, 9, 199
hallucination, 207
Hamming, 115, 224, 242
handheld, 86, 185, 187, 192, 194, 201, 203
hardware, 32, 36, 41, 42, 45, 100, 113, 115, 146, 157, 170, 171, 184, 223
harmonic, 11, 241, 242
Harris corner, 36
Heaviside, 21

heterogeneous, 133, 134, 138
high dynamic range, 184–192, 197, 198, 200–204
high frame-rate, 58, 65, 70, 71
high resolution, 57–60, 63, 66, 70, 71, 123, 125, 129, 131–134, 137, 139, 211, 212
high-speed monochrome, 133, 138
Hilbert space, 248
histogram, 228, 241, 255
histogram of oriented gradients, 244
homogeneous, 50, 51, 109, 110, 125, 129, 165
homographic, 60, 67
homography, 32, 34, 46, 50, 53, 75, 79, 81, 83, 85, 87, 97, 161, 162, 165, 166, 168–171, 175, 177–181, 193, 194
horizontal, 87, 154, 155
horizontally, 72
hot mirror, 60
HR-C camera, 124–126, 129, 130, 132–135, 138
HS-C camera, 124–127, 129, 132, 138
HS-C frame, 125
hybrid, 32, 55, 57–61, 63–65, 67, 70, 72, 73, 123–125, 130–133, 137, 138, 142, 178
hybrid imaging, 57–61, 63, 64, 67, 70, 72, 73
hyper-Laplacian, 9, 10, 12, 14, 41, 163
hysteresis, 23

identically distributed, 7, 143, 156
identity, 247, 248
ill-conditioned, 101
ill-posed, 31, 100, 101, 119, 142, 143, 150, 152, 185
illumination, 33, 110, 113, 253–256, 258, 259, 269, 273–278
image center, 110
image statistics, 8
implementation, 66
impulse, 9, 259–264, 268, 270, 272–278
in-plane, 35, 70, 79–82, 94, 150, 165, 175
inaccuracy, 64
incidental shake, 184
independent, 7, 10, 15, 65, 66, 103, 104, 106, 110, 143, 156, 162, 164, 166, 173, 233
independent and identically distributed, 66
indicator, 152
indirect, 105
indoor, 133, 136, 137, 258, 274, 277, 278
inertial, 36–38, 40, 42, 43, 45, 72, 73, 100, 106, 112, 116, 119, 141, 164, 170, 259
inertial sensor, 73
infinitesimal, 164
inflections, 231, 241–243
initial estimate, 73
inlier, 155
integrate, 15, 16, 33, 37, 40, 41, 124, 141, 143, 144, 153, 156, 159, 161, 163, 165, 179, 211, 266, 267
integration, 57, 61, 62, 68
inter-frame, 141
inter-pixel, 244
interference, 142
interior point, 18
interleaving, 159
intermediate, 16, 18, 19, 172
internal, 79
interpolate, 82, 83, 96, 108, 127, 168, 181, 195
interpolation, 62
intersect, 126, 127, 248
interval, 57, 62, 142
intra-frame, 159
intrinsic, 34, 38, 41, 43, 105, 109–111, 119, 129
inverse, 32, 34, 90, 101, 102, 106, 141–143, 148, 151, 152, 155, 159, 168, 187, 189, 191, 192, 196, 198, 199, 207, 225, 227, 236, 242
invertibility, 151, 153, 154
invertible, 144, 152
iris, 222–225, 236, 239, 240, 244
irradiance, 184–199, 201, 203, 204
irregular, 143
ISO, 178
isotropic, 10, 235
iteration, 63, 66
iterative, 6, 18, 24, 25, 51, 61, 63, 69, 84, 87–90, 104, 107, 113, 124, 126, 127, 167–173, 180, 215, 216

Jensen's inequality, 91

Kalman filter, 2
kernel, 4, 7, 15, 18–28, 31, 36, 62, 64, 68–70, 72, 76, 81–86, 88, 94–96, 101, 102, 104, 106–109, 125, 127, 128, 130–132, 137, 161, 163, 164, 166, 175, 177–179, 181, 212, 247–251, 253, 258, 260, 266, 267, 276
Kronecker delta, 194
Kuhn–Tucker condition, 167
Kullback–Leibler divergence, 16

Landweber method, 191
Laplacian, 8, 9, 12, 14, 19, 21, 23, 103, 188
lasso, 85
latent, 2–4, 7, 8, 12–21, 23, 25, 31–33, 35, 36, 40, 47–49, 75, 84, 85, 87, 90, 91, 103, 107, 113, 185, 186, 194, 199, 203, 204, 222, 223, 225, 231, 233–235, 237, 239, 241
lateral motion, 240, 242
layer, 72
LBP, 253–255
learned, 248, 253
least square, 2, 3, 9, 18, 25, 84, 85, 87, 92, 195, 236–239
length, 62, 73, 142, 144–147, 149, 150, 153, 157–159
lens, 59, 60, 62, 73, 101, 277
lexicographical, 264
light, 57–61, 100, 109, 110, 141, 142, 153, 154, 156, 158, 162, 164, 184, 187, 192, 209, 225, 236, 258–266, 269, 272–274, 277, 278

likelihood, 2, 3, 7, 9, 13, 16, 34, 40, 41, 63, 65, 66, 166, 169
limit, 94, 141, 156, 179, 180, 209, 211
line-shaped, 266
linear, 58, 70, 82–85, 87, 88, 101, 108, 109, 142, 144, 146–148, 150, 156, 159, 195, 203, 210, 211, 213–215, 223, 224, 229, 233–235, 241, 244, 247, 249, 250, 258, 260, 261, 272, 274
linear motion, 261
linear system, 147, 156
local, 14, 15, 55, 75, 87, 88, 90, 96, 103, 105, 107, 136, 153, 186, 194, 195, 197, 203, 207, 212, 231, 248, 252, 254, 255
log, 34, 41, 164
lookup table, 10, 12
low bandwidth, 123
low dynamic range, 134, 188, 191
low frame-rate, 58
low light, 31, 123, 124, 132–138
low pass, 152
low resolution, 57–60, 65, 66, 71, 123, 125, 129–132, 134, 142, 150, 159, 170, 178
LPQ, 248, 252, 254, 255
Lucas–Kanade, 47
luminance, 191

m-sequence, 142
M-step, 16
machine vision, 156, 157, 269, 274
magnified, 67
magnitude, 69, 75, 79, 84, 88, 145, 146, 153, 154, 261, 266
mapping, 185, 187–189, 193
marginalize, 15, 103, 104
maritime, 246, 247
mask, 70, 71, 127, 187, 189, 196–199, 212, 213, 217, 263, 265, 275, 277
matching, 126, 135, 137, 223–225, 235, 236, 239–242
Matlab, 11, 16, 26, 28, 91
matting, 71, 72, 151–154
max–min, 228–230
maximize, 103, 153, 170
maximum a posteriori, 32, 34, 35, 43, 48, 49, 65, 84–88, 93–96, 103, 271
maximum likelihood, 7, 13, 48, 143, 270
mean, 15, 118, 228, 229, 235, 262, 268, 270–272, 274
measurement, 208, 210
mechanical, 32, 115, 225, 226
median filter, 126, 130–132
metric, 21, 224, 229, 230, 259, 260, 270–272, 274, 278
micro, 43, 124, 156, 187
microcontroller, 60, 61
mid-range, 181, 195
minimization, 2, 3, 9–11, 15, 16, 18, 20, 23–26, 34, 35, 37, 40, 41, 47, 49, 51, 66
minimum, 145, 146, 153, 154
misalignment, 189, 197, 201
mixture, 68
model, 60, 61, 63, 72, 76, 88, 92, 93, 95, 161, 162, 164–166, 168–171, 173–176, 179–181
modulate, 141–143, 145, 212, 213, 223, 225, 226, 228, 229, 231, 242–244, 261
moments, 16
monochromatic, 57, 59
monochrome, 133, 134, 138
monotone, 173, 189, 192
morphing, 109
motion, 1, 2, 12–14, 16, 17, 21, 24, 25, 57–68, 70–73, 75, 76, 78, 80, 81, 123–128, 130–138, 141–145, 147–153, 155, 156, 158, 159, 161, 162, 164, 165, 168–170, 172–180, 258–262, 264–269, 272–275, 277, 278
motion blur, 1, 13, 17, 21, 32, 37, 43, 46, 48, 51, 57, 61–65, 70–72, 123, 128, 132, 135, 137, 138, 141, 142, 144, 145, 147, 148, 150–153, 158, 159, 161–166, 168–180, 184, 185, 187, 189, 192, 195, 201, 203, 207–210, 212, 246–256, 258–261, 264–269, 273, 274
motion deblurring, 1, 12, 16, 21, 123–125, 127, 128, 131, 138, 222, 223, 242, 244, 258, 260–262, 264, 269, 272–275, 278
motion density function, 33–35, 107, 185
motion direction, 233, 234, 236
motion flow, 57, 61, 62, 125–127, 134, 135, 138
motion invariant, 261, 265–269, 274, 276
motion path, 127, 162, 164–166, 168, 170, 171, 175, 177–179, 181, 218, 266
motion sensor, 112, 118
motion vector, 61, 63
mount, 124, 132, 133
moving, 57, 70–72, 123, 124, 133, 135, 136, 141–145, 147–152, 162, 164, 180, 212, 254
moving object, 70–72
moving objects, 57, 72
MPEG, 214
multi-body, 217
multi-camera, 209
multi-frame, 32
multi-image deconvolution, 32, 47, 48, 51, 52
multi-scale, 7, 66, 69, 84, 94
multi-sensor, 123, 124, 137, 138
multi-spectral, 137
multi-view, 135, 137
multiple, 31, 52, 55, 82, 105, 106, 108, 119, 124–126, 137, 141, 156, 164, 165, 184, 185, 187, 189, 190, 208–210, 225, 259, 265
multiplexing, 260, 265, 277

narrowband, 277
natural image statistics, 103, 228, 229, 233, 244, 278

natural prior, 8, 34, 40, 231, 240, 241
near-optimal, 228, 244
neighborhood, 40, 153
Nelder–Mead simplex method, 41
noise, 1, 2, 4–7, 9, 12–14, 16, 23, 24, 33, 34, 37, 40, 42, 48, 49, 51, 95, 101, 103–106, 108, 111, 115, 118, 129, 132, 135–138, 141–144, 146, 147, 150, 153–158, 162, 163, 165, 171–174, 176, 179, 180, 184, 223, 225, 228, 233, 237, 239, 243, 244, 258–264, 266, 268, 269, 271, 272, 274–276, 278
noisy, 57
non-blind, 1–3, 5, 7, 9, 12, 13, 18, 23, 25, 31, 34, 75, 87, 90–92, 94, 95, 97, 105, 106, 135, 163
non-blurred, 57, 70, 71, 259
non-convex, 94, 246, 247, 249, 250, 255, 256
non-invertible, 142, 155
non-iterative, 90–92
non-negative, 85, 127, 143, 248, 249
non-overlapping, 32
non-parametric, 187
non-rigid, 217
non-smooth, 151, 159
non-uniform, 76, 77, 80, 81, 93–95, 185, 192, 201, 203
non-visible, 60
nonlinear, 36, 63, 81, 84, 85, 94, 143, 187, 192, 210, 216, 259, 271
nonlocal, 7
norm, 3, 8–10, 14, 20
normalize, 62, 63, 113, 127, 131, 132, 153
numerically stable, 225
Nyquist frequency, 244

object, 2, 16, 17, 20, 21, 25, 70–73, 124, 129, 132, 136, 141–145, 147, 150–153, 157, 159, 207, 244, 258, 259, 261, 265–268, 274
observation, 66, 71
occlusion, 31, 129, 132, 180, 216–218
occupancy, 71
off-diagonal, 157
off-the-shelf, 156, 157
offline, 10, 100, 106
on the fly, 146, 159
on-board, 209
on-chip, 156, 159
opaque, 16, 145, 147, 156
operator, 101, 102, 106–108, 110, 143, 229, 233, 234
optical, 57, 59–62, 65–68, 164, 170
optical aberration, 108, 113
optical axis, 79, 81, 107, 111
optical center, 43, 78–80
optical coding, 258, 274
optical defocus, 195
optical flow, 32, 47, 61, 65–68, 106, 146, 214, 215
optical mask, 187
optical path, 57, 61
optical transfer function, 223, 267, 268
optics, 60, 73
optimal, 3, 10, 11, 15, 21, 126, 142, 145, 151, 153, 229, 230, 234, 244, 263
optimization, 31, 35, 36, 40, 41, 49, 51, 64, 67, 69, 124, 211–213, 215, 216
optimized, 67
orientation, 76, 77, 79–84, 95, 110
orthogonal, 36, 63
orthographic, 44
out-of-focus, 101, 102, 108, 134
out-of-plane, 80, 81, 165, 180
outdoor, 64, 65, 133, 141, 258, 274
outlier, 113, 126, 185
overdetermined, 190
overlap, 88–90, 112, 115

padded, 142, 145–147
panorama, 32, 46, 47, 54, 164
parallax, 123, 124, 135, 137, 185, 186, 204
parallel, 141, 142, 148, 156, 223
parameter, 35, 36, 38, 40, 45, 46, 48, 75, 76, 79, 85, 88, 95, 161, 171, 173, 246, 247, 251, 252
parameterization, 62
parametric, 15, 31, 47, 60, 61, 67, 106, 110, 119
parametric motion, 60, 61, 67
Parseval's theorem, 23, 90
partial derivative, 3, 7, 8, 11
partial least squares, 254, 255
patch, 32, 88–91, 107, 108, 112, 115, 116, 118, 135, 136, 186, 195–197, 201–203
patch-based, 68
pattern, 21, 141–145, 158, 261
peak, 135
penalize, 7–9, 24, 87, 104, 119, 141
per-pixel, 37, 164, 168, 210, 212
periodic, 113, 210
periphery, 60
permute, 144, 228
person detection, 244
perspective, 44, 51, 53, 148
phase, 20, 25
photo, 141, 142, 144, 145, 148, 149, 155, 158, 159
photograph, 31, 75, 76, 95, 100, 141, 142, 159, 161, 164, 207, 222
photon, 260, 262, 269, 272–276
physical, 101, 102, 106, 179
piecewise, 8
piecewise linear, 46, 50
pinhole, 79, 81, 109, 277
pipeline, 125, 134
pitch, 32, 79
pixel, 57–61, 68, 69
planar, 32, 33, 44, 51, 55, 107–111, 126, 141, 142, 148, 153, 155, 161, 164, 165

Plancherel's theorem, 11
PLS, 254, 255
point source, 102, 265, 266
point spread function, 1, 2, 4, 5, 7, 12–21, 23–25, 31, 37, 40, 41, 44, 45, 57, 61–71, 73, 77, 82, 84, 85, 88, 94, 101–103, 105–118, 120, 123, 125, 127, 131, 135, 137–139, 141–146, 148, 150–157, 159, 161–168, 171, 176–181, 192, 194, 195, 197, 209, 223–236, 238, 239, 241–244, 258, 261, 264, 266, 267, 274
point trace, 110, 111
point-and-shoot, 32, 54, 207
Poisson, 7, 22, 166, 169, 173, 275, 276
Poisson statistics, 63
polar, 150
polarization, 211
polygon, 101, 152
polynomial, 188
pose, 75, 80, 253–256
position, 76, 77, 79, 102, 108–112, 116, 117
positivity, 103
power spectrum, 5, 233–235, 243, 244
predicted, 16, 19, 167, 168
primal, 223
primary, 57, 59–61, 63–65, 71
principal component analysis, 68, 69
principal point, 80, 81
prior, 1, 2, 6–12, 16, 22, 31, 34, 35, 37, 40, 41, 48, 49, 51, 63, 72, 103–105, 107, 113, 119, 147, 159, 170, 210, 214, 222, 231, 236, 237, 258, 270–272, 274, 278
probability, 2, 7, 14–16, 26, 34, 103, 166, 169, 170, 271
probe, 247, 248, 251, 254, 255
processing, 73
programmable pixel compressive camera, 210–212, 217
project, 82, 109, 110, 126, 127, 230, 233
projective, 79, 80, 83, 109, 115, 126, 144, 161–165, 168–176, 178–181, 185, 230, 233, 234, 266
prototype, 60, 61, 64, 211, 217
pseudo, 135, 142, 147
PSF, 57, 61–71, 73, 123, 125, 127, 131, 135, 137, 139, 141–146, 148, 150–157, 159
PSNR, 118
PTLens software, 81
pulse, 142, 156
pursuit, 20
pyramid, 15, 20, 23, 66, 104, 189

quadrant, 154
quadratic, 8, 9, 11, 13, 14, 16, 23, 90, 92, 94, 107
quality, 57, 73
quality metric, 271, 272, 274, 278
quantize, 2, 171
quantum efficiency, 208, 259, 273, 274

radial distortion, 81
radiance, 62, 188
Radon, 233, 235
RAID 0, 134
random, 102, 146, 212, 213, 228, 229, 233
range, 36, 48, 52, 130, 131
RANSAC, 107, 126
ratio-based, 143
read noise, 259, 262, 263, 269, 272–276, 278
recognition, 222, 223, 242, 244, 246–248, 251–255
reconstruction, 85, 90, 132, 143, 147, 150, 168, 181, 208, 210, 215–218, 270, 278
recovery, 16, 17, 20, 124
rectangle, 63
rectification, 148
recursive, 2
redundant, 96, 142, 214
reference, 195, 197
refinement, 7, 15, 20, 23–25, 66, 69
reflector, 61
registration, 188, 189, 197, 254, 255
regularization, 1–6, 8, 18, 20, 21, 24, 51, 72, 85–87, 91, 92, 94, 105–107, 161–163, 170–176, 178, 179, 186, 191, 199, 207, 209, 214, 215, 222, 237
relative, 125, 141, 161, 165, 180, 189, 193
rendering, 222
representation, 67–69, 102, 161, 165, 166, 179, 180
reprojected, 129–132
resample, 127
resampling, 33, 34
residual, 167–170, 201, 203
resizing, 114
resolution, 15, 20, 36, 57–61, 63–66, 70, 71, 94, 96, 106, 113, 123–126, 128–135, 137–139, 141, 142, 145, 150, 151, 158, 159, 173, 207–212, 215, 233
response, 94, 95, 142, 144–146, 185, 187, 191
restoration, 19, 20, 24, 25, 52, 53, 75, 141, 185, 186, 201, 246, 248
reverse, 161, 168
Richardson–Lucy, 1, 2, 6, 7, 12, 13, 63, 77, 107, 125, 127, 132, 143, 147, 148, 161–163, 165–169, 171–176, 178–180, 185, 236
ridge, 62
rig, 59, 60, 123
rigid, 38, 39, 59, 61, 67, 106
ringing, 1, 2, 7, 12, 18, 46, 52, 113, 115, 118, 120, 132, 163
RMS, 162, 171–174, 176, 177
RMSE, 157
robust, 7, 9, 12, 20, 22, 105, 107, 126, 138, 207, 222, 239, 240, 251, 252
robustness, 72, 73
ROC, 224, 239, 240
roll, 44, 79, 115, 116, 120, 136, 210

root, 171
rotation, 25, 32–35, 38–40, 42–46, 52, 53, 61, 67, 75–82, 85, 94, 97, 102, 106, 107, 110, 111, 115, 148, 150, 155, 161, 164–166, 171, 173, 175–178, 180, 185, 192–194, 197, 201, 218, 223
rotational, 70
rotational drift, 40
sample, 10, 36, 40–42, 44, 50–53, 83, 95, 112, 117, 118, 126, 127, 129, 153, 158, 165, 176–178, 180, 187, 194, 224
saturation, 162, 180, 184, 185, 187, 188, 190, 198, 201, 202, 222
scalar, 85
scale, 43, 73, 123, 184, 232, 234
scan, 101, 136, 144, 148, 261
scene, 40, 41, 43, 75–80, 97, 102, 105–110, 113, 119, 123, 124, 128–138, 141, 142, 145, 161–163, 165, 179, 180, 184–187, 189–192, 194, 199, 203, 204, 208, 231, 244, 259
search, 40, 41, 145, 146, 153, 154, 228–230, 242, 269
second-order, 85
secondary, 57, 59–61, 64, 65, 70, 71, 73
segment, 31, 32, 68, 161, 164, 180
segmentation, 70, 71
semi-blind, 100
semi-definite, 167
sensing, 141
sensitivity, 144, 162, 252, 253
sensor, 32, 40, 42, 43, 45, 47, 55, 60, 72, 73, 75, 77–81, 107, 123–125, 130, 132–135, 137, 138, 141, 142, 150, 159, 165, 173, 176, 184, 207–211, 258, 259, 265, 268, 269, 273, 274, 277, 278
sequence, 57, 61, 64, 65, 67, 70, 71
set, 246–251, 255, 256, 258, 259, 261, 268, 273, 274
setup, 212, 217
shake, 75, 86, 88, 91, 93, 94, 97
shape, 70, 101, 103, 104, 109, 167, 175, 225
sharp, 16, 17, 23, 75–77, 80, 82–85, 88, 90, 91, 94, 97, 100, 103, 104, 106, 107, 112, 115, 117, 118, 141–143, 146, 147, 149, 156, 158, 159, 163, 164, 178, 208–210, 216, 218, 222, 223, 225, 227, 239, 246–249, 251, 253–255
sharp image, 67, 70, 72
shift, 247, 249–253
shift invariance, 70
shift invariant, 1, 12, 61, 62, 64, 67, 123, 137, 223, 265, 276
shock filter, 19–22, 84
short-exposure, 259, 262, 264, 269, 273, 274
shot noise, 226
shrinkage, 10
shutter, 75, 77, 80, 86, 93, 109, 115, 141–143, 148, 156, 157, 223–226, 228–232, 236, 239–244
SIFT, 126
signal dependent, 260, 263
signal independent, 260, 263
signal-to-noise, 58
signal-to-noise ratio, 5, 123, 134, 184, 259, 262–264, 266, 269, 270, 272, 274–278
similarity metric, 271, 272
simulated annealing, 104
simulation, 70, 71
sinc, 143, 146, 225
single pixel camera, 210
single-image deconvolution, 51, 52
singular value, 144, 146
slice, 144, 267
SLR, 42, 43, 123, 124, 141, 156
smartphone, 37, 100, 106, 107, 111–120
smear, 143, 144, 150, 156, 157
smooth, 3, 9, 20, 35, 85, 88, 145, 151–154, 172, 173
smoothness, 62, 72
solution space, 13
space–time, 265, 266
space-invariant, 31, 33, 36, 37, 41, 44, 75, 76, 81–84, 87, 88, 90, 96, 97, 101, 102, 105, 112, 114, 115, 118, 119, 161, 164, 165, 171, 175–177, 185, 186, 196, 208, 210, 260, 261, 264
space-variant, 16, 25, 31, 32, 35–38, 40, 41, 43–45, 48, 54, 55, 75, 76, 79, 83, 85, 87–92, 94, 96, 97, 102, 110, 112, 115, 116, 118–120, 161, 162, 164, 168, 171, 177, 180, 185, 186, 197, 203, 207
span, 144, 150, 155
sparse, 1, 3, 14, 16, 20, 23–25, 34, 35, 43, 48, 82–85, 91, 94, 103, 104, 107, 113, 163, 195, 210–212, 214, 237, 254, 255
spatial, 57–59, 61, 68, 70
spatial frequency, 142, 143, 145, 146, 149, 158, 159, 223–225, 227, 229, 231, 244, 267
spatial resolution, 57–59
spatially-varying, 67–69
spatio-temporal, 59, 209–212, 214, 215, 217
spectrum, 55, 233, 235, 241, 243, 260, 268, 277
specularity, 218
speed, 62, 86, 87, 91, 93, 128, 130–133, 138, 141, 153, 264–267, 274
spline, 62
split, 60, 62
spread spectrum, 142
stability, 5, 15, 32, 69, 100, 141
stable, 66
standard deviation, 40, 118, 254, 259
static, 60, 76, 79, 97, 129, 131, 135, 141, 145, 147, 148, 150, 151, 161, 165, 184, 188, 192–194, 203

stationary, 40, 43, 70, 147
stereo, 57, 124
stitch, 46, 54
storage, 67
stream, 123, 133, 134
stroboscopic motion, 32
structure, 4–6, 9, 13, 21, 22, 24, 25, 94, 96, 270, 272
structure from motion, 42, 43
sub-pixel, 82, 83, 212, 215
sub-problems, 84, 92
subject, 207, 211, 222–224, 226, 228, 231, 233, 241, 246, 247, 250, 253, 254
subsampling, 211, 213
subset, 62
subspace, 248
super-resolution, 31, 106, 125, 128–130, 138, 139, 150, 209, 216
support, 24, 25
SVM, 246, 247, 251–253
switching, 69
symmetrical, 103
synchronize, 60, 61, 124, 133, 134
synthetic, 36, 51–53, 135, 136, 138, 152, 169, 171, 173, 176, 179

target, 60
template, 223, 236, 243, 244
temporal, 41, 48, 50, 52, 54, 81, 142–144, 208, 210, 211, 214, 215, 261, 266, 268
temporal resolution, 58, 59
terrestrial, 101
tessellation, 62, 63, 127
text, 131, 132, 135–137
texture, 21, 25, 94, 107, 135, 210, 216, 222, 223, 225, 231, 236, 240, 243
textureless, 88
thermal noise, 208
threshold, 19–24, 36, 84, 85, 154
throughput, 210
Tikhonov, 3–6
timestamp, 116
Toeplitz, 102, 261, 264
tone-map, 134–136, 188–191, 199–201
total variation, 8–10, 21, 135, 163, 173–176, 184, 186, 199, 270–272
trace, 109, 110, 157, 234, 262, 263
tracking, 31, 32, 37, 43, 70, 123, 278
traditional, 141, 144, 147, 148, 151–159
training, 248, 251–253, 256
trajectory, 81, 101, 126, 161, 170, 223, 258, 264, 265, 274
transform, 2, 95, 143–145, 148, 150, 161, 165, 166, 168, 171, 178, 186, 192–199, 203, 209–211, 214
transition, 153, 154, 156, 198
translation, 33–35, 38–40, 42–44, 49, 50, 53, 60, 61, 77–79, 82, 95, 106, 107, 110, 111, 165, 176, 185, 193, 194, 197, 241, 244
translational, 62, 72
transparency, 4, 5, 14, 16, 17, 19–23, 237
transparent, 156
transpose, 25
trapezoidal, 144
trigger, 42, 43
trimmed, 112
tripod, 64, 65, 124, 133, 175, 184, 199
Tukey window, 198
turbulence, 101

un-blurred, 1, 2, 5, 17, 161
unconstrained, 76, 254, 255
uncorrelated, 105
under-determined, 208, 211, 214
underexposed, 106, 108, 109
uniform, 1, 12, 16, 25, 26, 36, 76, 81, 86, 89, 93–95, 142, 148, 161, 162, 164, 171, 178, 179
uniformly, 65
uniformly distributed, 65
universal quality index, 270, 272
unknown, 71
unstable, 101
update, 10–12, 17, 23, 48, 66, 69, 90, 104, 167, 168, 170
update rules, 69
upper triangular, 110
upsample, 66, 96, 125, 129–132, 150, 151, 215, 217
user assisted, 72

vanishing, 148
van Cittert, 6
variable, 71
variable splitting, 9
variance, 7, 16, 103–105, 116, 135, 143, 144, 146, 157, 169, 173, 174, 188, 229, 235, 262, 268, 269, 278
variational, 16, 104, 163, 171, 175, 180, 181
velocity, 37–40, 43, 46, 51, 226, 231, 232, 234, 236, 242–244
video, 51, 207, 208, 212, 214
view-dependent, 33
vignetting, 149, 156
visual, 2, 7, 11–13, 26, 27, 44, 57, 172, 173, 176, 179, 217, 272
Voronoi, 62, 63, 127
voxel, 36, 211, 213

Walsh–Hadamard, 145
warp, 33, 66, 125, 126, 129–132, 138, 148, 181, 193, 194, 199
wavelength, 60
wavelet basis, 210
weight, 3, 4, 10, 13, 18, 25, 49, 66, 68, 75, 81–83, 87, 88, 97, 102, 127, 154, 155, 171–173, 188–190, 193, 194, 197, 198, 229, 230, 249

wide-angle lenses, 115
Wiener, 1, 2, 5, 6, 12, 112, 113, 115, 163
wiener filter, 63
window, 21, 69, 102, 109, 110, 117, 181

X-ray, 142

yaw, 79

zero mean, 135, 173, 262
zero-padding, 88, 145
zoom, 60, 73, 82, 86, 162, 178, 201–203, 218